Modular Design Automation of Intelligent Robot Systems

IEEE Press

445 Hoes Lane

Piscataway, NJ 08854

Modular Design Automation of Intelligent Robot Systems

Zhun Fan
Shenzhen Institute for Advanced Study, China
Shenzhen Loop Area Institute, China

Zhaojun Wang
Shantou University, China

Wenji Li
Shantou University, China

Guijie Zhu
Shantou University, China

Jiafan Zhuang
Shenzhen Institute for Advanced Study, China

IEEE*Press*
WILEY

Published by John Wiley & Sons, Inc., Hoboken, New Jersey.

Library of Congress Cataloging-in-Publication Data has been applied for:

Hardback ISBN: 9781394318490
ePDF ISBN: 9781394318513
ePub ISBN: 9781394318506
oBook ISBN: 9781394318520

Cover Design: Wiley
Cover Image: © Cavan Images/stock.adobe.com

Contents

Acronyms

AI	artificial intelligence
AUROC	area under the receiver operating characteristic
BG	bond graph
BGGP	bond graph and genetic programming
CDP	constraint dominance principle
CHT	constraint handling technique
CIC	convergence infill criterion
CIMOP	constraint imbalanced multi-objective optimization problem
CMOEA	constrained multi-objective evolutionary algorithm
CMOGP	constrained multi-objective genetic programming
CMOP	constrained multi-objective optimization problem
CNN	convolutional neural network
CRL	causal representation learning
CV	constraint violation
DE	differential evolution
DIC	diversity infill criterion
DRL	deep reinforcement learning
EA	evolutionary algorithm
ECMOP	expensive constrained multi-objective optimization problem
EDA	electronic design automation
FCM	functional causal model
FIC	feasibility infill criterion
FOV	field of view
FSA	finite state automaton
GA	genetic algorithm
GNN	graph neural network
GP	genetic programming

GPR	Gaussian process regression
GRN	gene regulatory network
HBG	hybrid bond graph
HFC	hierarchical fair competition
HV	hypervolume
HVI	hypervolume improvement
IGD	inverted generational distance
LLM	large language model
MAC	multiplication accumulation operation
M2M	multi-objective to multiobjective
MCDM	multi-criteria decision-making
MEMS	microelectromechanical systems
MODENA	modular design automation
MOEA	multi-objective evolutionary algorithm
MOP	multi-objective optimization problem
NAS	neural architecture search
PID	proportional-integral-derivative
PF	pareto front
POMDP	partially observable Markov decision process
PPS	push–pull search
PSO	particle swarm optimization
RBF	radial basis function
RU	resonator unit
SA-PPS	surrogate-assisted PPS
SE	sensitivity
SFS	structure fitness sharing
SP	specificity
SPL	success weighted by path length
SR	success rate
TEAs	trial-and-error approaches
UAV	unmanned aerial vehicle
VAE	variational autoencoder

1

Introduction to Design Automation and Intelligent Robotic Systems

1.1 Background of Design Automation

1.1.1 Historical Evolution

Design automation is an important branch of knowledge automation and is listed by McKinsey as one of the second largest and most disruptive technologies that will determine the future economy. It plays a crucial role in the design of intelligent robots, especially in the design of the robot's body, control system, and vision system. By utilizing technologies such as genetic programming (GP), evolutionary computing, deep learning, reinforcement learning, and causal inference, design automation can significantly improve design efficiency, avoid the repetitive trial-and-error process in traditional design methods, and thus promote the automatic discovery of innovative designs.

For example, famous robot companies such as ABB attach great importance to research in the field of modular design automation (MODENA) for intelligent robots. ABB's robot research and development team has established a long-term cooperative relationship with Professor Peter Krus's team at Linköping University in Sweden and is committed to in-depth research on the automatic design optimization problem of robot systems. In their publicly published papers, they clearly define the concept of "Design Automation" [1]. In their cooperative research, design automation of geometric graph models is at the core, and the classic multi-objective evolutionary algorithm (MOEA) is mainly used to optimize the design of robot systems [1]. However, since the classical MOEA is relatively simple in the constraint-handling mechanism, and it is difficult to effectively solve large-scale and complex constrained multi-objective optimization problems (CMOPs), systems designed using this method are often limited in scale expansion.

Another important representative work in MODENA for intelligent robots comes from Professor Hod Lipson of Columbia University in the United States.

They used evolutionary computing methods to automatically design a movable robot system on a computer and physically realized it through 3D printing technology, creating what is claimed to be the world's first robot system automatically designed and manufactured by a computer [2]. Professor Hod Lipson's team has continued to make many new progress in this field [3, 4].

The group led by Professor Erik Goodman, the founding director of the BEACON National Center for Scientific and Technological Innovation at Michigan State University, has also carried out pioneering research in the field of electromechanical system design automation. They mainly adopted a method combining GP and bond graph and genetic programming (BGGP) and successfully achieved the automatic design of a class of electromechanical systems [5]. Subsequently, the team of Professor Clarence D. Silva, a famous authority in the field of mechatronics at the University of British Columbia in Canada, completed another doctoral dissertation and extended this research from the field of linear systems to the field of nonlinear systems. At the same time, they also proposed an important concept of mechatronic design quotient [6]. This concept can integrate multiple design goals in engineering design into one goal, so that a single-objective optimization algorithm can be used for automatic design. Subsequently, Fan et al. [7] extended the automatic design of electromechanical systems from continuous systems to hybrid electromechanical systems containing continuous dynamic systems and discrete events, and realized the parallel design of controllers and controlled objects, further promoting research in this field. In conclusion, research in the field of electromechanical system design automation has formed a certain inheritance internationally [5–8]. As a typical representative of electromechanical systems, research on MODENA for intelligent robots [9–12] combined with electromechanical system design automation [5–8] is forming a broader research direction.

For example, in the aspect of robot body design, Fan et al. [13] used the algorithm based on the push-pull search (PPS) framework [14] to conduct design optimization on a six-degree-of-freedom teaching robotic arm and obtained a design scheme better than that designed by human engineers. In the aspect of robot vision system design, Fan et al. embedded the proposed lightweight codec network model [15] into a composite mobile robot and developed a road crack detection robot for open environments. In the aspect of automatic design of robot body and controller, the team of Professor Li Fei-Fei from Stanford University proposed a new computing framework – deep evolutionary reinforcement learning [16]. Based on this framework, embodied agents can perform multiple tasks in multiple complex environments. In addition, this research also demonstrated for the first time through morphological learning the Darwin–Baldwin effect in evolutionary biology. In the aspect of automatic morphological design of swarm robots, Fan et al. [17] proposed a design automation framework for automatic generation of swarm robot morphologies. The gene regulatory network (GRN) generated by this framework not only has a simpler structure but also has better performance than the model designed by human experts. In addition, the GRN model generated by this framework can generate irregular swarm patterns in dynamic environments and thus flexibly surround and capture targets.

Although many achievements have been made in the MODENA for intelligent robots, a series of challenges still remain. For example, how to effectively integrate and optimize the integrated design of the robot's "morphology (body) – brain (controller) – eyes (vision)," and how to deal with the complex coupling relationships between various modules in multi-energy domain complex systems. Future research on design automation needs to make further breakthroughs in the following aspects [18]: (1) multi-angle unified modeling of intelligent robots; (2) expensive constrained multi-objective optimization method based on surrogate models; (3) robot design automation based on Digital Twins; (4) domain adaptation and generalization in design automation; (5) design decision control method integrating large models and domain agents; and (6) design of embodied intelligent robot systems based on MODENA.

1.1.2 Importance in Modern Engineering

The design of intelligent robots is inherently complex, and traditional methods often rely on tedious trial-and-error processes. This approach results in time inefficiencies and hampers innovation. Design automation effectively addresses these challenges. It represents a significant branch of knowledge automation, or the automation of knowledge work. Unlike traditional automation, which focuses on enhancing human physical capabilities, knowledge automation leverages mental activities such as cognition, reasoning, and imagination. According to a McKinsey report, knowledge automation ranks as the second most disruptive technology. The advent of technologies like ChatGPT and other generative artificial intelligence (AI) tools marks the beginning of the knowledge automation era, where tasks such as essay writing and code programming are increasingly automated. Design automation plays a crucial role in knowledge automation because it encompasses a major part of human mental and innovative activities. Consequently, design automation has become an indispensable part of modern engineering.

1.2 Intelligent Robotic Systems

1.2.1 Definition and Components

Intelligent robotic systems are the specific application of intelligent machines in the field of robotics. They are complex, computer-controlled robotic systems equipped with multiple visual and non-visual sensors, possessing the ability to make decisions and solve problems within their operational domain. Their design and operation are based on the theory of hierarchical intelligent control systems and general systems theory, utilizing both analytical and heuristic methods for system modeling and control.

The components of intelligent robotic systems are divided into three hierarchical levels. The organizational level is basically structured after a knowledge-based system [19]. It is primarily responsible for functions such as reasoning, planning, decision-making, feedback, and long-term memory exchange. At this level, information processing resembles certain functions of the human brain, such as the processing and application of knowledge. For instance, when planning a task, it starts with high-level information, considering the goals of the task, available resources, and various constraints to generate a series of possible action plans. The second level is the coordination level, which serves as a crucial interface between the organizational and execution levels. It processes real-time information and generates specific sub-tasks based on the plans developed by the organizational level. This layer consists of multiple coordinators, each with a fixed structure and specific function. For example, the visual system coordinator is responsible for processing visual information, and the sensor system coordinator handles information received from various sensors. Communication between these coordinators is managed by a scheduler, whose structure adapts based on instructions from the organizational level. The final level is the execution level, responsible for carrying out specific control functions and translating decisions into actual actions. It includes hardware controllers that directly drive various parts of the robotic system, such as the movement of robotic arms or the processing of image and sensor data.

1.2.2 Challenges in Design and Optimization

In modern engineering and technology, designing and optimizing robotic systems are crucial for achieving efficient and stable operations. However, this process presents several challenges that need urgent attention:

1. **Defining Functional Boundaries**: Accurately defining the functional boundaries of modules is challenging, especially in complex task scenarios. It is difficult to determine which functions should be integrated into a single module and which should be distributed across different modules to optimize performance and scalability. Varying task requirements can lead to differences in module partitioning, and there is a lack of unified standards or methods to guide the rational division of module functions. This can result in redundancy or the absence of essential functionalities.
2. **Designing Module Interfaces**: Creating universal and efficient module interfaces is another significant challenge. Modules need to exchange data, energy, and physical connections, so ensuring compatibility, reliability, and

high-speed transmission is crucial. As technology evolves and module functionalities update, upgrading interfaces while maintaining compatibility becomes increasingly complex. For example, a new sensor module may need to adapt to existing control modules, requiring interface design to be forward-looking and flexible.

3. **Establishing System Architecture**: Developing an effective system architecture model to integrate modular components is complex. This requires consideration of the distributed nature of modules, parallel processing capabilities, and overall system stability and fault tolerance. Existing control theories often fall short in addressing the dynamic changes and uncertainties of modular systems. For instance, when a module fails or a new module is added, rapidly adjusting the control strategy to maintain normal system operation remains an urgent issue to resolve.
4. **Efficient Optimization Algorithm Design**: The design optimization of robotic systems is often a complex, CMOP with mixed variables and costly evaluations. Traditional methods cannot effectively solve these problems, making constrained multi-objective evolutionary algorithms (CMOEAs) almost the only choice. Additionally, suitable surrogate models need to be designed to address the expensive nature of these problems. Therefore, developing efficient optimization algorithms for the integrated parallel optimization of intelligent robotic systems is a key challenge.
5. **Design Knowledge Extraction and Intelligent Feedback**: Traditional design approaches overlook the vast amount of data generated during the optimization of intelligent robots. Mining this data can effectively guide algorithm searches and update optimization models, forming a closed-loop iterative feedback optimization system. Thus, integrating deep learning and causal inference to achieve knowledge extraction and intelligent feedback in the design of intelligent robotic systems is also a key challenge.

1.3 Overview of MODENA

Robots are increasingly used in sectors such as manufacturing, agriculture, service industries, and defense to handle tasks that are repetitive, physically demanding, or hazardous [20]. However, in more complex and dynamic environments, robots lacking intelligence face challenges in adapting swiftly and effectively. This challenge has led to significant research on enhancing robots with intelligence [21, 22]. Intelligent robots integrate AI technologies into robotic systems, allowing them to function autonomously by learning and adapting to changes in both tasks and the environment. This is achieved through techniques such as machine learning, image recognition, and target detection, as opposed to simply following preset commands. Key elements that influence a robot's intelligent behavior include its structure (morphology), control system (brain), and visual perception system (eyes). In this regard, our research within the MODENA framework is centered on creating an automated approach for the design of the "Body-Brain-Eye" components in intelligent robots.

Advancements in technologies like deep learning, machine learning, evolutionary computing, intelligent control, and robotics have spurred considerable academic interest in the automation of intelligent robot design [23, 24]. This area is also seen as a vital part of knowledge-worker automation [25]. MODENA is a method that breaks down the design process of intelligent robotic systems into several simpler, standalone functional modules. These modules are modeled as unified graphs, which simplifies the design optimization. This modular strategy allows for the automatic design and integration of system components. Specifically, for intelligent robots, MODENA separates the morphology (body) [26], control system (brain) [27], and vision system (eyes) [28] into distinct, interpretable graphical modules within a digital twin framework. By leveraging AI techniques such as GP [29], evolutionary computation [30], deep learning [31], reinforcement learning [32], and causal reasoning [33], the modular components are automatically integrated and optimized, with evolving rules for combination. This approach automates the design of robots, enabling new discoveries to be transformed into modular units that can be added to the module library, continuously improving the system. In comparison to traditional methods, the MODENA approach enhances design efficiency and performance, generating innovative solutions that are not constrained by the limitations of human expertise or conventional trial-and-error methods.

1.4 Objectives and Significance of the Book

This book introduces a groundbreaking design paradigm, MODENA, aimed at revolutionizing the design of intelligent robots. MODENA enhances efficiency and fosters innovative designs that are often beyond the reach of human designers. This automated design framework is applicable across various aspects of robotic systems, including morphology (body), vision systems (eyes), and control and coordination (brain). Each of these components is explored through detailed case studies within the book. Furthermore, it addresses current challenges in the field and envisions the future of robot design.

The book delves into the use of computational models to construct the bodies, brains, and eyes of robots. It explains how MODENA accelerates and improves the design process compared to traditional methods. The integration of AI techniques, including CMOEAs, GP, causal discovery, and reinforcement learning, is emphasized as a core part of this innovative approach. A standout feature of this book is its comprehensive approach, which not only delves into the technical intricacies of robot design but also provides a broad perspective on current challenges and future trends. This holistic coverage of the integrated design process for intelligent robots sets it apart from other resources.

For readers interested in robotics, this book serves as an essential guide to understanding and applying MODENA techniques. It is invaluable for students, researchers, and professionals seeking to stay abreast of the latest advancements in robotics and design automation. By offering a clear overview of the design process and tackling key challenges, the book empowers readers to create more efficient and innovative robotic systems.

2

Foundations of MODENA

2.1 Principles of Modular Design Automation

Modular design automation (MODENA) is a design approach that divides the overall design process of intelligent robotic systems into several simpler, independent functional modules. These modules are represented as unified graph models, which enable design optimization. This approach facilitates the automated generation and integration of these modules. In intelligent robotics, MODENA breaks down the robot's morphology (body) [26], controller (brain) [27], and vision system (eyes) [28] into distinct, interpretable graph-based modules, all organized within a digital twin framework. Leveraging artificial intelligence (AI) techniques such as genetic programming (GP) [29], evolutionary computation [30], deep learning [31], reinforcement learning [32], and reasoning [33], these modules are autonomously integrated and evolve following predefined rules. This automation simplifies the design process of intelligent robots. Moreover, throughout the design, new modules can be identified and added to the module library, allowing their inclusion in a closed-loop system for ongoing enhancement of performance. Compared to traditional approaches, MODENA significantly boosts the efficiency and effectiveness of intelligent robot design by fostering innovative solutions that are not bound by human intuition and experience, and avoiding the iterative trial-and-error methods commonly seen in conventional design practices [2].

The MODENA approach presented here (refer to Figure 2.1) has attracted increasing interest within academic circles in recent years. This method utilizes constrained multi-objective GP techniques to automatically generate and refine the topologies and parameters of various graph models, including bond graphs (BGs), finite state machines, gene regulatory networks (GRNs), deep neural networks, and Bayesian networks [18]. Through this process, design rules

Modular Design Automation of Intelligent Robot Systems, First Edition. Zhun Fan et al.

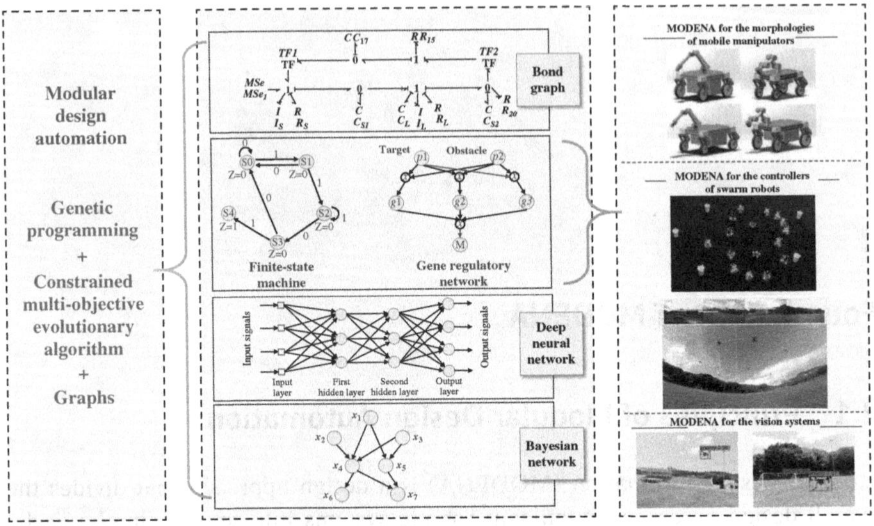

Figure 2.1 Key components in MODENA. [18] / Springer Nature / CC BY 4.0.

for high-performance intelligent robots are developed. To address multi-objective optimization problems (MOPs), two primary techniques, constrained multi-objective evolutionary algorithms (CMOEAs) and GP, are applied in tandem. CMOEAs are effective in managing conflicting objectives while handling constraints and both discrete and continuous variables. Meanwhile, GP is used to optimize the topologies and internal parameters of graph models, producing innovative and high-performing structures. Based on the characteristics of the target system, distinct graph models are applied to various subsystems of the intelligent robot, such as the morphology, controller, and vision system. For the morphology and controller subsystems, BGs are specifically used to model multi-domain physical and control systems, facilitating a thorough dynamic analysis.

In the modular design of the robotic system's body, morphological design is primarily achieved through constrained multi-objective GP and other isogeometric topology optimization, with module units including robotic arms, joints, sensors, and actuators. These modules can be flexibly assembled through parameterization and topology optimization to adapt to complex and dynamic environments. The control system module includes the optimization of single-robot and swarm-robot controllers. For single-robot controllers, neural networks and genetic algorithms (GAs) are widely used to optimize proportional-integral-derivative (PID) controllers, achieving efficient control of complex tasks [18]. For swarm-robot controllers, finite state machines and GRNs are employed to abstract and coordinate swarm behavior. For instance, finite state machines can decompose behavior into several states, enabling mobile robots to switch flexibly between different tasks, while GRNs design adaptable and robust collaborative strategies by simulating the interaction patterns of biological cells.

The vision module extensively utilizes deep neural networks and Bayesian networks. Deep neural networks can learn hierarchical representations of visual data, thereby achieving advanced perceptual capabilities for text, images, and sound [31]. Bayesian networks model uncertain causal relationships between variables, aiding robots in making real-time decisions in complex environments and providing robust perception of external information [3, 4, 34].

In the swarm control strategy module, researchers analyze the information interactions between individual robots to establish a logic relationship library, and generate behavior control models adapted to complex task environments through evolutionary algorithms (EAs). For example, through modular methods, robot swarms can adjust their behavioral patterns according to changes in objectives and environments, thereby achieving dynamic task decomposition and efficient collaboration [35].

In summary, current approaches to design automation typically involve the creation of a foundational library of modules using graphical modeling tools. These modules are then optimized through methods such as evolutionary computation to determine the optimal solution. For mechatronic system design, BGs are commonly utilized as the modeling language [36, 37]. For vision systems, deep neural networks are generally employed for representation [38, 39]. When designing behaviors for swarm robots, languages like finite state machines and GRNs are frequently used [17, 40, 41]. These modeling languages are modular, parameterizable, and consistently represented using graphical models.

2.2 Evolutionary Computation in MODENA

2.2.1 Fundamentals

2.2.1.1 Evolutionary Algorithms

EAs are optimization techniques grounded in the principles of natural selection and genetic mechanisms. They are crucial in the field of computational intelligence, particularly in AI. These algorithms operate with a population-based strategy, using trial-and-error to discover solutions, emulating the "survival of the fittest" principle. The process begins by randomly generating a population of candidate solutions, known as "individuals." A fitness function evaluates how well each individual solves the problem, with those exhibiting higher fitness being more likely to reproduce. Crossover combines the traits of individuals to produce new solutions, while mutation introduces random changes to maintain diversity. This iterative process continues until a predefined stopping condition is met.

Unlike traditional gradient-based optimization methods, EAs are global optimization techniques with broad applicability. They are self-organizing, self-adaptive, and self-learning, making them effective for solving complex problems that are non-continuous, non-differentiable, or non-convex, without being limited by the properties of the problem [42].

2.2.1.2 Genetic Programming

GP is an important subset of EAs that mimics biological evolution to automatically create computer programs aimed at solving specific problems. In GP, each solution is represented as a tree-like structure, where function nodes (such as arithmetic operations) and terminal nodes (including variables or constants) form the components. The system refines these programs through a process that draws inspiration from natural selection.

At the start, a population of randomly generated programs is created, and each program is assessed based on its ability to solve the problem. Programs that perform better have a higher probability of being chosen for reproduction. Reproduction involves operations such as selection, crossover (swapping subtrees between programs), and mutation to create new programs. Over successive generations, the population's performance improves, ultimately producing a well-performing program. GP is highly versatile and adaptive, applicable in areas like symbolic regression, pattern recognition, and control system design.

Each individual in GP is an independent computer program, typically represented using a tree structure. This representation naturally expresses the hierarchical relationships and computational processes within programs. For example, an algebraic expression can be represented as a syntax tree (Figure 2.2), where internal nodes denote operations (like addition and multiplication) and leaf nodes

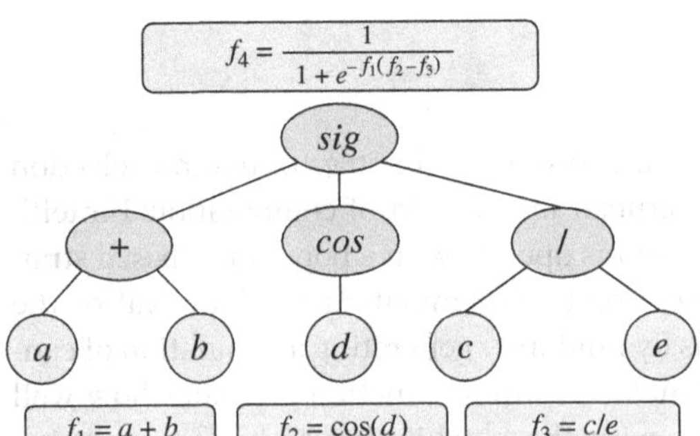

Figure 2.2 Example of a GP tree.

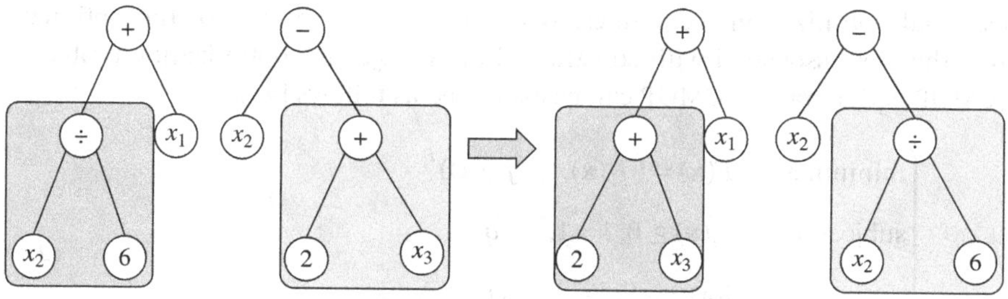

Figure 2.3 Example of crossover operators in GP.

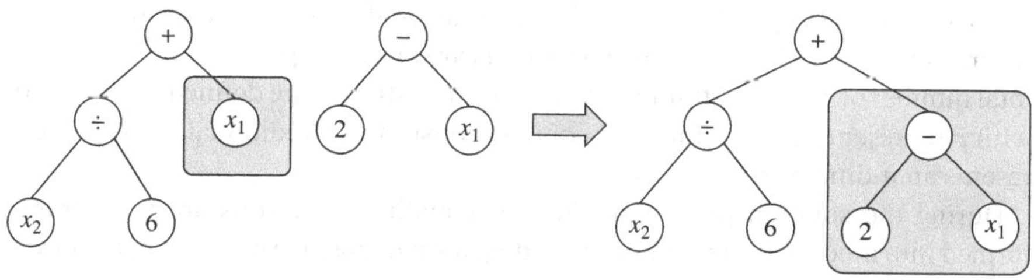

Figure 2.4 Example of mutation operators in GP.

represent variables or constants. This tree structure makes the program's architecture clear and facilitates genetic operations.

In GP, the design of evolutionary operators must consider the unique tree-structured representation of programs. Crossover in GP involves exchanging subtrees between two parent programs. More precisely, the system randomly chooses a crossover point in each parent tree and exchanges the subtrees at those points, as shown in Figure 2.3.

As shown in Figure 2.4, mutation in GP generally consists of randomly selecting a node in the program tree, removing the subtree that is rooted at that node, and replacing it with a newly generated random subtree.

This section introduces the basic principles of EAs and GP. As a key branch of EAs, GP applies evolutionary concepts to automatically generate programs. Its unique tree-based representation and evolutionary operations offer a novel approach to solving complex program synthesis problems.

2.2.1.3 Constrained Multi-objective EAs

MODENA for robots typically involves optimizing multiple conflicting objectives while ensuring that various constraints are met. In this context, CMOEAs are essential optimization techniques. Below is a brief overview of the optimization theories associated with constrained multi-objective optimization problems (CMOPs). Generally, a CMOP can be expressed as follows [43]:

$$\begin{cases} \text{minimize} & F(\mathbf{x}) = (f_1(\mathbf{x}), \ldots, f_m(\mathbf{x}))^T \\ \text{subject to} & g_i(\mathbf{x}) \geq 0, i = 1, \ldots, q \\ & h_j(\mathbf{x}) = 0, j = 1, \ldots, p \\ & \mathbf{x} \in \mathbb{R}^n \end{cases} \tag{2.1}$$

Let $F(\mathbf{x}) = (f_1(\mathbf{x}), f_2(\mathbf{x}), \ldots, f_m(\mathbf{x}))^T$ denote an m-dimensional objective vector, where $F(\mathbf{x}) \in \mathbb{R}^m$. The inequality constraints are given by $g_i(\mathbf{x}) \geq 0$, where q is the total number of such constraints. The equality constraints are defined by $h_j(\mathbf{x}) = 0$, with p representing the number of equality constraints. Additionally, $\mathbf{x} \in \mathbb{R}^n$ represents an n-dimensional decision vector.

During the solution process of CMOPs, equality constraints are often transformed into inequality constraints by adding a small positive constant δ, as shown below:

$$h_j(\mathbf{x})' \equiv \delta - |h_j(\mathbf{x})| \geq 0 \tag{2.2}$$

The total constraint violation is typically used to manage multiple constraints in the following manner:

$$\phi(\mathbf{x}) = \sum_{i=1}^{q} |\min(g_i(\mathbf{x}), 0)| + \sum_{j=1}^{p} |\min(h_j(\mathbf{x})', 0)| \tag{2.3}$$

A solution $\mathbf{x}^k \in \mathbb{R}^n$ is considered feasible if $\phi(\mathbf{x}^k) = 0$. For two feasible solutions, $\mathbf{x}^a$ and $\mathbf{x}^b$, $\mathbf{x}^a$ is said to dominate $\mathbf{x}^b$ if $f_i(\mathbf{x}^a) \leq f_i(\mathbf{x}^b)$ for all $i \in 1, \ldots, m$ and $f_j(\mathbf{x}^a) < f_j(\mathbf{x}^b)$ for at least one $j \in 1, \ldots, m$. The set of Pareto optimal solutions is referred to as the Pareto optimal set (*PS*), and their corresponding values in the objective space form the Pareto optimal front (*PF*).

Both multi-objective optimization and constraint handling are essential for addressing CMOPs. One common approach for multi-objective optimization is multi-objective evolutionary algorithms (MOEAs), which are stochastic, population-based meta-heuristic algorithms designed to solve multiple objectives simultaneously. These algorithms simulate natural selection, progressively steering solutions toward the Pareto front (PF).

MOEAs are generally classified into three primary approaches:

1. **Pareto Dominance-based Approaches [44]:** These methods enhance selection pressure through Pareto dominance relationships and variants, while employing diversity maintenance strategies such as crowding distance

or grid partitioning to evaluate the diversity of elite archives. They naturally identify superior solutions and maintain a uniform distribution on the PF. However, maintaining elite archives can be complex in dynamic environments or MOPs. Representative algorithms include NSGA-II [45], PESA-II [46], and SPEA2 [47].

2. **Decomposition-based Approaches [48]:** These approaches decompose the original MOP into multiple subproblems, which are then solved in a collaborative manner. This strategy enhances solution accuracy by tackling each subproblem individually. They may face challenges in archive maintenance. Representative algorithms include MOEA/D [49], C-MOGA [50], and MOEA/D-M2M [51].
3. **Indicator-based Approaches [52]:** These methods use performance indicators as selection criteria to screen elite individuals, making the selection process quantifiable. However, algorithm performance heavily depends on the accuracy and relevance of the chosen indicators. Representative algorithms include IBEA [53], SMS-EMOA [54], HypE [55], and MOMBI-II [56].

To effectively address constraints in optimization, it is crucial to balance minimizing objectives with satisfying constraints. Current research emphasizes developing constraint-handling techniques (CHTs) to achieve this balance. According to their design, CHTs can be classified into the following types [57]:

1. **Penalty Function-based Methods:** These methods use a penalty factor λ to incorporate constraints into the objective functions, balancing them accordingly [58]. The main challenge is setting appropriate penalty coefficients. This category consists of the Static Method [59], Dynamic Method [60], and Adaptive Method [61–63].
2. **Objective and Constraint Separation Methods:** These methods avoid the need for penalty factors by separately evaluating objectives and constraint violations. Techniques include the constraint dominance principle (CDP) [45], epsilon-constraint approach [64], and stochastic ranking technique [65].
3. **Multi-objective Methods:** These techniques convert constraints into extra objectives, transforming the CMOP into a higher-dimensional, unconstrained problem that can be addressed by MOEAs or many-objective evolutionary algorithms [66, 67]. However, this can increase computational complexity when many constraints are involved.
4. **Methods Converting CMOPs to Other Problems:** These approaches transform CMOPs into forms that are better suited for existing optimization methods, striving for an improved balance between objectives and constraints. Notable examples are multi-population cooperative optimization [68], which divides the problem into different populations focusing on various aspects, and two-phase optimization [14], which splits the process into two distinct phases to enhance efficiency.

2.2.1.4 Expensive Constrained Multi-objective Optimization

While the algorithms and constraint-handling techniques previously mentioned are effective for solving CMOPs, they face difficulties when applied to scenarios where evaluating objectives and constraints is computationally expensive. These scenarios, known as expensive constrained multi-objective optimization problems (ECMOPs), are common in domains like engineering design, where simulations or experiments for function evaluations involve substantial computational effort. Solving ECMOPs requires not only minimizing the number of evaluations but also overcoming issues such as small feasible regions, diverse PFs, and the risk of premature convergence to local optima.

To overcome these difficulties, surrogate-assisted CMOEAs have emerged as a promising solution. These techniques leverage surrogate models, including Gaussian processes [69, 70], radial basis functions (RBFs) [71], and support vector machines [72, 73], to approximate costly evaluations, thereby substantially cutting down on computational overhead. Recent research has proposed novel strategies for solving ECMOPs. For instance, Mai et al. introduced a Pareto set learning approach embedded within a push-pull search (PPS) framework. This method employs neural networks to predict Pareto optimal solutions, enhancing convergence and avoiding local optima [74]. Similarly, Li et al. introduced the SEACoEA algorithm, which uses a surrogate-ensemble model combining RBF and Kriging for coevolutionary optimization, effectively balancing exploration and exploitation [75]. Additionally, the SA-PPS algorithm integrates Bayesian active learning with adaptive infill criteria, optimizing performance across diverse problem settings [76]. These algorithms represent significant advancements in handling the complexity of ECMOPs, paving the way for efficient and robust solutions under constrained evaluation budgets.

2.2.1.5 Performance Metrics for CMOEAs

Evaluating the performance of CMOEAs requires a thorough assessment of both the convergence and diversity of the non-dominated solution set produced by the algorithm. In addition, the feasibility of the solutions should be taken into account. Common metrics for evaluation include success rate (SR), hypervolume (HV) [77], and inverted generational distance (IGD) [78].

1. **SR:**

A fundamental requirement for CMOEAs is to find feasible solutions. Researchers typically run algorithms many times independently and calculate the SR as the proportion of runs that successfully find feasible solutions:

$$SR = \frac{N_{success}}{N_{total}} \times 100\% \tag{2.4}$$

where N_{total} is the total number of algorithm runs, and $N_{success}$ is the number of runs where the algorithm successfully found feasible solutions. Here, "success" is defined as finding at least one solution that satisfies all constraints in a single run. The SR directly reflects the algorithm's capability to handle constraints.

2. **HV:**

In practical scenarios, simply identifying feasible solutions is not enough; the quality of these solutions plays a significant role. The HV metric assesses the performance of an algorithm by measuring the volume of the multi-dimensional space formed by the non-dominated solution set and a reference point:

$$HV = \lambda(\cup_{i=1}^{|P|} v_i) \tag{2.5}$$

where λ denotes the Lebesgue measure, P refers to the non-dominated solution set, and v_i represents the HV associated with solution i and the reference point. A higher HV value signifies superior overall performance of the algorithm. The reference point is generally chosen as the worst point, which is a vector consisting of the maximum values across all dimensions of the non-dominated set. As illustrated in Figure 2.5, for two-dimensional problems, the space defined by the non-dominated set and the reference point corresponds to the shaded gray area.

The HV metric has the significant advantage of simultaneously considering both the convergence and diversity of solution sets. The HV value increases when solutions are closer to the true PF or more evenly distributed. Additionally, the HV calculation does not rely on prior knowledge of the PF, making it highly practical. However, HV computation can be time-consuming and is sensitive to the selection of the reference point.

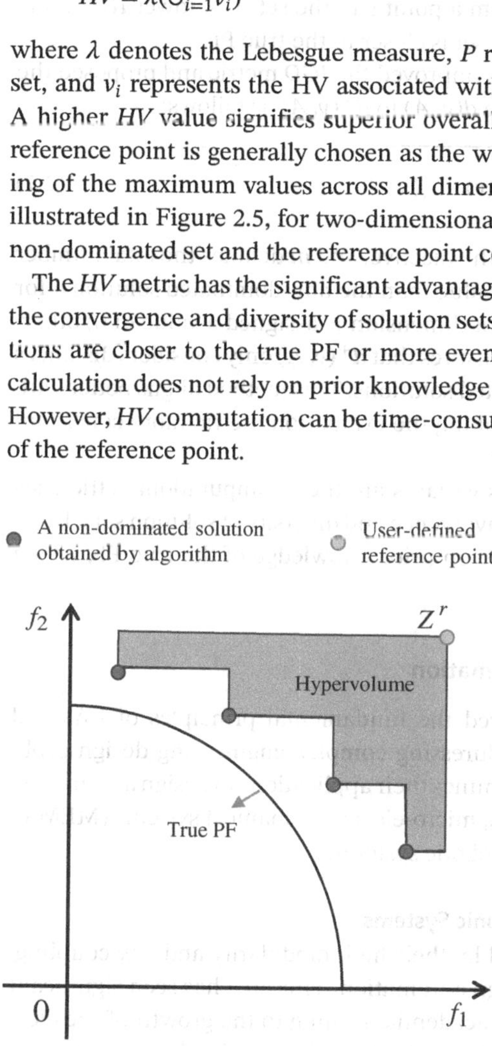

Figure 2.5 Illustration of the HV indicator.

3. IGD:

Another widely used performance metric is *IGD*, which is designed to evaluate how close the non-dominated solution set produced by the algorithm is to the true PF. The formula for its calculation is as follows:

$$IGD(A, P^*) = \frac{\sum_{v \in P^*} d(v, A)}{|P^*|} \tag{2.6}$$

where A denotes the non-dominated solution set produced by the algorithm, P^* is a collection of uniformly distributed points on the true PF, and $d(v, A)$ represents the minimum Euclidean distance from a point v to the set A. A lower *IGD* value signifies that the algorithm's solution set is closer to the true PF.

To better handle MOPs, researchers improved the *IGD* metric and proposed the IGD^+ metric. This involves modifying $d(v, A)$ to $d^+(v, A)$ as follows:

$$d^+(v, A) = \min_{a \in A} \sqrt{\sum_{i=1}^{m} \max(0, f_i(v) - f_i(a))^2} \tag{2.7}$$

where f_i represents the value of the i-th objective, and m denotes the total number of objectives. If point v has a smaller value than the non-dominated solution a for the i-th objective, the distance for that dimension is assigned a value of 0; otherwise, the distance is computed. This ensures that $d^+(v, A)$ only considers distances when point v dominates non-dominated solutions. This method aligns better with the essence of Pareto dominance, making the evaluation metric more consistent with Pareto dominance relationships.

The main strengths of *IGD* and its variants are their computational efficiency and their ability to assess both the convergence and diversity of solution sets. However, a limitation is their dependence on prior knowledge of the true PF, which restricts their applicability.

2.2.2 Application in Design Automation

In the previous sections, we explored the fundamental principles of EAs and GP, which are powerful tools for addressing complex engineering design problems. This section systematically examines their applications in design automation across four levels: electronic systems, micro-electromechanical systems (MEMS), complex mechatronic systems, and robotic systems.

2.2.2.1 Application of EAs for Electronic Systems

Electronic systems are characterized by their high modularity and low coupling, which makes them suitable for design automation. This area has seen significant advancements in both industry and academia, leading to the growth of the electronic design automation (EDA) industry [79]. Within academic research, EAs have found widespread applications in the design automation of electronic systems [80–85]. For instance, Zhang et al. [86] utilized GAs for power electronic system design, employing an adaptive search method to manage the design space exploration between controllers and controlled systems. However, this approach is limited to optimizing parameters in electrical systems with fixed topologies. Castejón et al. [87] introduced an automatic analog circuit synthesis technique based on grammatical evolution. They utilized Backus–Naur Form grammar to decode chromosomes into analog circuit netlists and applied GAs to explore the solution space, thereby enabling the automatic design of optimal circuit topology and scale for seven benchmark circuits. Rojec et al. [88] presented a global parameter optimization method for analog circuit design, capable of performing both structural topology searches and parameter optimization for given circuits. Koza et al. [89] employed GP to automatically design analog circuits, achieving simultaneous topology searches and parameter optimization. They successfully designed eight different analog circuits with varying topological structures, although their method did not account for the concurrent design of controllers and controlled systems. Building on the work of Koza et al. [89], Mattiussi et al. [90] proposed an analog gene-encoding method for evolutionary circuit design and reverse engineering of networks, such as analog electronic circuits, neural networks, and GRNs. As highlighted, EDA has made significant strides in both industry and academia, offering valuable insights for the automation of complex mechatronic system design.

2.2.2.2 Application of EAs for MEMS

As special types of electromechanical systems, MEMS are microsystems that tightly integrate microelectronic devices with micromechanical components. Currently, EAs have also made significant progress in design automation for MEMS [91–95]. For instance, Nabavi et al. [96] introduced an automated design and optimization technique for MEMS-based piezoelectric energy harvesters, in which GAs were employed to adjust the key physical parameters of the harvesters, thereby improving their output power. Through finite element analysis and prototype testing of the optimized harvesters, they demonstrated GAs' high optimization efficiency in increasing voltage amplitude and reducing computation time. Subsequently, they proposed a GA-based design automation technology [97], using GA to optimize the physical dimensions of energy harvesters to adjust harvesting frequency and improve energy collection efficiency, proving GA's effectiveness through prototype testing. To address the geometric uncertainties in MEMS design caused by manufacturing processes, Fan et al. [98] introduced an automated robust design method based on enhanced differential evolution (DE) algorithms, validating its effectiveness through the layout design of a comb-drive micro-resonator.

Generally, complex MEMS consist of thousands of micro-electromechanical components, making their design process and optimization quite complex. Many scholars have adopted hierarchical design thinking combined with EAs to optimize MEMS design. Related works include: Fedder et al. [99] borrowed hierarchical design concepts from VLSI and analog circuit design to propose a hierarchical design method for MEMS. Mukherjee et al. [100] used this hierarchical method to study the layout design of comb-drive micro-resonators, automatically designing and optimizing geometric parameters of functional components. However, their method cannot simultaneously optimize MEMS topology structure and parameters. Fan et al. [101] developed a hierarchical design automation approach for MEMS, initially leveraging the topology search capability of GP and the unified modeling power of BG to optimize MEMS system-level behavioral models automatically, followed by using constrained GAs to optimize the geometric parameters of the components. Using this hierarchical design automation method, they designed a filter composed of multiple coupled micro-resonators.

Besides the above works, another representative line of research focuses on simultaneous design optimization of both MEMS structure and parameters. For example, Kamalian [102] introduced an interactive EA that integrates human design expertise and subjective assessments of candidate designs, enabling the parallel optimization of MEMS topology and parameters. Zhang [103] further developed this work and proposed an interactive hybrid EA that integrates a MOGA and a gradient-based local optimization method, providing more flexibility in helping designers identify promising design topologies before optimizing their parameters.

2.2.2.3 Application of EAs for Complex Mechatronic Systems

A mechatronic system is fundamentally a multi-domain physical system that incorporates control systems [104]. It represents a key phase in the development of modern products, combining elements from diverse engineering disciplines, including mechanical, electrical, hydraulic, and control engineering. Within these systems, the controller functions as the central control unit (brain), with its performance having a direct impact on the overall system quality. This section explores the topic from three viewpoints: the design of mechatronic system structures, the design of controllers, and the integrated design of both controllers and controlled objects.

1. Design of Mechatronic System Morphologies

In the development of mechatronic products, the conceptual design of the structural morphology is a critical step before detailed design and product development, as it generates solutions that fulfill the functional requirements of the design challenges. To effectively design the body structure, a modeling tool that supports a unified approach for multi-domain systems is essential. The BG, a modeling language capable of simultaneously representing physical subsystems and controllers, has been widely utilized in complex systems like robots [105, 106], electric vehicles [107], and more. Especially in open-ended topology search designs, the BG's ability to provide a unified representation makes it an ideal tool. Figure 2.6 presents an example of a single BG model, which can consistently represent resonator units (RUs) across three different domains: mechanical, electrical, and micro-electro-mechanical systems.

Tay et al. [108] applied BG to automate the design process of mechatronic systems, ensuring they met specified requirements, by utilizing a GA to explore the design space. Seo et al. [109] proposed a methodology called BG/GP for the automatic design of mechatronic systems. This approach combines the BG model's ability to represent multi-domain physical systems with the optimization capabilities of GP, allowing for the exploration of open-topology design spaces and the optimization of both system topologies and parameters. In the BG/GP method, BG is used to model multi-domain systems, while GP is employed to automatically search the design space. Figure 2.7 illustrates the genotype-to-phenotype mapping in BG/GP, demonstrated through an example of circuit design optimization in the electronic domain. In this approach, BG acts as a bridge between the GP tree and the final physical implementation, reflecting the genotype-to-phenotype mapping process. Typically, the process begins with an initial "embryo" structure, which grows topologically through operations (such as Insert_J0, Insert_J1), and node operations (like add_I, add_C, and add_R) to form a GP tree, as seen in Figure 2.7d. This GP tree is then converted into a BG model (Figure 2.7e), where ":" represents the growth zone of the embryo (topological growth points). The resulting topological structure includes various nodes (0-node and 1-node) and BG

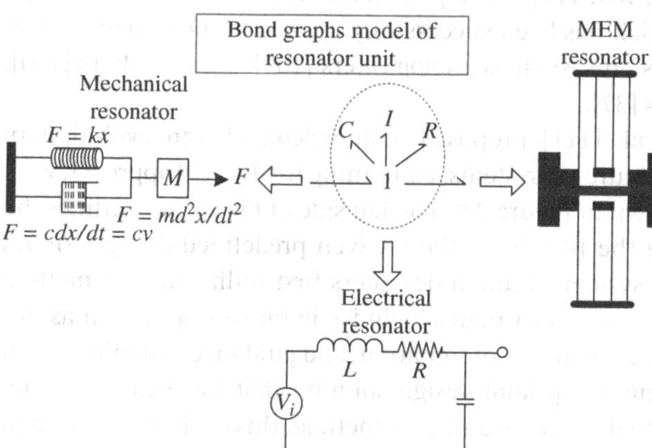

Figure 2.6 A single BG representing resonators across various application domains.

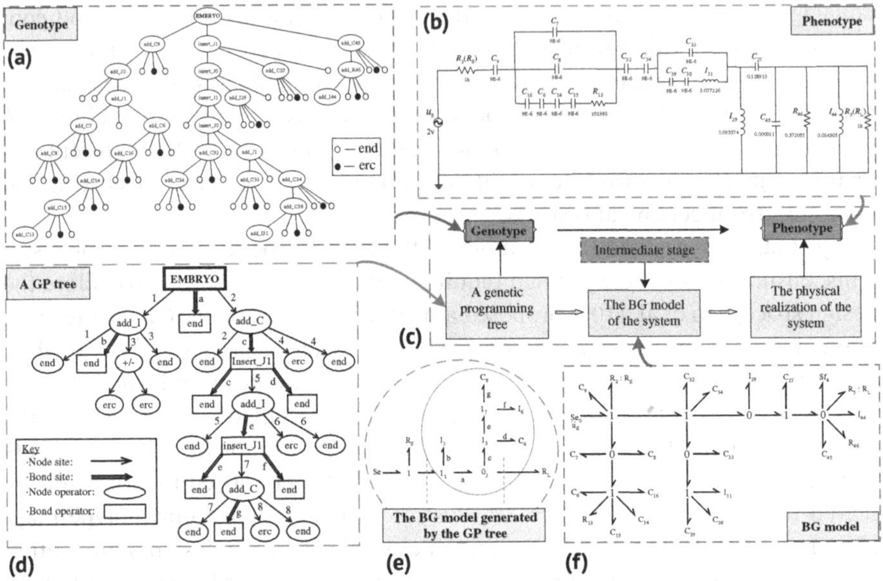

Figure 2.7 An example of genotype-to-phenotype mapping.

components with specific parameters (such as I-components, C-components, and, in some cases, R-components), all connected through keys. Figure 2.7a, b, and f depicts the evolution of the genotype from the circuit embryo to its final circuit diagram (phenotype) via the BG model. BG/GP not only automates the search for open-ended topologies but also allows for simultaneous optimization of system parameters. Fan et al. [37, 109, 110] developed an automatic design method that integrates BG and GP, which has been successfully applied to the design of electrical and mechatronic systems, such as analog filters [109], electric filters [110], and printer driver systems [37].

Additionally, Wang et al. [111] proposed a knowledge-driven evolutionary design framework for mechatronic systems, combining the BG/GP approach with human knowledge, as shown in Figure 2.8. The left side of Figure 2.8 outlines the fundamental workflow of the BG/GP method. Given predefined design criteria for complex mechatronic systems, human designers first utilize the BG method to model the system. The design procedure includes iterative stages such as offspring creation, fitness assessment, recombination, and guidance, combined with GP techniques, to generate an optimal design solution that satisfies the predefined criteria. In contrast to the standard BG/GP method, this evolutionary design process in Wang's framework incorporates a significant interaction with a human knowledge base, which plays out in two main ways: (1) knowledge extraction,

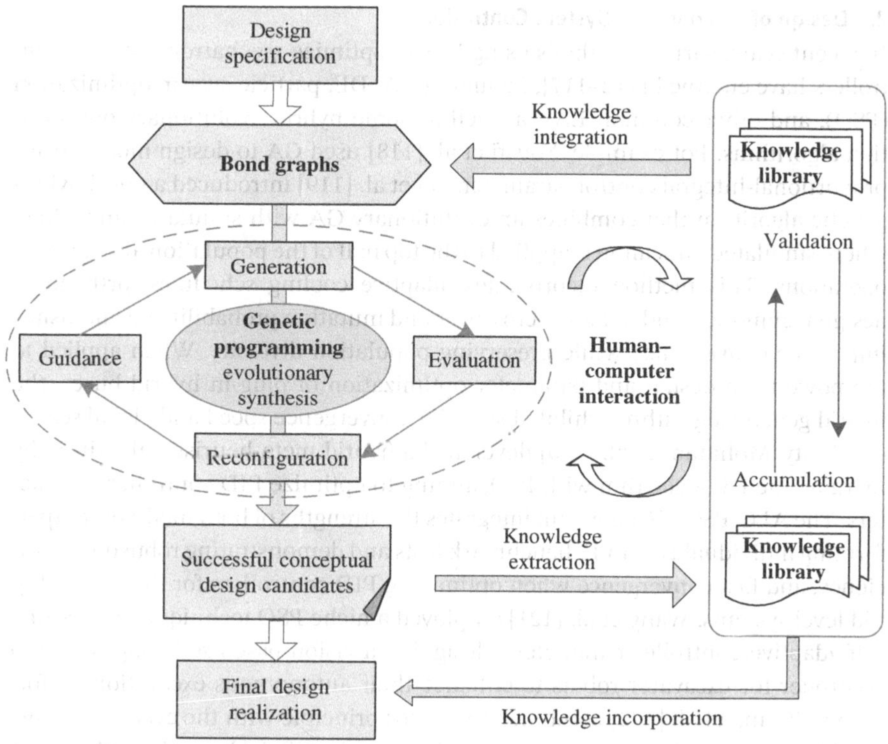

Figure 2.8 The framework for evolutionary synthesis of complex mechatronic systems.

where BG modules are derived from successful candidate designs during the evolutionary cycle. These modules are stored in a knowledge base as a library of basic building blocks, which can later be reintegrated into the design process to enhance modularity, promote design reuse, and reduce the search space in GP, thereby accelerating the discovery of optimal solutions. (2) Integration of design knowledge, where knowledge gained from human expertise or from the evolutionary process itself is fed back into the design loop. After undergoing iterative validation and recombination, this knowledge enhances the ability of GP to conduct open-topology searches, leading to more optimal design outcomes. This framework is demonstrated in applications such as the synthesis of a quarter-car suspension control system and the design of a MEMS bandpass filter. These examples highlight how the automated BG/GP-based design method introduces innovative approaches to mechatronic system design, with the potential to enhance the performance of existing design solutions.

2. Design of Mechatronic System Controllers

In recent years, various methods using EAs to optimize mechatronic system controllers have emerged [112–117], including GA, DE, particle swarm optimization (PSO), and other common EAs, as well as some hybrid evolutionary optimization algorithms. For example, Yousfi et al. [118] used GA to design motor vector proportional-integral control parameters. Li et al. [119] introduced a novel hybrid genetic algorithm that combines an evolutionary GA with simulated annealing, where simulated annealing is applied to the top half of the population after genetic operations. This method incorporates adaptive cooling schedules, orthogonal design techniques, and dynamic crossover and mutation probability mechanisms, improving convergence while preserving population diversity. When applied to the powertrain design and parameter optimization of plug-in hybrid buses, the hybrid genetic algorithm exhibited superior convergence speed and global search capability. Moharam et al. [120] developed a hybrid meta-heuristic algorithm by merging the DE algorithm with PSO, aiming to optimize PID controller parameters. The ALC-PSODE algorithm integrates the strengths of both methods, outperforming individual EAs in 12 benchmark tests and demonstrating robustness, efficiency, and fast convergence when optimizing PID controllers for three-tank liquid level systems. Wang et al. [121] employed a niche PSO technique to fine-tune self-adaptive controller parameters, designing a vision-based self-adaptive servo controller for excavator robots to enhance their autonomous excavation performance. Zhang et al. [122] combined the chaos principle with the cyclic exchange neighborhood mechanism to propose an improved artificial bee colony algorithm for optimizing fractional-order PID controllers of automatic voltage regulation systems. They used chaos traversal to prevent the search from falling into local optima and applied a cyclic exchange neighborhood mechanism to expand the search range, thereby accelerating the discovery of optimal controller parameters. Zheng et al. [123] proposed a parameter tuning optimization method for permanent magnet synchronous motor fractional order proportional-integral speed controllers based on a DE algorithm. Further, Zheng et al. [124] optimized controller parameters using a DE algorithm, considering time-domain and frequency-domain requirement specifications, and obtained $PI^{\lambda}D^{\mu}$ controllers with good robustness and optimal dynamic response performance.

The above methods all optimize controller parameters under fixed controller structures, which significantly limits controller performance. To improve controller performance, researchers have also studied the simultaneous optimization of controller parameters and structures. Koza et al. [125–128] conducted extensive research and experiments in controller design automation using GP methods. A GP-based automatic design method [125] uses GP's program trees to represent the controller's topological structure and parameters. GP self-determines the modules for signal processing that compose the controller, module types, module connections, and control parameters. Building on Koza et al.'s research, Bruijnen et al. [129] further proposed a hybrid optimization algorithm combining GAs and GP, where GAs optimize controller parameters and GP optimizes the controller structure. Lapa et al. [130] proposed a flexible fuzzy PID controller design method combining GA and extended PSO, where GA selects the controller structure while ePSO selects structural parameters. Simulation results showed that this controller structure design process can be automated. In 2017, Zhan et al. [131] proposed an enhanced DE algorithm for the joint optimization of controller structure and parameters. This method addressed the performance constraints imposed by the controller structure and improved the automation level in the design process. Through simulations and tests on a magnetic levitation ball control system, the controller exhibited superior disturbance rejection capability when subjected to various disturbances, confirming the effectiveness and advantages of the approach. In 2019, Zhan et al. [132] introduced a universal framework for the design of structurally optimized disturbance-rejecting controllers, as illustrated in Figure 2.9. This framework integrates the search and evolution abilities of EAs, allowing for the automatic determination of both controller structure and parameters for different control systems. The research elaborates on the

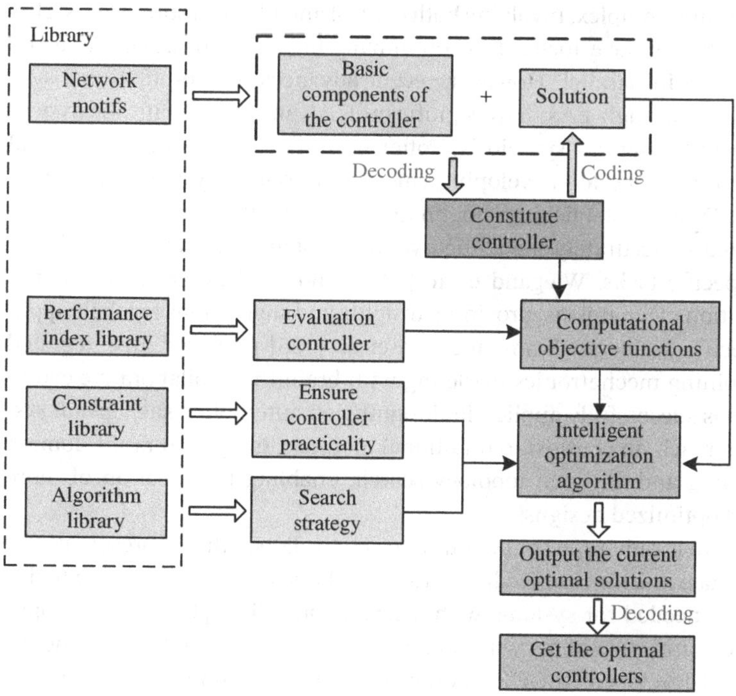

Figure 2.9 The framework of disturbance rejection controllers with optimized structures. Adapted from [132].

composition and expression of controller structures and establishes a set of generation rules for these structures. Efficient coding and decoding strategies were applied to reduce the solution space and enhance the algorithm's search efficiency. Using this framework, robust controllers tailored for typical nonlinear systems with diverse characteristics were designed, with simulations and experiments validating their feasibility, advancement, and versatility of the intelligent optimization algorithms.

Control systems are ubiquitous in production and daily life, and AI-based technologies are increasingly becoming a hot topic in controller optimization. As discussed, evolutionary computation, an AI technique, offers new approaches for optimizing the design of mechatronic system controllers.

3. Integrated Design of Controllers and Controlled Objects

Mechatronic systems typically consist of various subsystems from different engineering domains. Since each discipline often requires its own specific design language, it becomes challenging to establish a unified approach for integrating multi-domain system designs. Moreover, traditional design methods often fail to guarantee optimal performance for the overall system. Mechatronic systems are inherently complex, involving both control and physical subsystems. Historically, few studies have focused on optimizing both the controllers and the controlled objects in parallel. However, recent advances in computational synthesis, especially through EAs, have significantly advanced mechatronic system design. Inspired by natural symbiosis, Potter et al. [133] introduced a general co-evolution framework for developing rule-based control systems for autonomous robots. Despite establishing a foundational co-evolution architecture, challenges persisted in accurately modeling dynamic systems and selecting suitable species for specific tasks. Wiegand et al. [134] conducted experiments to analyze collaboration mechanisms, providing insights and suggestions for their application in specific problems. Wang and Fan et al. [135] proposed an integrated method combining mechatronics modeling with brain-limb collaborative evolution, which was successfully applied in designing an automobile suspension system. Their approach, in contrast to traditional methods, integrated multi-domain system modeling and an open topology search, enabling the creation of more advanced and optimized designs.

Since BG models only represent continuous controllers, other representations, such as finite state machines or feedforward controllers (a type of model predictive controller), are needed for systems with discrete controllers [136–138]. To optimize hybrid dynamical systems with both continuous and discrete components, Dupuis et al. [7] developed HBGGP, a design automation technique that combines hybrid bond graph (HBG) with GP. In this approach, HBG is utilized to represent systems involving both continuous and discrete dynamics, while GP explores the design space of HBGs to optimize the topologies and parameters of DC–DC converters. Subsequent work by Dupuis et al. [139] focused on the evolutionary design of controllers for hybrid mechatronic systems, employing Finite State Automata to model discrete controllers. A case study involving a two-tank system demonstrated the success of their approach in designing a finite state automaton (FSA) controller for hybrid mechatronic systems. These case studies underline how the BG/GP-based automated design methodology can provide fresh insights into mechatronic system design, potentially improving upon traditional approaches.

In conclusion, the broad application of EAs in design automation illustrates their distinct advantages. Notably, GP has attracted considerable interest from researchers because it can simultaneously conduct open-ended searches for topology and optimize design parameters.

2.2.2.4 Application of EAs for Robotic Systems

The design of a robotic system with autonomy and adaptability is particularly challenging, as it represents a complex mechatronic system that integrates mechanics, power electronics, software, and control. However, evolutionary computing methods offer significant potential for the automated design of robotic systems, garnering great attention from both industry and academia. For example, the well-known robotics company ABB has maintained a long-term partnership with Professor Peter Krus's research group at Linköping University in Sweden, focusing on the development of methods for automating the design of robotic systems. Tarkian [140] provided a clear definition of "design automation" and applied traditional MOEAs to optimize robot design.

In 2000, Lipson et al. [141] successfully designed the first computer-generated robotic system using evolutionary computing and realized its physical construction through 3D printing technology. This research achievement was published in *Nature*. Subsequently, Professor Lipson's team achieved a series of significant breakthroughs in the field of soft robotics and modular self-configurable robots. For instance, in their work outlined in [142], they proposed a method for the design automation of soft robots, combining EAs with a representation method using Gaussian mixture models to achieve an open-topology search for soft robots. This method facilitated the creation of soft robots with advanced functional requirements, flexible material distribution, and diverse configurations.

In the field of robot morphology structure design, scholars have made significant achievements using evolutionary computing methods, demonstrating notable advantages in design optimization problems concerning mechanical arms [143–145], grippers [146, 147], and transmission devices [148, 149]. For instance, West et al. [150] studied the dynamic modeling and parameter evaluation of a seven-degree-of-freedom robotic arm. They devised a mechanism utilizing GAs to estimate and update robotic arm model parameters under conditions of noise and operational uncertainty. To enhance the performance of robotic arms, Xiao et al. [143] applied NSGA-II to optimize key features such as the geometric length of the arm, motors, and gearboxes, grounded in robotic kinematics and dynamics. Experimental findings revealed that the algorithm produced a mechanical arm that was both lighter and more efficient than the UR5 model. Hassan et al. [151] undertook extensive studies on gripper modeling and optimization, creating detailed geometric, kinematic, and dynamic models. By incorporating functional requirements along with geometric and force constraints, they structured the gripper design as a multi-objective problem and utilized NSGA-II to determine the optimal configuration. Additionally, Fan et al. [30] employed the PPS algorithm [152] to optimize a six-degree-of-freedom teaching robot arm, achieving a design superior to those produced by human engineers. Experimental validations highlighted the substantial advantages of the PPS-MOEA/D method, demonstrating that the adoption of the PPS framework notably enhanced algorithm performance in solving robotic optimization design challenges.

In the domain of design automation for robotic systems, EAs have been utilized to optimize not only morphological structures under various geometric constraints but also robot controllers, achieving notable advancements, especially in PID controller optimization. For instance, Sharma et al. [153] employed the cuckoo search algorithm to fine-tune the parameters of a fractional-order fuzzy PID controller, addressing the trajectory tracking challenge of a two-link manipulator. Experimental outcomes confirmed that the optimized controller effectively ensured precise trajectory tracking while enhancing robustness against model uncertainties, noise, and external disturbances. For the trajectory tracking of autonomous mobile robots, Ali et al. [154] leveraged the artificial bee colony and GA to adjust PID controller parameters, validating the controller's performance through predefined test trajectories. Results indicated that PID controllers optimized using the artificial bee colony algorithm outperformed those optimized with the GA. Taherkhorsandi et al. [155] applied a multi-objective GA to optimize a biped robot's controller parameters, designing an adaptive robust hybrid controller that integrated PID control and sliding mode control. This controller successfully enabled the biped robot to walk smoothly on inclined surfaces. Son et al. [156] combined an improved DE algorithm with a neural network to propose an adaptive feedforward neural network-PID controller combination for controlling the joint angle position of a SCARA parallel robot. Experimental results demonstrated that the optimized adaptive feedforward neural network PID controller had adaptive self-tuning capabilities, enabling precise and robust position control of the SCARA parallel robot. Furthermore, for the issue of posture deformation in mobile robots, Hai et al. [157] introduced an evolutionary game theory-based pigeon optimization algorithm for automatically optimizing the parameters of a self-disturbance controller to control the posture deformation of the robot. Simulation experiments indicated that this optimization method effectively enhanced the robustness of the self-disturbance controller. The application of EAs in robot controllers is not limited to the aforementioned types of robots but can also be applied to the design optimization of wheeled robots [158], soft robots [159], underwater robots [160], excavator robots [16], and surgical robots [161].

In summary, using EAs for the design optimization of robot morphologies and controller parameters can create robot systems that meet specific performance requirements. However, in the realm of robot system design automation, robots are viewed as complex systems comprising both a morphology structure ("body") and a controller ("brain"). Focusing solely on optimizing either the morphology or the controller may not yield the best overall system performance. Therefore, integrating the design of both the morphology and the controller is essential for developing high-performance robot systems. Traditionally, the approach has been to manually design the morphology first, followed by the design and optimization of control strategies for robots with fixed structures [162]. Recently, researchers have applied biological evolution concepts to integrate the design of morphological structures and controllers. For instance, Endo et al. [163] used a GA to co-evolve the structure and controller of a bipedal humanoid robot, achieving a refined robot structure and walking pattern. Marbach et al. [164] combined GP and GAs for the co-evolutionary design of a homogeneous modular robot's morphology and controller, overcoming limitations of manual design. They took advantage of GP's ability to search open topologies, allowing for the autonomous configuration of modules, which enabled the robots to adapt to different tasks and environments. The GA-optimized controller parameters produced robust nonlinear oscillation controllers for exploring different motion strategies. Villarreal-Cervantes et al. [165] proposed a hybrid gradient EA for the coordinated design optimization of a planar parallel robot's structure and control system, achieving optimal parameters for both. Experimental results showed that this coordinated design method surpassed the sequential design method in global search capability and efficiency, optimizing both geometric and controller parameters simultaneously. Valdez et al. [135] used three heuristic algorithms to optimize two serial robotic arms, comparing parallel and sequential optimization designs. They found that parallel optimization offered superior global search capability and efficiency, achieving optimal geometric and controller parameters concurrently. For a comprehensive overview of using EAs for the co-evolution of robot bodies and controllers, refer to [139].

2.3 Neural Architecture Search

In the realm of AI, neural architecture search (NAS) has emerged as a pivotal area of research. It is dedicated to the autonomous discovery of efficient neural network configurations within a predefined framework, thereby diminishing the need for human intervention during the design phase. This field is of paramount importance for enhancing the visual capabilities of autonomous robots, which is essential for establishing the foundation of automated design for such intelligent systems. NAS facilitates the optimization of neural network structures tailored to visual perception tasks, such as object detection and scene interpretation. This leads to a marked enhancement in the efficacy and adaptability of the visual components within robotic systems. This section delves into the theoretical underpinnings, algorithmic approaches, and the profound influence of NAS on the automation of intelligent robot design.

2.3.1 Theory and Algorithms

NAS represents an automated approach to discovering the most effective neural network configurations, playing a pivotal role in contemporary AI research. From the perspective of automated machine learning and hyperparameter optimization, traditional machine learning relies heavily on manual intervention for model selection, feature engineering, and hyperparameter tuning. This becomes increasingly difficult as the amount of data and model complexity increase. As a part of automated machine learning, NAS is dedicated to automating these processes. In neural network models, hyperparameters are essential, dictating aspects like the depth of layers, neuron quantity, and the rate at which the model learns. They cannot be obtained automatically through training and need to be determined by search algorithms to find the optimal value. The task of NAS is to find the best combination of hyperparameters and network architecture in a given search space to improve model performance. The definition of the search space is crucial. It covers aspects such as network layer types, layer parameters, connection methods, and activation functions. Different network layer types are suitable for different tasks. For example, convolutional layers are well-suited for processing image data, and fully connected layers are often used in classification tasks. Layer parameters determine the specific calculation method and output feature size of the layer. Connection methods can be simple sequential connections or complex jump connections, which have an important impact on model performance. Activation functions determine the output characteristics of neurons and affect the nonlinear representational ability of the model.

In terms of algorithms, there are algorithms based on reinforcement learning, EAs, and gradients. The algorithm based on reinforcement learning models the search process as the interaction between an agent and the environment. The controller generates a network architecture and obtains a reward through training and evaluation. It continuously adjusts the strategy to find the optimal architecture. It has strong exploration ability, but there are problems of low optimization efficiency, unstable training, and high computational cost. The EA draws on the idea of biological evolution and searches for the optimal architecture through processes such as natural selection and genetic mutation. It is a global search algorithm with strong versatility, but the convergence speed is slow and the computational cost is high. The algorithm based on gradients represents the search space as a continuous function space and quickly finds a better solution through gradient optimization. However, for complex search spaces, it may be difficult to represent, and it is prone to falling into local optima.

2.3.2 Application in Design Automation

NAS plays a pivotal role in the automation of designing visual systems, facilitating the automatic exploration of neural networks that yield enhanced performance in diverse visual tasks. In this part, we explore the realm of visual system design automation through the lens of prevalent visual tasks in modern robotics, which include object detection, image segmentation, and applications for embedded systems.

2.3.2.1 NAS for Object Detection

Robotic systems that employ object detection can pinpoint and ascertain the location of specific objects within visual data, thereby facilitating actions like grasping [166, 167], object tracking [168, 169], and other tasks. The object detection network framework mainly consists of three segments: the backbone, the connector, and the front-end. The backbone extracts image traits, the connector blends features, and the front-end classifies and locates objects. Presently, there are two main approaches to object detection architectures: (1) a comprehensive search for network configurations [170]; and (2) a targeted search within the network structure while utilizing other existing framework components. [171]. A range of approaches have been crafted to delve into and enhance object detection network architectures, tailored to the nuances of object detection tasks and the intrinsic properties of network structures.

For example, Chen et al. [172] introduced DetNAS, an approach designed to address the loss of object location details that can occur when using an image classification network as the foundation. With ShuffleNetV2 as the search space, it searches the whole network. The algorithm undergoes pre-training on ImageNet datasets and fine-tuning on object detection task datasets, also employing an EA to identify optimal sub-network. Wang et al. [171] proposed NAS-FCOS, a rapid NAS technique. It utilizes established image classification networks as the backbone and extends the architecture with a feature pyramid network and a detection module. The NAS-FCOS method confines its search to the optimization of the feature pyramid network and the detection module, exploring these components across separate search domains. Employing a long short-term memory network as an agent and and leveraging a reinforcement learning-driven search strategy, the system constructs neural networks. Structural-to-Modular NAS [173] uses a two-stage strategy to identify appropriate architectures for object detection. First, it combines various existing networks to find the optimal mix concerning inference speed and accuracy. Subsequently, it scrutinizes all network structures within the Pareto solution set across individual modules.

2.3.2.2 NAS for Image Segmentation

Image segmentation is a technique that assigns a class label to every pixel within an image, representing a detailed predictive challenge. This technology is instrumental for robots in executing diverse tasks, such as identifying manufacturing defects and quantifying measurements [174], as well as in the field of medical diagnostics [175, 176]. Presently, the exploration of image segmentation architectures is bifurcated into two primary approaches: one that seeks optimal module configurations within a predefined network structure, and another that endeavors to optimize both the overarching network design and the specific module configurations in tandem.

For example, Liu et al. [177] proposed Auto-DeepLab, which was the first study on the application of NAS to image segmentation. It adopts architecture-level and unit-level search methods to study the overall architecture and unit structure, respectively. They frame the architecture search problem as a differentiable problem and use gradient-based optimization methods to solve it. Nekrasov et al. [178] developed a compact semantic segmentation network for mobile use, treating the existing network as the encoder and focusing the search efforts on the decoder within an encoder-decoder setup. Wei et al. [179] introduced Genetic U-Net for the segmentation of retinal vessels, utilizing a U-shaped architecture and evolutionary strategies to refine the network's cellular structure. Employing binary encoding for network architecture representation, the Genetic U-Net assesses model fitness based on performance against a test dataset. By applying GAs, it evolves more capable candidate models and selects the most effective network configurations. This approach surpasses current methods in segmentation accuracy and minimizes model complexity in various benchmark datasets. Additionally, the Genetic U-Net serves as a flexible platform that can be adjusted to a variety of visual tasks, tailoring optimal models according to the particular traits of the training data, as illustrated in Figure 2.10.

2.3.2.3 NAS for Embedded Devices

Existing NAS techniques can produce models with high accuracy, but they often have unacceptable computing latency and are not suitable for real-world intelligent robots deployed on embedded devices with limited memory and computing

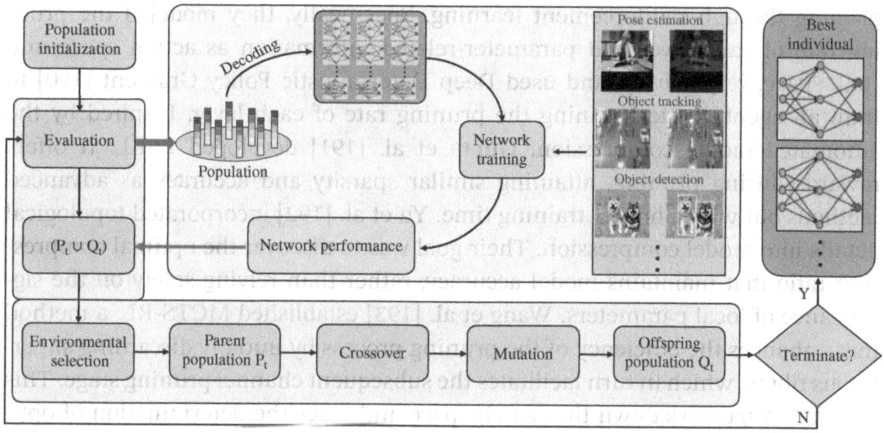

Figure 2.10 The framework of Genetic U-Net [179]. In the framework, if the dataset is pose estimation [180], object tracking [181], or object detection [182], the framework can automatically generate the corresponding optimal neural network model. [18] / Springer Nature / CC BY 4.0.

resources. Therefore, designing suitable NAS methods for embedded devices has become an urgent and critical research focus.

In the domain of embedded systems, model efficiency is paramount, with a focus on reducing computational delay and memory usage. Cai et al. [183] introduced ProxylessNAS, which formulates model latency as a continuous variable and integrates it into the optimization process as a regularization term to yield models with minimal delay. Wu et al. [184] created DANS to predict the latency of the overall model using module-level delay estimation and employed reward functions for delay optimization to guide search strategies. López et al. [185] proposed E-DNAS, a method that uses a multi-objective, differentiable loss function to balance classification accuracy and minimum delay on feature maps. Luo et al. [186] proposed LightNAS, a two-stage approach: an initial extensive search for latency-compliant models, followed by an iterative refinement to select the most accurate candidate. In addressing the memory consumption problem, Cassinmon et al. [187] suggested adding soft constraints on caching and performance, and hard constraints on memory usage and latency to the reward function to steer the search toward a model that satisfies resource constraints. In addition, Wan et al. [188] introduced DMaskingNAS, which employs an efficient feature reuse mechanism and shape propagation, significantly broadening the search space by enabling exploration in both spatial and channel dimensions.

Besides constructing models from scratch, another direction worthy of consideration is the automatic compression of existing large models. He et al. [189] initially presented automated model compression, which realizes automatic model pruning through reinforcement learning. Specifically, they modeled the pruning rate of each layer and parameter-related information as action space and state space, respectively, and used Deep Deterministic Policy Gradient [190] to train an agent for determining the pruning rate of each layer. Inspired by the automated model compression, Gupta et al. [191] developed PuRL. It offers rewards during pruning, attaining similar sparsity and accuracy as advanced methods but with a briefer training time. Yu et al. [192] incorporated topological details into model compression. Their goal was to discover the optimal compression ratio that maintains model accuracy, rather than relying solely on the significance of local parameters. Wang et al. [193] established MCTS-RL, a method that enhances the efficiency of the pruning process by initially discarding superfluous filters, which in turn facilitates the subsequent channel pruning stage. This innovation narrows down the search space and eases the determination of optimal channel pruning ratios. Beyond this, other compression strategies like tensor decomposition [194], data quantization [195], and knowledge distillation [196] have proven their merit. Yet, these approaches necessitate human expertise and are not central to the automation of design processes, which is why they are not elaborated upon in this context. For a more in-depth exploration of these topics, readers are directed to the referenced studies [197–200] for further research.

To summarize, NAS has emerged as a cutting-edge approach within the realm of automated vision system design, facilitating the autonomous exploration of neural network configurations that yield superior results in a multitude of visual recognition endeavors. Within this segment, we present a succinct analysis from multiple viewpoints to clarify how the various techniques examined herein intersect and diverge, with a visual representation provided in Figure 2.11.

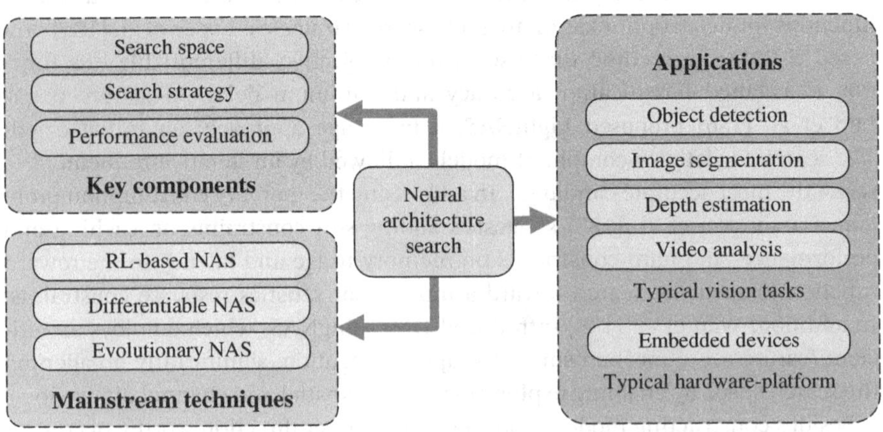

Figure 2.11 A summary of design automation for vision systems with NAS.

Current research in NAS predominantly concentrates on three critical elements: (1) the search space [172, 179, 188, 189, 193, 201–203], encompassing all potential network architectures to be considered; (2) the search strategy [171, 173, 177, 178, 204–209], which dictates the process of identifying a suitable candidate that aligns with specific criteria from the search space; and (3) the performance evaluation [183–187, 192, 210–212], responsible for creating performance metrics for a candidate and offering feedback to inform the search strategy.

In terms of practical techniques, the current body of research is predominantly divided into three areas: NAS that utilizes reinforcement learning-based NAS [213–216], NAS that is based on differentiable approaches [217, 218], and NAS that employs evolutionary strategies [219–221]. While reinforcement learning-based NAS is capable of delivering exceptional performance, it often requires thousands of GPUs in a matter of days to train even medium-size datasets. Compared to reinforcement learning-based NAS, NAS based on differentiable methods tends to be more efficient. However, it frequently results in poorly conditioned architectures due to flawed gradient-based optimization processes. Evolutionary NAS, on the other hand, is not affected by local minima and does not rely on gradient information, which has demonstrated its effectiveness in addressing intricate non-convex optimization challenges [222], even in scenarios where the mathematical form of the objective function is not known [223].

In an applied context, current research typically either integrates specific prior knowledge into the development of NAS methods or addresses unique challenges in specific visual tasks. Taking the field of binocular depth estimation as an example, some studies [201, 207, 212] suggest that the network architecture should be constructed as a stereoscopic matching network, and then the optimal internal configuration should be found.

2.4 Causal Discovery

In MODENA, understanding the relationships between different modules and factors is crucial. We need to clearly understand how various factors interact and influence the final design outcome. By uncovering causal relationships, we can ensure that design results are interpretable, providing a solid foundation for decision-making. In causal discovery algorithms, there are typically five categories of methods: constraint-based methods, score-based methods, function-based methods, methods based on the Granger causality test, and methods based on deep learning.

1. Constraint-based Methods

The constraint-based method, also referred to as the conditional independence-based method, is a key approach in causal discovery [224]. This method primarily leverages the Causal Markov Assumption and the Causal Faithfulness Assumption as constraints. It determines the existence of causal relationships among variables based on the outcomes of conditional independence tests, thereby constructing a causal structure. The process of this method can be broadly divided into three stages:

1. The first stage involves constructing an undirected complete graph among variables, implying that all variables are pairwise undirected and connected.
2. The second stage is dedicated to constructing a skeleton graph. This is achieved by utilizing the conditional independence test to ascertain whether variables are independent of each other. If they are found to be independent, the corresponding edges are eliminated.
3. The third and final stage involves determining the directions of the edges in the skeleton graph based on specific orientation rules [225]. This results in the desired causal graph.

Prominent constraint-based causal discovery methods include the IC algorithm [226], the PC algorithm [224], the FCI algorithm [227], among others.

2. Score-based Methods

The score-based method for causal discovery relies on three fundamental assumptions about the observed variables: (1) the Causal Markov Assumption; (2) the Causal Faithfulness Assumption; and (3) the Causal Sufficiency Assumption. This approach primarily identifies the causal graph with the highest score by evaluating the fit between the data and the graph [228].

More specifically, the score-based method considers all possible causal graphs obtainable through a search as the search space and uses a scoring function as the criterion to assess the quality of a causal graph. Consequently, the challenge of identifying the optimal causal graph is reframed as the task of finding the optimal scoring function within the search space [229].

This can be formally expressed as follows: Given a dataset $D = \{d_1, d_2, \ldots, d_n\}$, the aim is to identify the causal graph G^* such that

$$G^* = \arg\max_{G \in G_n} S(D, G) \tag{2.8}$$

where D denotes the observed data, G_n represents the search space, and the possible values of the variable G include all directed acyclic graphs with the observed variables as nodes [230]. S is the scoring function [231]. Therefore, the pivotal aspects of the score-based method are the selection of the scoring function and the design of the graph search algorithm.

3. Function-based Methods

The Functional Causal Model (FCM), also referred to as Structural Causal Models [225], is a framework for depicting causal relationships. As pointed out by Qiao et al. [232] and Zhang et al. [233], in any set of variables with causal relationships, one variable can always be expressed as a function of other variables, along with independent noise. Thus, the FCM characterizes a causal system by formulating a set of equations: $Y = f(X, e)$, where Y is the outcome variable, X is the causal variable, and e is the independent noise. Here, X and e are assumed to be independent. The Linear Non-Gaussian Acyclic Model [234] and the Additive Noise Model [235] are prominent models based on the FCM framework.

4. Methods Based on Granger Causality Test

In 1969, Granger introduced the concept of a testable causal relationship, which later became known as Granger Causality [236, 237]. This definition is primarily based on two principles: (1) the cause precedes the effect; and (2) the cause provides unique information that can predict the future of the effect, which is typically modeled using Vector Autoregression. Despite ongoing debates about its effectiveness, the application of Granger Causality continues to draw considerable interest [238–242].

5. Methods Based on Deep Learning

The array of causal discovery methods discussed previously primarily determines the existence of causal relationships among variables based on observed data, with the goal of generating relatively accurate causal graphs. Essentially, the problem of causal discovery is transformed into a combinatorial optimization problem, where the accuracy of the solution is of paramount importance. However, as the number of nodes in the problem increases, the complexity of the combinatorial optimization problem grows exponentially. This leads to a significant reduction in the speed of computation, severely limiting the scale of the problem that can be feasibly solved [243].

In contrast to combinatorial optimization methods, causal discovery methods based on deep learning employ continuous optimization techniques to train neural networks. They transform the combinatorial graph search problem into a continuous optimization problem [244]. The objective is to identify an adjacency matrix A that minimizes a specific scoring function $S(A)$, under the constraint that the function $h(A)$ is equal to zero, thereby ensuring the acyclicity of the graph [245].

Compared to the previously described causal discovery methods, deep learning-based methods have the capacity to learn extremely complex functions. They possess superior fitting capabilities and more robust approximation data generation mechanisms.

2.5 Comparison with Traditional and LLM-based Methods

This section compares the MODENA method with traditional trial-and-error approaches (TEAs) and large language model (LLM) methods. Through comparative analysis, it further demonstrates the advantages and potential applications of the MODENA method.

2.5.1 Contrast with Traditional Approaches

In the design of intelligent robotic systems, MODENA and TEAs present different innovative pathways. MODENA emphasizes enhancing efficiency through modularization and automation, whereas TEAs rely on experience and intuition, continuously adjusting to refine the design. These two methods differ in terms of design process, efficiency, performance, adaptability, and innovation, as detailed below:

1. Design Process

The design process in TEAs follows an iterative cycle of hypothesis, prototyping, testing, and modification. Although this approach is flexible, it is also labor-intensive and time-consuming. In complex systems such as robotic design, engineers must manually adjust parameters and components to achieve the desired performance. This manual adjustment increases the risk of overlooking optimal solutions and repeating inefficient tasks [246]. In contrast, MODENA leverages advanced computational techniques, including GP, deep learning, and EA, to automate the optimization of the design process through a unified graph modeling approach. For instance, in a robotic system, the representations include: (1) BGs for the robot's body; (2) BGs and finite state machines for continuous and discrete controllers; (3) deep learning networks for the machine vision system; and (4) GRNs or finite state machines for the behavior of swarm robots. The commonality among these elements is their graphical representation, which can be generated and evolved using GP [18]. By constructing hierarchical modular designs and automating parameter and topology optimization and integration, MODENA significantly reduces the manual labor required in TEAs and systematically explores innovative designs.

2. Efficiency and Performance

The TEAh in traditional methods is time-consuming and often requires numerous iterations to achieve satisfactory results [247]. In multivariable systems, the interactions between modules further complicate optimization. Manual adjustments might also fail to uncover innovative solutions, potentially leading to stagnation in performance improvements [248]. In contrast, MODENA employs algorithms to quickly evaluate a vast number of configurations, thereby enhancing efficiency by rapidly identifying optimal solutions [248]. For example, GP is particularly effective at exploring the design topology space and performing parallel parameter optimization, making it ideal for generating and evolving various graphical representations. By incorporating CMOEAs, researchers can utilize these tools to tackle practical challenges across different levels and domains in robotic system design automation [18]. This systematic approach not only reduces development time but also results in high-performance designs.

3. Adaptability and Innovation

TEAs heavily rely on the expertise and creativity of designers, particularly in interdisciplinary fields. While this approach can swiftly generate initial concepts, its success often hinges on personal experience and struggles to manage complexity effectively [249]. In contrast, MODENA excels in adaptability and innovation through its use of modularization and automated design optimization. It possesses strong knowledge extraction capabilities [18]. For instance, during the design optimization of various robotic systems, valuable design rules can be mined from the optimal solution sets obtained using constrained multi-objective GP methods. These rules help human designers enhance their design capabilities and expertise. As designers improve their skills, they can embed more domain-specific knowledge into the GP methods, further boosting algorithmic performance. This creates a feedback loop of human–machine interaction, where both human and algorithmic capabilities improve interactively. This dynamic process resembles an upward spiral, achieving the goal of human–machine hybrid augmented intelligence. Moreover, the MODENA approach can swiftly adapt to changes in design requirements. By combining modern EAs with generative models, it can uncover innovative designs that traditional methods might miss, reducing the reliance on human intuition and enabling breakthroughs in intelligent robotic system design [248]. In summary, MODENA represents a paradigm shift in intelligent robotic system design. It replaces the time-consuming and labor-intensive trial-and-error method with an efficient, systematic, and scalable approach that facilitates the automatic discovery of innovative designs. While the TEA remains valuable for exploratory tasks, MODENA's principles of automated optimization and modular design make it the preferred choice for developing complex, high-performance systems.

2.5.2 Differences from LLM-based Methods

Unlike GPT and other generative AI methods that rely on LLMs, MODENA offers superior explainability. In MODENA, GP algorithms generate graphical structural representations, providing excellent interpretability for the structures produced [18]. This approach is fundamentally different from the popular LLMs based on deep neural networks. By defining GP functions, the generation process of graphical structures can be thoroughly traced, allowing for the discovery of generation rules within the optimal solution set. Additionally, MODENA incorporates causal discovery algorithms to effectively mine causal relationships and rule knowledge among various factors. This facilitates the extraction of rule knowledge and creates an interactive feedback loop with the algorithm's optimization process, enhancing both human and algorithmic capabilities in a spiraling manner. This approach achieves human–machine hybrid augmented intelligence in the automation of robotic system design.

Generative AI based on LLMs uses deep learning techniques and is trained on vast amounts of language data. These models, built on the Transformer architecture, generate coherent and fluent text by learning the statistical relationships among words and phrases in the corpus [248]. Although LLMs exhibit flexibility and adaptability in handling unstructured text data, their training demands extensive data and computational resources, resulting in high development and maintenance costs. In contrast, MODENA's modular design approach emphasizes resource efficiency and system stability, making it more advantageous for intelligent robotic system design applications.

We believe that MODENA and LLMs will be combined in highly constructive ways to complement each other's strengths in the future. For instance, the 2024 Nobel Prize in Chemistry was awarded to American scientist David Baker for his contributions to "computational protein design," while the other half went to Demis Hassabis and John M. Jumper from Google DeepMind for their "contributions to protein structure prediction." MODENA is a method that involves decomposing complex systems, such as robots or potentially biological molecules like proteins, into modular components. These components are then independently optimized to achieve optimal overall performance. In the context of protein design, this involves deconstructing proteins into independent domains or modules. Computational methods and predictive models are then used to optimize and predict the behavior of these modules, ensuring they achieve the desired functions and performance.

3

Optimization Methods in MODENA

The design automation of intelligent robot systems is a critical application within the broader framework of knowledge automation. Its primary objective is to enhance design efficiency and system performance through the use of intelligent algorithms. A key scientific challenge in this process lies in the development of high-performance algorithms capable of solving complex robot design optimization problems. In real-world engineering applications, such problems are typically modeled as multi-objective optimization problems (MOPs), characterized by the following technical challenges:

1. Mixed Design Variables: Robot design optimization involves both continuous and discrete variables. Continuous variables, such as mechanical dimensions and speed parameters, coexist with discrete variables like material types and module configurations. This mix of variable types renders traditional single-point search methods, such as gradient descent or linear programming, ineffective because these approaches rely on variable continuity and differentiability, which are not guaranteed in such problems.
2. Diverse and Conflicting Optimization Objectives: Robot design requires the simultaneous optimization of multiple performance metrics, including weight, energy consumption, manufacturing cost, stability, and operational efficiency. These objectives are often in conflict. For instance, minimizing weight may weaken structural integrity or increase production costs. Achieving a balance among these competing objectives and identifying a set of high-quality non-dominated solutions Pareto front (PF) remains a fundamental challenge.

3. Complex Constraints: Robot design problems are subject to numerous physical, performance, and spatial constraints that are often nonlinear and nonconvex. Examples include mechanical strength limits, motor power output constraints, and obstacle avoidance requirements in path planning. Such constraints contribute to a complex and irregular feasible solution space, making it difficult for existing constrained multi-objective optimization algorithms to effectively identify feasible solutions. This challenge is particularly significant in high-dimensional or intricate scenarios, where the search for feasible solutions may fail entirely.

This chapter systematically explores the role of constrained multi-objective evolutionary algorithms (CMOEAs) in addressing key challenges in robot design optimization and examines their applications across various scenarios. The content is structured as follows: Section 3.1 introduces the push-pull search (PPS) mechanism, which offers an innovative approach to efficiently identify feasible regions in MOPs. Section 3.2 explains the integration of the PPS mechanism with the MOEA/D framework, highlighting its role in enhancing optimization performance. Section 3.3 examines the combination of the PPS mechanism with the NSGA-II framework and its benefits in solving complex optimization problems. Section 3.4 discusses the application of PPS in tackling expensive constrained optimization problems commonly encountered in real-world engineering scenarios. Section 3.5 proposes an evolutionary neural architecture search method to optimize neural network structures, specifically for intelligent robot vision system design. Section 3.6 details the use of genetic programming (GP) to automate the design of intelligent robot control systems, streamlining the process of control strategy development.

In summary, this chapter highlights the advantages and future potential of CMOEAs in automating intelligent robot design. It provides a detailed analysis of the PPS mechanism, its integration with mainstream evolutionary algorithm frameworks, and its application to expensive constrained optimization problems. Additionally, it explores methods such as evolutionary neural architecture search and GP. Together, these approaches offer effective solutions for addressing complex constraints and conflicting objectives in robot design optimization, advancing robot system design toward improved performance and enhanced intelligence.

3.1 General Framework of PPS

This section introduces the PPS mechanism [14], an innovative solution for efficiently identifying feasible regions in constrained multi-objective optimization problems (CMOPs). Unlike traditional methods that treat all infeasible regions equally, PPS focuses on regions that intersect with the optimal solution set.

This targeted approach reduces unnecessary exploration, enhances convergence efficiency, and strengthens global search capabilities. By accelerating the discovery of feasible solutions and improving solution quality, PPS effectively identifies the optimal solution set.

Various types of constraints create infeasible regions with different shapes in the decision space, which, in certain cases, can impact the PFs in the objective space. The impact of these infeasible regions on the PF can be classified into three distinct scenarios. Figures 3.1–3.3 illustrate the search behavior of the PPS mechanism for each scenario. These illustrations highlight how PPS efficiently adapts its search strategy under different constraint conditions [14].

Figure 3.1 shows that the unconstrained PF is the same as the constrained PF. In this case, infeasible regions can obstruct the search path toward the PF, as demonstrated in Figure 3.1a. In the push search stage of PPS, constraints are disregarded, enabling the population to explore the objective space extensively without relying on constraint-handling mechanisms. This approach allows the population to traverse infeasible regions efficiently and rapidly approach the vicinity of the unconstrained PF, as illustrated in Figure 3.1a–e. Since the constrained PF and the unconstrained PF are identical, the pull search stage does not further influence the population, as depicted in Figure 3.1f.

In this case, the unconstrained PF is entirely covered by infeasible regions, making all points on it infeasible. As shown in Figure 3.2a, the constrained Pareto optimal points lie along constraint boundaries. To address this, the PPS mechanism

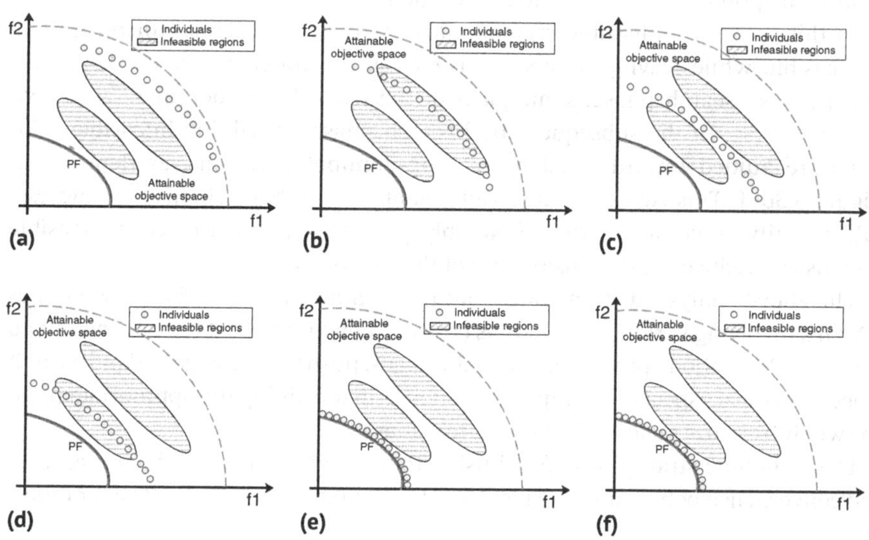

Figure 3.1 The constrained PF coincides with the unconstrained PF.

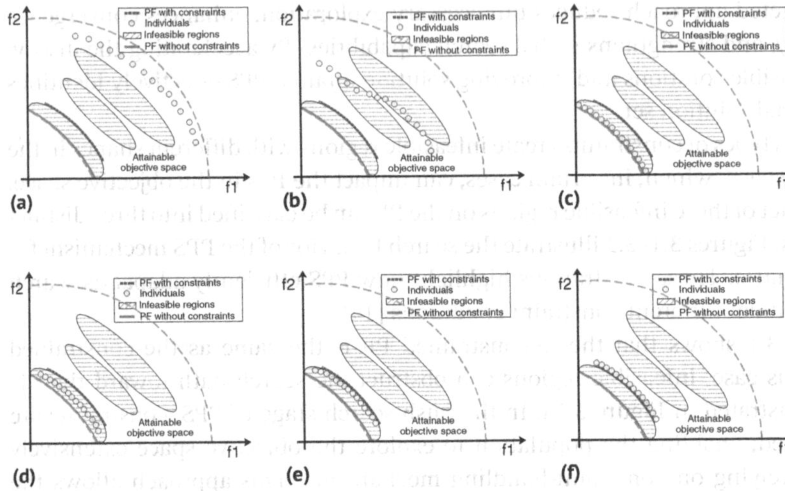

Figure 3.2 The entire unconstrained PF is fully infeasible, with the true PF located along specific constraint boundaries.

first uses the push strategy to guide the working population toward the vicinity of the unconstrained PF, as illustrated in Figure 3.2a–c. In Figure 3.2d–f, once the population reaches this region, the pull strategy is employed to refine the search, pulling the population toward the constrained PF.

In this scenario, infeasible regions make parts of the original unconstrained PF feasible while leaving others infeasible, as illustrated in Figure 3.3a. During the push search phase, some parts of the true PF are identified, as shown in Figure 3.3c. In the subsequent pull search phase, remaining infeasible solutions are guided toward feasible and non-dominated regions, as depicted in Figure 3.3d–f. This two-stage approach enables the PPS mechanism to successfully identify the complete true PF. Notably, in this case, the existence of infeasible regions can reduce the dimensionality of the PF.

The above analysis demonstrates that the PPS mechanism effectively handles CMOPs involving various interactions between constraints and the unconstrained PF. The PPS method operates in two main stages: push search stage and pull search stage. A critical aspect of its implementation is determining the appropriate point to switch from the push to the pull search phase.

The switching strategy is defined using the rate of change r_k, which measures the convergence behavior of the ideal and nadir points over the last l generations:

$$r_k \equiv \max\{rz_k, rn_k\} \leq \epsilon \tag{3.1}$$

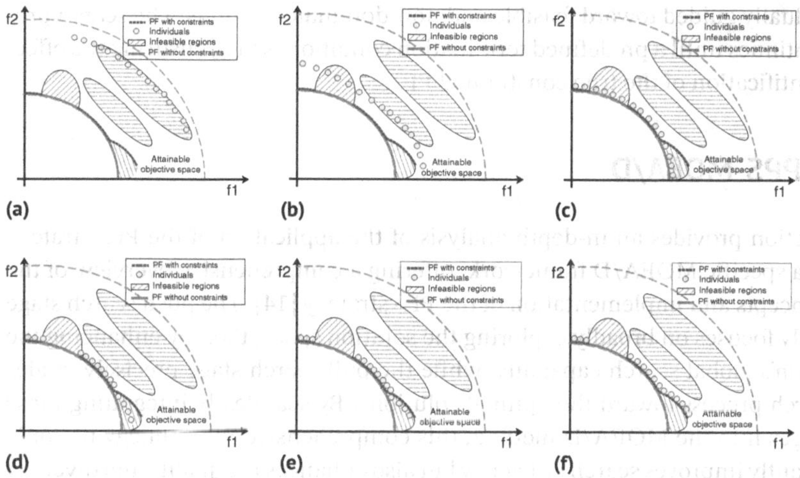

Figure 3.3 Infeasible regions make parts of the original unconstrained PF feasible while others remain inaccessible.

Here, r_k represents the maximum rate of change, with ϵ as a user-defined threshold (set to $1e-3$ in this study). The rates of change for the ideal and nadir points, rz_k and rn_k, are computed as follows:

$$rz_k = \max_{i=1,\ldots,m}\left\{\frac{|z_i^k - z_i^{k-l}|}{\max\{|z_i^{k-l}|, \Delta\}}\right\} \tag{3.2}$$

$$rn_k = \max_{i=1,\ldots,m}\left\{\frac{|n_i^k - n_i^{k-l}|}{\max\{|n_i^{k-l}|, \Delta\}}\right\} \tag{3.3}$$

In Equations (3.2) and (3.3), z^k and n^k represent the ideal and nadir points in generation k, while z^{k-l} and n^{k-l} correspond to the points in generation $k-l$. N denotes the population size, and Δ (set to $1e-6$) is a small positive number, ensuring the denominators do not become zero. Initially, r_k is set to 1.0. At each generation, r_k is updated based on the above equations. Once r_k falls below or equals the predefined threshold ϵ, the search process transitions from the push phase to the pull phase.

In summary, the PPS mechanism operates in two sequential stages: the push search and the pull search. During the push search phase, constraints are ignored to efficiently approximate the unconstrained PF. Once the convergence criterion defined by Equation (3.1) is satisfied, the algorithm transitions to the pull search phase, where constraints are fully incorporated, and infeasible solutions are gradually guided toward feasible and non-dominated regions. The search process continues until a predefined termination condition is met, ensuring the effective identification of the true constrained PF.

3.2 PPS-MOEA/D

This section provides an in-depth analysis of the application of the PPS strategy within a specific MOEA/D framework, offering a comprehensive overview of the core concepts and implementation of the PPS strategy [14]. The push search stage primarily focuses on broadly exploring the solution space, thereby enhancing the algorithm's global search capability, while the pull search stage precisely guides the search process toward the optimal solutions. By seamlessly integrating these two stages into the MOEA/D method, this comprehensive PPS strategy not only significantly improves search efficiency but also enhances the quality and diversity of the obtained solutions, offering an efficient and reliable pathway for solving optimization problems.

3.2.1 Basic Description of PPS-MOEA/D

1. Push Search Stage

During the push search stage, the MOEA/D algorithm operates in an unconstrained manner, focusing solely on identifying non-dominated solutions while disregarding any constraints. The primary objective of this phase is to thoroughly explore the solution space, providing high-quality initial solutions and greater solution diversity for subsequent optimization stages.

To address MOPs, MOEA/D simplifies the process by breaking down the MOP into multiple single-objective optimization subproblems. These subproblems are associated with decomposition functions through weight vectors λ^i, where each weight vector represents a specific trade-off among objectives. This ensures that the subproblems can individually focus on optimizing specific directions, thereby covering the entire solution space. For MOEA/D, three classic decomposition methods are commonly used: the Tchebycheff method [250], the boundary intersection method [251], and the weighted sum method [250]. This study adopts the Tchebycheff method as the decomposition strategy due to its superiority in effectively guiding the search process toward Pareto optimal solutions and its high flexibility in handling trade-offs between multiple objectives. The specific implementation details of the Tchebycheff method are as follows:

$$g^{te}(\mathbf{x}, \lambda^i, z^*) = \max_{j=1,\dots,m} \frac{1}{\lambda_j^i}(|f_j(\mathbf{x}) - z_j^*|) \tag{3.4}$$

where the weight vector λ^i is a key parameter, ensuring non-negativity and normalization. The term z^* is defined as the ideal point, which represents the minimum objective value achievable for each objective j among all solutions $\mathbf{x}^k$ in the population. This method evaluates the quality of a solution $\mathbf{x}$ by determining the maximum deviation, weighted by the reciprocal of λ_j^i, from the ideal point across all objectives.

In the push search stage, the evaluation and selection of newly generated solutions for the next generation are carried out by utilizing the g^{te} value as the primary criterion. This process is implemented following the steps described in detail in Algorithm 3.8.

Algorithm 3.1: Push Subproblem.

Function *result* = *PushSubproblems*($\mathbf{x}^j$,$\mathbf{y}^i$,z^*)
  result = *false*
  if $g^{te}(\mathbf{y}^i|\lambda^j, z^*) \leq g^{te}(\mathbf{x}^j|\lambda^j, z^*)$ **then**
    $\mathbf{x}^j = \mathbf{y}^i$
    result = *true*
  end
  return *rcsult*
end

2. Pull Search Stage

In this search stage, PPS-MOEA/D adopts a specific mechanism to gradually guide the solutions on the unconstrained PF toward feasible and non-dominated regions, thereby improving the solution quality and accelerating the convergence of the optimization process. To achieve this goal, a constraint-handling mechanism specifically designed for infeasible solutions is introduced during the pull search phase. This mechanism applies appropriate penalties to infeasible solutions, encouraging the population to move toward better feasible solutions. On this basis, this chapter further proposes an improved epsilon constraint-handling method to handle complex constraints more efficiently, with its specific mathematical formulation as follows:

$$\varepsilon(k) = \begin{cases} (1-\tau)\varepsilon(k-1), \text{ if } rf_k < \alpha \\ \varepsilon(0)\left(1 - \dfrac{k}{T_c}\right)^{cp}, \text{ if } rf_k \geq \alpha \end{cases} \tag{3.5}$$

where rf_k represents the ratio of the number of feasible solutions to the number of infeasible solutions in the k-th generation. This metric is used to dynamically evaluate the feasibility distribution of the current population. The parameter τ determines the rate at which constraint relaxation decreases when $rf_k < \alpha$, with its range being $\tau \in [0, 1]$, allowing for flexible control of the relaxation speed. The parameter α is used to balance the algorithm's search preference between the feasible and infeasible solution regions, with its range also being $\alpha \in [0, 1]$. Additionally, cp controls the rate of constraint relaxation reduction when $rf_k \geq \alpha$, providing further refinement to the dynamic behavior of the algorithm. The value of $\varepsilon(k)$ is updated according to the above rules until the generation counter k reaches the predefined control generation T_c. Notably, $\varepsilon(0)$ is initialized as the maximum overall constraint violation of the population at the end of the push search phase, ensuring sufficient flexibility in the early stage.

Compared to the ε-setting method proposed in [252], the improvement of Equation (3.5) lies in its use of an exponentially decreasing strategy to dynamically adjust the value of $\varepsilon(k)$ when $rf_k < \alpha$. This fast-decreasing feature enables the algorithm to more efficiently find feasible solutions that satisfy the constraints, significantly improving the convergence speed of feasible solutions in the early stages. Meanwhile, when $rf_k \geq \alpha$, Equation (3.5) is consistent with the ε-setting method in [252], ensuring stability and accuracy (ACC) of the algorithm in the later stages.

Algorithm 3.2: Pull Subproblem.

```
Function Result: result = PullSubproblems(x^j, y^i, ε(k), z*)
result = false
if φ(y^i) ≤ ε(k) and φ(x^j) ≤ ε(k) then
    if g^te(y^i|λ^j, z*) ≤ g^te(x^j|λ^j, z*) then
        x^j = y^i;
        result = true;
    end
else if φ(y^i) == φ(x^j) then
    if g^te(y^i|λ^j, z*) ≤ g^te(x^j|λ^j, z*) then
        x^j = y^i;
        result = true;
    end
else if φ(y^i) < φ(x^j) then
    x^j = y^i;
    result = true;
end
return result;
```

3. Process Flow of PPS-MOEA/D

The pseudocode for PPS-MOEA/D is thoroughly outlined in Algorithm 3.3. Within this framework, the process of initialization is detailed in lines 2–4 of the algorithm. In line 5, the algorithm also initializes the ideal point (z^*), which is used to record the objective values of the current optimal solution set. In line 6, the generational counter (k) tracks the evolutionary progress of the algorithm. In line 7, the parameter r_k is initialized to 1.0. This parameter controls the maximum rate of change for the ideal and nadir points, thereby limiting fluctuations in the objective space and ensuring the stability of the optimization process. At the same time, the search stage is marked as the "push phase" (*PushStage* = *true*), indicating that the algorithm is currently exploring the objective space while temporarily ignoring constraints. Additionally, the variable *maxViolation* is initialized to −1, which is used to record the maximum overall constraint violation encountered so far. This initial value signifies that no constraint violation has occurred at this stage. Subsequently, in line 8, the algorithm updates the ideal and nadir points based on the solution set of the k-th generation. These two key parameters define the optimal boundaries of the current objective space, and their specific calculation method can be referred to in Algorithm 3.4. Line 9 updates *maxViolation*, which monitors the maximum constraint violation encountered in the current stage. The details of this update process can be found in Algorithm 3.5.

Subsequently, PPS-MOEA/D repeatedly executes lines 11–45 until the predefined maximum generation $T_{\max}$ is reached. Among these, lines 12–23 are responsible for setting the value of the dynamic constraint relaxation parameter $\varepsilon(k)$ and determining the search strategy. In line 12, the algorithm calculates the maximum rate of change of the ideal and nadir points, r_k, which reflects the fluctuations in the objective space. The specific details of this calculation can be found in Algorithm 3.6. Immediately following this, in line 16, $\varepsilon(k)$ and its initial value $\varepsilon(0)$ are formally initialized, serving as the core parameters for the constraint relaxation strategy. Lines 24–42 provide a detailed description of the subproblem updating process. In line 25, S represents the neighborhood index set of the current solution $\mathbf{x}^i$, where the definition of the neighborhood plays a crucial role in the algorithm's design. Subsequently, in lines 26–27, the algorithm employs a differential evolution operator to create a new candidate solution, denoted as $\overline{\mathbf{y}}$. Subsequently, in line 28, polynomial mutation is applied to $\overline{\mathbf{y}}$ to produce a new candidate solution $\mathbf{y}^i$. To ensure the feasibility of the candidate solution, PPS-MOEA/D performs boundary correction at this stage: if any component of $\mathbf{y}^i$ falls below its specified lower boundary, it is adjusted and set equal to the lower boundary. Conversely, if any component exceeds its defined upper boundary, it is corrected and assigned the value of the upper boundary. This process effectively ensures the validity and feasibility of the solution. z^* is updated in line 29, and the mechanism for updating the ideal point is of great importance for ensuring the dynamic adjustment of the optimization objectives. The detailed process can be referred to in Algorithm 3.7. Subsequently, in line 30, *maxViolation* is updated to record the maximum constraint violation value in the current generation. Lines 32–41 describe the application of different search strategies. When *PushStage* = *true*, the algorithm adopts a push search strategy (line 35), which focuses on exploring the objective space and expanding the non-dominated solution set. In contrast, when *PushStage* = *false*, a pull search strategy is used (line 37), aiming to pull the solution set toward the feasible region defined by the constraints. In line 43, the algorithm increments the generation counter k and recalculates both the ideal and nadir points to align with the solution set of the k-th generation. The detailed method for setting these points can be found in Algorithm 3.4. Finally, in line 44, the algorithm updates the solution set *NS* based on the non-dominated sorting method, filtering out the current feasible and non-dominated solution set. This updating process ensures the diversity and superiority of the solution set, with the specific method detailed in [45].

The steps required to determine the ideal and nadir points for the k-th generation are comprehensively outlined in Algorithm 3.4, which provides a detailed breakdown of the procedure. In lines 2–4, the ideal point z^k and nadir point n^k are first initialized. This step serves as the foundation for the subsequent update process, ensuring that the ideal point accurately reflects the optimal objective values of the current generation, while the nadir point represents the worst boundaries of the objective space. Subsequently, in lines 7–9, each element of the ideal point z^k is updated iteratively. During this process, the algorithm dynamically adjusts the ideal point based on the objective values in the current solution set, ensuring that it consistently reflects the optimal objective value set of the current generation. In lines 10–12, each element of the nadir point n^k is updated iteratively as well. The update process for the nadir point is similarly based on the current solution set, defining the boundaries of the objective space by selecting the worst-case objective values.

Algorithm 3.5 provides a detailed explanation of the process for updating the maximum overall constraint violation. When a newly generated solution $\mathbf{y}^i$ is introduced, the algorithm first calculates its overall constraint violation value $\phi(\mathbf{y}^i)$. This value is then compared with the currently recorded maximum overall constraint violation, *maxViolation*. If the violation value of the new solution, $\phi(\mathbf{y}^i)$, exceeds the current *maxViolation*, it indicates that the new solution represents a more severe constraint violation. In such a case, *maxViolation* is updated to $\phi(\mathbf{y}^i)$ to ensure that the algorithm dynamically tracks the worst-case constraint violation.

Algorithm 3.6 provides a detailed explanation of the specific pseudocode steps for calculating r_k. In lines 2–3, the algorithm calculates rz_k and rn_k based on

Algorithm 3.3: PPS-MOEA/D.

Input: N: the number of subproblems. $T_{\max}$: the maximum generation. N weight vectors: $\lambda^1, \ldots, \lambda^N$. T: the size of the neighborhood. δ: the probability of selection from neighbors. n_r: the maximal number of solutions replaced by a child. T_c: the control generation for $\varepsilon(k)$.

Output: *NS*: a set of feasible non-dominated solutions

Initialization:
Decompose a CMOP into N subproblems associated with weight vectors.
Generate a population $P = \{\mathbf{x}^1, \ldots, \mathbf{x}^N\}$ randomly.
For each $i = 1, \ldots, N$, set $B(i) = \{i_1, \ldots, i_T\}$, where $\lambda^{i_1}, \ldots, \lambda^{i_T}$ are the T closest weight vectors to λ^i.
Set the ideal point $z_j^* = \min_{i=1,\ldots,N} f_j(\mathbf{x}^i)$.
Set $k = 1$.
Set $r_k = 1.0$, *PushStage* = *true*, *maxViolation* = -1.
SetIdealNadirPoints(P, k).
For each $i = 1, \ldots, N$, *UpdateMaxViolation*($\mathbf{x}^i$,*maxViolation*).
Main Loop:
while $k \le T_{\max}$ **do**
    if $k \ge T_c$ **then** Set r_k = *CalcMaxChange*(k) ;
    if $k < T_c$ **then**
        if $r_k \le \epsilon$ *and PushStage* == *true* **then**
            PushStage = *false*;
            $\varepsilon(k) = \varepsilon(0)$ = *maxViolation*;
        end
        if *PushStage* == *false* **then**
            Update $\varepsilon(k)$ according to Equation (3.5);
        end
    else
        $\varepsilon(k) = 0$;
    end
    for $i \leftarrow 1$ **to** N **do**
        if *rand* $< \delta$ **then** $S = B(i)$ **else** $S = \{1, \ldots, N\}$;
        Set $r^1 = i$ and select two indexes r^2 and r^3 from S randomly, and $r^2 \ne r^3$.
        Generate a new solution $\bar{\mathbf{y}}$ from $\mathbf{x}^{r^1}$, $\mathbf{x}^{r^2}$ and $\mathbf{x}^{r^3}$ by a DE operator.
        Perform a polynomial mutation [253] on $\bar{\mathbf{y}}$ to generate a new solution $\mathbf{y}^i$, and repair $\mathbf{y}^i$.
        UpdateIdealPoint$(\mathbf{y}^i, z^*)$;
        UpdateMaxViolation$(\mathbf{y}^i$, maxViolation);
        Set $c = 0$.
        while $c \ne n_r$ *or* $S \ne \emptyset$ **do**
            select an element j from S randomly.
            if *PushStage* == *true* **then**
                result = *PushSubproblems*$(\mathbf{x}^j, \mathbf{y}^i, z^*)$;
            else
                result = *PullSubproblems*$(\mathbf{x}^j, \mathbf{y}^i, \varepsilon(k), z^*)$;
            end
            if *result* == *true* **then** $c = c + 1$;
            $S = S \backslash \{j\}$;
        end
    end
    $k = k + 1$; *SetIdealNadirPoints*(P, k);
    NS = *NDSelect*(*NS* $\cup$ *P*, *N*);
end

Algorithm 3.4: Set Ideal and Nadir Points at k-th Generation.

```
Function SetIdealNadirPoints(P,k)
    for j ← 1 to m do
        z_j^k = 1e30; n_j^k = −1e30;
    end
    for i ← 1 to N do
        for j ← 1 to m do
            if f_j(P(i)) < z_j^k then
                z_j^k = f_j(P(i))
            end
            if f_j(P(i)) > n_j^k then
                n_j^k = f_j(P(i))
            end
        end
    end
end
```

Algorithm 3.5: Update the Maximum Constraint Violation.

```
Function UpdateMaxViolation(y^i,maxViolation)
    if φ(y^i) > maxViolation then
        maxViolation = φ(y^i) [14];
    end
end
```

Equations (3.2) and (3.3), respectively, which represent the current rates of change of the ideal and nadir points. Subsequently, in line 4, the algorithm computes r_k using Equation (3.1), which represents the maximum rate of change of the ideal and nadir points. By calculating r_k, the algorithm can better understand the dynamic trends of changes in the objective space.

Algorithm 3.7 provides a detailed explanation of the specific pseudocode steps for updating the ideal point z^*. The ideal point z^* in multi-objective optimization represents the optimal values across all objectives, making its updating process critically important. When a newly generated solution $\mathbf{y}^i$ is introduced, the algorithm sequentially checks the value of each objective function $f_j(\mathbf{y}^i)$. If the j-th objective value $f_j(\mathbf{y}^i)$ of $\mathbf{y}^i$ is smaller than the current value of the ideal point z_j^*, it indicates that a better solution has been found for this objective dimension. In such a case, the algorithm updates z_j^* to $f_j(\mathbf{y}^i)$.

Algorithm 3.6: Calculate the Maximum Rate of Change of Ideal and Nadir Points r_k.

```
Function r_k = CalcMaxChange(k)
    Calculate rz_k according to Equation (3.2).
    Calculate rn_k according to Equation (3.3).
    r_k = max{rz_k, rn_k}, according to Equation (3.1).
end
```

Algorithm 3.7: Update the Ideal Point.

```
Function UpdateIdealPoint(y^i, z*)
    for j ← 1 to m do
        if f_j(y^i) < z*_j then
            z*_j = f_j(y^i)
        end
    end
end
```

3.2.2 Experimental Study

3.2.2.1 Experimental Settings

To comprehensively evaluate the effectiveness of PPS-MOEA/D, some classical algorithms are selected for comparison experiments, including MOEA/D-Epsilon [254], NSGA-II-CDP [45], C-MOEA/D [255], MOEA/D-SR [256], MOEA/D-IEpsilon [257], and MOEA/D-CDP [256]. These algorithms are tested on the LIR-CMOP1-14 problem set [257], which is characterized by having a large number of infeasible regions. This design allows for a thorough examination of the algorithms' performance in solving complex constrained optimization problems. The specific experimental parameter settings are provided in [14].

In addition, to objectively measure the optimization capability and effectiveness of PPS-MOEA/D, two widely used and important evaluation metrics are adopted: IGD [78] and hypervolume (HV) [77]. IGD evaluates the proximity of the solution set generated by the algorithm to the true PF, where smaller values indicate that the solution set is closer to the ideal PF. On the other hand, HV measures the size of the objective space covered by the solution set, with larger values indicating better diversity and coverage of the solution set. These two metrics provide a comprehensive analysis of the algorithms' performance in terms of both ACC and diversity of solutions.

3.2.2.2 Experimental Discussion

Table 3.1 provides the statistical results of the inverted generational distance (IGD) values obtained by PPS-MOEA/D and six other CMOEAs on the LIR-CMOP1-14 problems after 30 independent runs. Additionally, Table 3.1 includes the results of the Wilcoxon rank sum test conducted at a significance level of 0.05 to compare PPS-MOEA/D with the six other CMOEAs. In the table, the symbol † indicates that the corresponding algorithm performs significantly worse than PPS-MOEA/D, while ‡ indicates that it performs significantly better. The Wilcoxon test results clearly show that, in terms of the IGD metric, PPS-MOEA/D significantly outperforms six other CMOEAs across all 14 test problems. This demonstrates that PPS-MOEA/D exhibits superior performance and stronger robustness when addressing complex CMOPs.

Table 3.2 presents the statistical results of the HV values obtained by seven algorithms on the LIR-CMOP1-14 problems after 30 independent runs. Additionally, Table 3.2 includes the Wilcoxon rank sum test results conducted at a significance level of 0.05 to compare PPS-MOEA/D with the other six CMOEAs. In the table, the symbol † indicates that the corresponding algorithm performs significantly worse than PPS-MOEA/D, while ‡ indicates that it performs significantly better. The statistical results clearly show that, in terms of the HV metric, PPS-MOEA/D significantly outperforms C-MOEA/D, MOEA/D-SR, NSGA-II-CDP, MOEA/D-CDP, and MOEA/D-Epsilon across all test instances. This demonstrates that PPS-MOEA/D exhibits excellent convergence and diversity when handling complex CMOPs. For LIR-CMOP4, LIR-CMOP9-10, and LIR-CMOP13, there are no statistically significant differences between PPS-MOEA/D and MOEA/D-IEpsilon, indicating that both algorithms perform comparably on these problems and are capable of generating high-quality solution sets. However, for the remaining 10 test instances, PPS-MOEA/D demonstrates significant advantages in the HV metric and performs significantly better than MOEA/D-IEpsilon.

To comprehensively analyze the performance advantages of PPS-MOEA/D, the populations corresponding to the median HV values obtained from 30 independent runs of each tested CMOEAs on LIR-CMOP6, LIR-CMOP7, and LIR-CMOP11 are plotted in Figures 3.5–3.7, respectively. These distribution plots visually illustrate the solution sets of different algorithms, providing a more intuitive basis for performance comparison. The selection is based on their representation of three typical patterns of infeasible regions influencing the PFs. Specifically, LIR-CMOP6 represents the first situation, where the constrained PF completely overlaps with the unconstrained PF. In this case, the algorithm must exhibit strong exploration capabilities to efficiently approximate this common front. LIR-CMOP7 reflects the second typical situation, where the entire unconstrained PF is infeasible, and the true PF lies on specific constraint boundaries. This requires the algorithm to accurately identify and capture feasible regions. LIR-CMOP11

Table 3.1 IGD results of PPS-MOEA/D and six other CMOEAs on LIR-CMOP1-14.

Test instance		PPS	IEpsilon	Epsilon	CDP	SR	C-MOEA/D	NSGA-II-CDP
LIRCMOP1	mean	**6.41E−03**	7.97E−03	5.74E−02 †	1.11E−01†	1.31E−02†	1.26E−01†	3.23E−01†
	std	1.94E−03	3.55E−03	2.89E−02	5.04E−02	1.66E−02	7.03E−02	7.33E−02
LIRCMOP2	mean	**4.67E−03**	5.23E−03†	5.39E−02†	1.43E−01†	9.63E−03†	1.40E−01†	3.03E−01†
	std	7.84E−04	1.01E−03	2.13E−02	5.55E−02	7.23E−03	5.44E−02	7.24E−02
LIRCMOP3	mean	**8.55E−03**	1.13E−02†	8.81E−02†	2.61E−01†	1.78E−01†	2.80E−01†	4.08E−01†
	std	5.18E−03	6.42E−03	4.36E−02	4.33E−02	7.20E−02	4.21E−02	1.15E−01
LIRCMOP4	mean	**4.68E−03**	4.85E−03	6.51E−02†	2.53E−01†	1.95E−01†	2.59E−01†	3.85E−01†
	std	1.12E−03	2.05E−03	3.01E−02	4.34E−02	6.40E−02	3.51E−02	1.35E−01
LIRCMOP5	mean	**1.84E−03**	2.13E−03†	1.15E+00†	1.05E+00†	1.04E+00†	1.10E+00†	5.53E−01†
	std	9.26E−05	3.79E−04	1.98E−01	3.63E−01	3.66E−01	2.99E−01	6.88E−01
LIRCMOP6	mean	**2.49E−03**	2.33E−01†	1.27E+00†	1.09E+00†	9.43E−01†	1.31E+00†	5.74E−01†
	std	3.40E−04	5.06E−01	2.95E−01	5.20E−01	5.90E−01	2.08E−01	4.21E−01
LIRCMOP7	mean	**2.80E−03**	3.73E−02†	1.51E+00†	1.46E+00†	1.08E+00†	1.56E+00†	2.38E−01†
	std	9.85E−05	5.41E−02	5.09E−01	5.58E−01	7.58E−01	4.24E−01	4.06E−01
LIRCMOP8	mean	**2.78E−03**	2.75E−02†	1.62E+00†	1.38E+00†	1.01E+00†	1.58E+00†	6.02E−01†
	std	7.56E−05	5.92E−02	3.05E−01	6.15E−01	7.24E−01	3.71E−01	7.39E−01
LIRCMOP9	mean	9.94E−02	**4.98E−03**	4.90E−01†	4.81E−01†	4.85E−01†	4.81E−01†	6.44E−01†
	std	1.52E−01	1.37E−02	4.22E−02	5.24E−02	4.78E−02	5.24E−02	1.60E−02
LIRCMOP10	mean	**2.11E−03**	2.11E−03	2.13E−01†	2.16E−01†	1.92E−01†	2.13E−01†	5.97E−01†
	std	7.75E−05	7.11E−05	5.32E−02	6.81E−02	6.81E−02	4.63E−02	3.20E−02

(Continued)

Table 3.1 (Continued)

Test instance		PPS	IEpsilon	Epsilon	CDP	SR	C-MOEA/D	NSGA-II-CDP
LIRCMOP11	mean	**2.83E−03**	5.81E−02†	3.47E−01†	3.42E−01†	3.16E−01†	3.81E−01†	4.87E−01†
	std	1.36E−03	5.79E−02	9.28E−02	9.22E−02	7.49E−02	8.95E−02	1.05E−02
LIRCMOP12	mean	**2.70E−02**	3.36E−02†	2.52E−01†	2.69E−01†	2.06E−01†	2.50E−01†	5.80E−01†
	std	5.00E−02	5.18E−02	8.98E−02	9.06E−02	5.61E−02	9.63E−02	1.17E−01
LIRCMOP13	mean	**6.46E−02**	6.46E−02	1.20E+00†	1.21E+00†	8.86E−01†	1.18E+00†	1.39E+01†
	std	2.18E−03	1.64E−03	3.06E−01	3.17E−01	5.76E−01	3.78E−01	2.26E+00
LIRCMOP14	mean	**6.42E−02**	6.54E−02†	1.02E+00†	1.11E+00†	1.03E+00†	1.25E+00†	1.36E+01†
	std	1.69E−03	2.04E−03	4.86E−01	3.98E−01	4.70E−01	5.30E−02	2.17E+00
Wilcoxon-Test (S-D-I)		–	9-5-0	14-0-0	14-0-0	14-0-0	14-0-0	14-0-0

PPS, IEpsilon, Epsilon, CDP, and SR are abbreviations for PPS-MOEA/D, MOEA/D-IEpsilon, MOEA/D-Epsilon, MOEA/D-CDP, and MOEA/D-SR, respectively. "S-D-I" represents the performance relationship: S means PPS is significantly better, D means no significant difference, and I means PPS is significantly worse.

† Indicates that the corresponding algorithm performs significantly worse than PPS-MOEA/D.

Table 3.2 HV results of PPS-MOEA/D and six other CMOEAs on LIR-CMOP1-14.

Test instance		PPS	IEpsilon	Epsilon	CDP	SR	C-MOEA/D	NSGA-II-CDP
LIRCMOP1	mean	**1.02E+00**	1.01E+00†	9.59E−01†	7.54E−01†	9.96E−01†	7.41E−01†	5.16E−01†
	std	1.58E−03	2.43E−03	3.28E−02	8.95E−02	2.91E−02	1.22E−01	5.57E−02
LIRCMOP2	mean	**1.35E+00**	1.35E+00†	1.28E+00†	1.06E+00†	1.34E+00†	1.07E+00†	8.24E−01†
	std	1.01E−03	1.32E−03	2.88E−02	1.08E−01	1.47E−02	9.10E−02	1.15E−01
LIRCMOP3	mean	**8.70E−01**	8.68E−01†	7.98E−01†	4.86E−01†	5.91E−01†	4.71E−01†	4.08E−01†
	std	2.65E−03	3.92E−03	3.93E−02	4.31E−02	1.07E−01	4.09E−02	2.88E−02
LIRCMOP4	mean	**1.09E+00**	1.09E+00	1.02E+00	7.35E−01†	8.15E−01†	7.31E−01†	6.17E−01†
	std	2.47E−03	2.46E−03	4.19E−02	5.44E−02	8.70E−02	5.16E−02	1.06E−01
LIRCMOP5	mean	**1.46E+00**	1.46E+00†	4.30E−02†	1.63E−01†	1.82E−01†	9.72E−02†	9.39E−01†
	std	2.92E−04	1.33E−03	2.35E−01	4.43E−01	4.39E−01	3.70E−01	3.21E−01
LIRCMOP6	mean	**1.13E+00**	9.26E−01†	5.40E−02†	1.88E−01†	3.02E−01†	2.33E−02†	4.13E−01†
	std	1.77E−04	4.23E−01	2.21E−01	3.87E−01	4.62E−01	1.28E−01	1.89E−01
LIRCMOP7	mean	**3.02E+00**	2.86E+00†	3.03E−01†	3.74E−01†	9.88E−01†	2.04E−01†	2.40E+00†
	std	2.66E−03	1.96E−01	9.07E−01	9.58E−01	1.27E+00	7.52E−01	6.52E−01
LIRCMOP8	mean	**3.02E+00**	2.94E+00†	1.06E−01†	5.17E−01†	1.10E+00†	1.66E−01†	1.90E+00†
	std	1.14E−03	1.86E−01	5.49E−01	1.05E+00	1.20E+00	6.11E−01	7.56E−01
LIRCMOP9	mean	3.57E+00	**3.71E+00**	2.74E+00†	2.77E+00†	2.75E+00†	2.77E+00†	2.06E+00†
	std	2.24E−01	1.88E−02	1.48E−01	1.84E−01	1.64E−01	1.84E−01	1.08E−02
LIRCMOP10	mean	**3.24E+00**	3.24E+00	2.89E+00†	2.88E+00†	2.93E+00†	2.89E+00†	2.04E+00†
	std	3.08E−04	2.48E−04	1.02E−01	1.36E−01	1.35E−01	9.77E−02	4.45E−02

(Continued)

Table 3.2 (Continued)

Test instance		PPS	IEpsilon	Epsilon	CDP	SR	C-MOEA/D	NSGA-II-CDP
LIRCMOP11	mean	**4.39E+00**	4.23E+00†	3.34E+00†	3.35E+00†	3.38E+00†	3.24E+00†	3.11E+00†
	std	2.22E−04	1.84E−01	2.57E−01	2.57E−01	2.90E−01	2.55E−01	1.54E−02
LIRCMOP12	mean	**5.61E+00**	5.59E+00†	4.88E+00†	4.83E+00†	5.03E+00†	4.89E+00†	3.28E+00†
	std	1.53E−01	1.58E−01	3.17E−01	3.28E−01	1.75E−01	3.45E−01	3.61E−01
LIRCMOP13	mean	**5.71E+00**	5.71E+00	4.55E−01†	4.63E−01†	1.89E+00†	6.29E−01†	0.00E+00†
	std	1.27E−02	1.30E−02	1.30E+00	1.42E+00	2.57E+00	1.71E+00	0.00E+00
LIRCMOP14	mean	**6.19E+00**	6.18E+00†	1.33E+00†	8.81E−01†	1.27E+00†	1.80E−01†	0.00E+00†
	std	1.31E−02	1.09E−02	2.45E+00	1.97E+00	2.29E+00	2.60E−01	0.00E+00
Wilcoxon-Test (S-D-I)		–	10-4-0	14-0-0	14-0-0	14-0-0	14-0-0	14-0-0

PPS, IEpsilon, Epsilon, CDP, and SR are abbreviations for PPS-MOEA/D, MOEA/D-IEpsilon, MOEA/D-Epsilon, MOEA/D-CDP, and MOEA/D-SR, respectively. "S-D-I" represents the performance relationship: S means PPS is significantly better, D means no significant difference, and I means PPS is significantly worse.

† Indicates that the corresponding algorithm performs significantly worse than PPS-MOEA/D.

represents the third and most complex situation, where infeasible regions almost entirely overlap with the PF, leaving only narrow feasible regions in the objective space. This scenario poses significant challenges for the algorithm's convergence and local search capabilities.

For the LIR-CMOP6 problem, there are two large infeasible regions in front of the PF, and the unconstrained PF completely overlaps with the constrained PF, as shown in Figure 3.4a. This complex layout of infeasible regions poses significant challenges to the convergence of optimization algorithms. As illustrated in Figure 3.5a, b, PPS-MOEA/D and MOEA/D-IEpsilon can successfully cross these large infeasible regions and approach the target PF. In sharp contrast, other five algorithms fail to effectively overcome the obstacles posed by the infeasible regions, resulting in their populations being trapped at the boundaries of the infeasible regions, as shown in Figure 3.5c–g. The fundamental reason for this difference lies in the design mechanisms of the algorithms. PPS-MOEA/D employs a push search stage, during which the constraints are ignored, allowing it to freely explore infeasible regions and overcome these obstacles. This mechanism is particularly effective in the early stages of optimization, as it enables the algorithm to focus on finding regions close to the optimal solutions without being overly restricted by constraints. In addition, MOEA/D-IEpsilon adopts an improved epsilon constraint-handling method, dynamically adjusting the epsilon level to allow a certain degree of infeasible solutions to enter the population. This strategy effectively alleviates the obstruction caused by the infeasible regions to the search path of the population, enabling the algorithm to successfully cross large infeasible regions and ultimately achieve convergence. In contrast, the other five CMOEAs lack mechanisms in their algorithm design to handle large-scale infeasible regions (as shown in Figure 3.4a). These algorithms fail to effectively overcome the obstacles posed by the infeasible regions, resulting in their populations being trapped at the boundaries of the infeasible regions during the process of approaching the PF. As a consequence, their solution sets cannot further

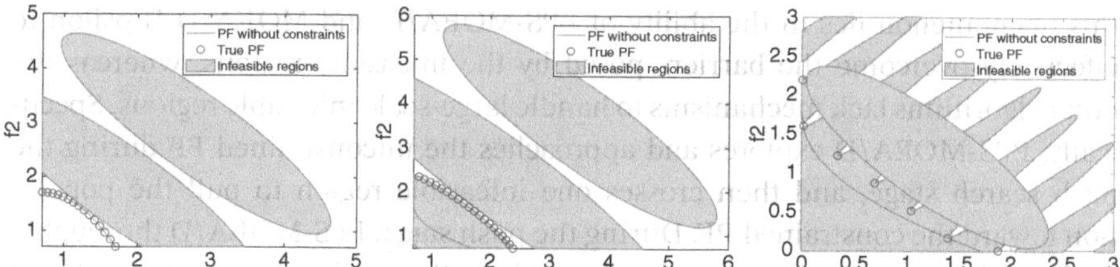

(a) LIR-CMOP6 (b) LIR-CMOP7 (c) LIR-CMOP11

Figure 3.4 Illustrations of the feasible and infeasible regions of LIR-CMOP6, LIR-CMOP7, and LIR-CMOP11.

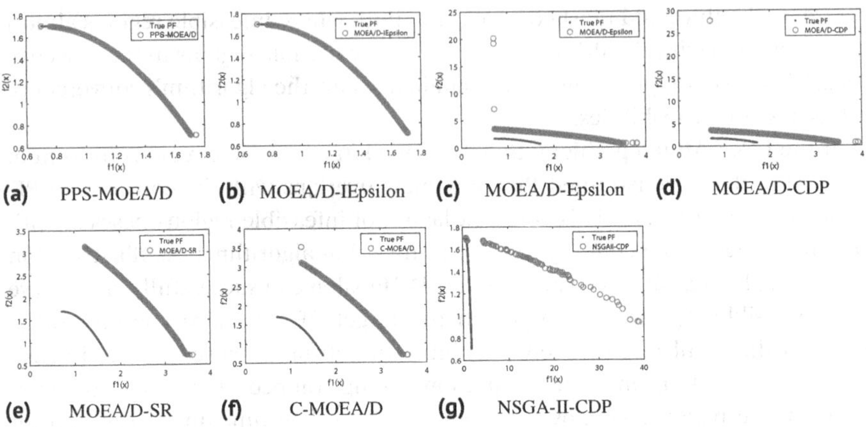

Figure 3.5 The non-dominated solutions achieved by each algorithm on LIR-CMOP6 with the median HV values.

converge toward the target front. This phenomenon is clearly demonstrated in Figure 3.5c–g, which vividly shows the stagnation of populations at the boundaries of the infeasible regions. In conclusion, PPS-MOEA/D, with its unique mechanism, successfully overcomes the limitations posed by the infeasible regions in LIR-CMOP6, demonstrating strong search capabilities and convergence. In contrast, other algorithms, lacking similar strategies, have their populations restricted by the infeasible regions, failing to fully realize their optimization potential.

For the LIR-CMOP7 problem, there are three large infeasible regions in front of the PF, and the unconstrained PF is completely covered by the infeasible regions, making it entirely infeasible, as shown in Figure 3.4b. This complex layout of infeasible regions poses significant challenges to the convergence of optimization algorithms. As can be clearly observed in Figure 3.6a, b, PPS-MOEA/D and MOEA/D-IEpsilon successfully converge to the true PF, while the other five CMOEAs fail to converge to the true PF, as shown in Figure 3.6c–g. The root cause of this phenomenon lies in the ability of PPS-MOEA/D and MOEA/D-IEpsilon to effectively overcome the barriers posed by the infeasible regions, whereas the other algorithms lack mechanisms to handle large-scale infeasible regions. Specifically, PPS-MOEA/D explores and approaches the unconstrained PF during the push search stage, and then crosses one infeasible region to pull the population toward the constrained PF. During the push stage, PPS-MOEA/D thoroughly explores the constraint landscape and calculates the maximum overall constraint violation. This information is then utilized in the pull search stage to guide the proper setting of the ε parameter in the constraint-handling method. This combined push-pull strategy enables PPS-MOEA/D to reliably overcome infeasible regions and efficiently converge to the true PF. In contrast, MOEA/D-IEpsilon,

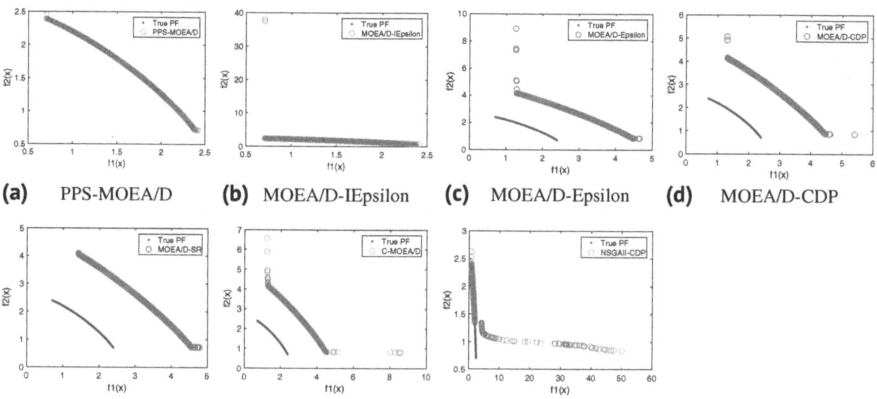

(a) PPS-MOEA/D **(b)** MOEA/D-IEpsilon **(c)** MOEA/D-Epsilon **(d)** MOEA/D-CDP

(e) MOEA/D-SR **(f)** C-MOEA/D **(g)** NSGA-II-CDP

Figure 3.6 The non-dominated solutions achieved by each algorithm on LIR-CMOP7 with the median HV values.

although occasionally capable of crossing the two large infeasible regions, often fails to achieve full convergence to the true PF, with some individuals in the population unable to reach the target, as shown in Figure 3.6b. This limitation arises because MOEA/D-IEpsilon lacks a mechanism to pre-explore the constraint landscape, preventing it from effectively using constraint information to dynamically adjust the ε parameter, as PPS-MOEA/D does. For the other five algorithms, their algorithm designs lack specific mechanisms to cross large-scale infeasible regions. As a result, the populations of these algorithms are restricted by the infeasible regions during the convergence process and fail to effectively approach the constrained PF. Ultimately, their populations are trapped at the boundaries of the infeasible regions, a phenomenon clearly demonstrated in Figure 3.6c–g. In conclusion, PPS-MOEA/D, with its unique push-pull mechanism, reliably identifies the true PF of LIR-CMOP7, exhibiting strong adaptability and convergence capabilities. Although MOEA/D-IEpsilon occasionally finds the true PF, its overall performance is limited. The other five CMOEAs, constrained by the barriers of the infeasible regions, fail to achieve effective convergence. This further validates the importance of mechanisms designed to overcome large-scale infeasible regions.

For the LIR-CMOP11 problem, the infeasible regions make part of the original unconstrained PF feasible, as shown in Figure 3.4c. The PF of LIR-CMOP11 is discontinuous and consists of seven Pareto optimal solutions, two of which are located on the unconstrained PF, while the other five are distributed in specific regions outside the infeasible areas. PPS-MOEA/D successfully identifies all seven Pareto optimal solutions, as shown in Figure 3.7a. In contrast, the other six CMOEAs fail to find all seven Pareto optimal solutions, as illustrated in

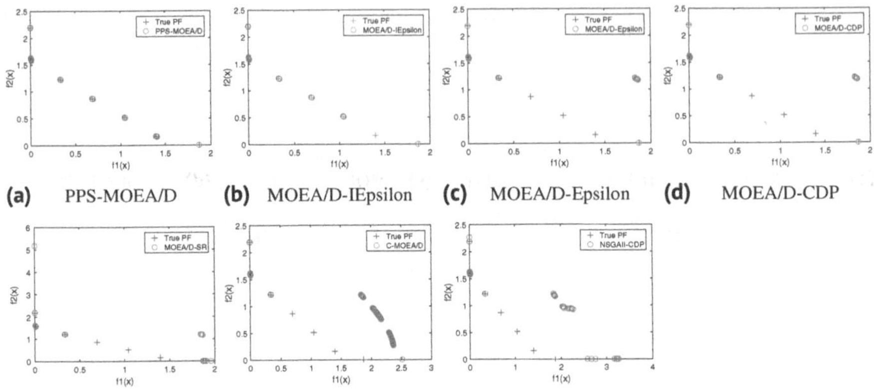

(a) PPS-MOEA/D **(b)** MOEA/D-IEpsilon **(c)** MOEA/D-Epsilon **(d)** MOEA/D-CDP

(e) MOEA/D-SR **(f)** C-MOEA/D **(g)** NSGA-II-CDP

Figure 3.7 The non-dominated solutions achieved by each algorithm on LIR-CMOP11 with the median HV values.

Figure 3.7b–g. The main reason for this phenomenon lies in the ability of PPS-MOEA/D to effectively ignore the barriers posed by the infeasible regions during the push search stage, allowing its population to efficiently approach the unconstrained PF. During the push stage, the entire unconstrained PF of LIR-CMOP11 can be fully explored. Since two of the Pareto optimal solutions are located on the unconstrained PF, PPS-MOEA/D quickly identifies these two solutions during the push stage. Furthermore, PPS-MOEA/D thoroughly explores the constraint landscape in the push stage, providing critical information for further searching in the pull search stage. In the pull stage, the population of PPS-MOEA/D needs to cross only one infeasible region to locate the remaining five Pareto optimal solutions. This strategy significantly enhances the algorithm's efficiency and convergence performance. In contrast, the other six CMOEAs lack a similar push-pull combined mechanism in their algorithm design, requiring their populations to traverse multiple infeasible and overlapping regions to find the Pareto optimal solutions. Additionally, these algorithms fail to sufficiently explore the constraint landscape during the search process, making it difficult to properly set constraint-handling parameters. This deficiency directly limits their ability to search effectively and to identify Pareto optimal solutions efficiently. Therefore, the proposed PPS-MOEA/D, with its unique push-pull combined mechanism, reliably finds all seven discrete Pareto optimal solutions, demonstrating strong adaptability and convergence performance. The other six algorithms, lacking similar mechanisms, can only occasionally locate some Pareto optimal solutions, failing to fully realize their optimization potential. This further validates the effectiveness and superiority of PPS-MOEA/D in addressing complex CMOPs.

3.2.3 Conclusion

The proposed PPS-MOEA/D skillfully integrates the strategies of the push stage and the pull stage, effectively addressing the challenge of population stagnation caused by infeasible regions through a phased optimization mechanism. During the push stage, PPS-MOEA/D ignores the constraint conditions, allowing the population to freely explore the objective space and quickly approach the unconstrained PF. The primary goal of this stage is to establish a solid population foundation for the subsequent pull stage while gaining deeper insights into the constraint distribution of the problem by exploring infeasible regions. The push stage not only enables the algorithm to efficiently traverse infeasible regions but also collects critical information, which serves as valuable input for parameter tuning during the pull stage. In the pull stage, PPS-MOEA/D makes full use of the constraint information gathered in the push stage to dynamically adjust the parameters of the constraint-handling mechanism, guiding the population more precisely and pulling the solution set toward the true constrained PF. Through this approach, PPS-MOEA/D effectively handles complex constraints, particularly in scenarios with large-scale infeasible regions, where the population can quickly converge to the target area. Experimental results demonstrate that on the LIR-CMOP benchmark problems, PPS-MOEA/D exhibits significant advantages and performs markedly better than the other six CMOEAs. By combining the push and pull strategies, the PPS framework introduces a novel approach to optimization algorithms, showcasing outstanding performance in addressing MOPs with complex constraint distributions.

3.3 PPS-M2M

When addressing CMOPs, a critical challenge for multi-objective evolutionary algorithms (MOEAs) is maintaining a balance between convergence and diversity. Many CMOEAs struggle with this balance, often causing the population to cluster around certain regions of the PFs, leading to poor diversity and incomplete convergence. To overcome this issue, this section introduces a new method, referred to as PPS-M2M [258]. In this approach, the CMOP is decomposed into some simpler CMOPs, each assigned to a subpopulation. These subpopulations work collaboratively to solve their respective problems. Specifically, each subpopulation applies a two-stage strategy for constraint handling: constraints are ignored during the push stage to allow traversal across infeasible regions, and they are fully considered in the pull stage to guide solutions toward feasible and non-dominated regions. This cooperative strategy enables the working populations to efficiently explore the search space while balancing convergence and diversity.

3.3.1 Basic Description of M2M Strategy

The core idea behind the multi-objective to multiobjective (M2M) decomposition strategy is to distribute the optimization workload, enabling the algorithm to search the solution space more effectively. By breaking the problem into parts, each subpopulation focuses on a specific segment of the objective space, which enhances convergence while maintaining solution diversity. Specially, to illustrate the decomposition of the objective space, we first divide it into K distinct and non-overlapping subregions. This is achieved by systematically and uniformly distributing K directional vectors within the first octant of a unit sphere. These vectors are carefully arranged to ensure even coverage across the region, where each individual vector v^k acts as the central reference point for the corresponding k-th subregion. As a result, the objective space is partitioned into K mutually exclusive subregions. In this chapter, these subregions are denoted as $\Omega_1, \ldots, \Omega_k$. For clarity, consider the specific case where there are $m = 2$ objectives, and the function space is divided into $K = 6$ regions. Figure 3.8 provides a visual representation, showing six directional vectors ($v^1, \ldots, v^6$) that are uniformly distributed across the first octant [258].

3.3.2 PPS Search Strategy

In PPS-MOEA/D, the integration of PPS with MOEA/D is described in detail. In the context of CMOEAs, constraint-handling methods based on dominance relations represent another major approach. Therefore, in this section, we provide a detailed discussion on how PPS can be integrated with NSGA-II.

In the first search stage of PPS, the subpopulation explores the objective space using the NSGA-II algorithm without incorporating any constraint-handling mechanism. At this stage, the primary objective is to comprehensively explore the

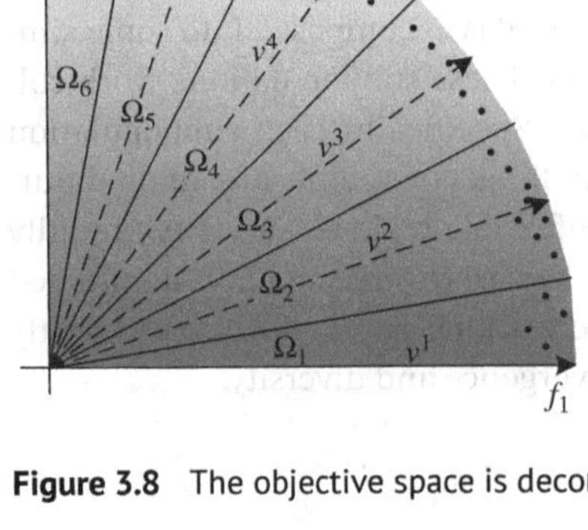

Figure 3.8 The objective space is decomposed by M2M strategy.

solution space without being limited by the complexities or restrictions introduced by constraints. By disregarding constraints temporarily, the algorithm can focus on identifying a wide range of non-dominated solutions, ensuring better coverage of the search space and enhancing solution diversity. During this phase, high-quality offspring individuals are carefully selected using two key criteria: non-dominated sorting and crowding distance. These two mechanisms help maintain both solution quality and diversity. The selection process for the next generation involves comparing two individuals, *i* and *j*, using the following criteria:

1. If *j* cannot dominate *i*, then *i* is preferred.
2. If *i* and *j* are mutually non-dominated, the one with the greater crowding distance is chosen.

To conclude, if two individuals fall into distinct non-domination ranks, priority is given to the one with the lower rank, as it represents a more optimal solution. On the other hand, if the individuals share the same rank, the decision is based on their crowding distance: the individual with a larger crowding distance, which signifies greater diversity and better distribution within the population, is selected to be part of the next generation.

During the push search stage, the proposed algorithm focuses on optimizing an unconstrained optimization problem while temporarily ignoring any constraints. The step-by-step procedure for this stage is outlined in Algorithm 3.8. Initially, in line 2, the subpopulation P_k undergoes non-dominated sorting. Following this, from lines 3–6, solutions are incrementally selected and added to $P_k^{'}$ until the total number of solutions in $P_k^{'}$ exceeds S_k. Next, in lines 7–9, if there are still unoccupied slots in $P_k^{'}$ (specifically, when the number of selected solutions $|P_k^{'}|$ is less than the required size S_k), the remaining slots are filled by selecting additional solutions from F_i. In the final step, as described in line 10, the existing subpopulation P_k undergoes an update process. Specifically, it is replaced with the newly constructed subpopulation $P_k^{'}$, which has been carefully generated and refined in the preceding steps.

In the second search stage of the PPS, the optimization process transitions from unconstrained exploration to a more focused search that incorporates constraints. This stage is essential for refining the previously discovered solutions, ensuring that they not only dominate other solutions but also meet the required feasibility criteria. The specific steps involved in this phase are outlined in Algorithm 3.9. A key challenge during this stage is to maintain a proper balance between exploring both feasible and infeasible regions, as over-concentration in one area may result in the population becoming stuck in local optima. To mitigate this issue, an enhanced epsilon constraint-handling method, introduced by [257], is employed. This advanced mechanism improves the algorithm's ability to effectively navigate the complex constraint landscape, ensuring that the search process remains diverse and capable of finding global optima while still adhering to the problem's constraints.

Algorithm 3.8: Push Subproblems.

```
Function PushSubproblems(P_k,S_k)
    F = nondominated-sort(P_k), F = (F_1, F_2, ...);
    P'_k = ∅ and i = 1;
    While(|P'_k| + |F_i| ≤ S_k)
        P'_k = P'_k ∪ F_i;
        i = i + 1;
    calculate crowding-distance in F_i;
    sort solutions in F_i by crowding-distance in a descending order
    P'_k = P'_k ∪ F_i[1 : (S_k − |P'_k|)];
    P_k = P'_k;
end
```

Algorithm 3.9: Pull Subproblem.

```
Function result = PullSubproblems(M_k,gen,T_max)
    //T_max: the maximum generation.
    result = false;
    if gen ≤ 0.9T_max then
        for k ← 1 to K do
            An improved epsilon constraint-handling mechanism is used
             to search for non-dominated and feasible solutions in M_k;
        end
    else
        Merge M_1, M_2, ..., M_K into a single population M_s.
        An improved epsilon constraint-handling mechanism is employed
         on M_s to search non-dominated and feasible solutions;
    end
    return result;
end
```

3.3.3 Combination of PPS with M2M

In Algorithm 3.10, PPS-M2M begins with initialization steps in lines 1–3. Specifically, in line 2, the population is partitioned into K subpopulations, each containing S individuals. The maximum change rate (r_k) for the ideal and nadir points is set to 1.0 in line 3. Line 3 also sets the search stage flag to "push" (*PushStage=true*). The main iterative loop runs from lines 4 to 38. A series of new individuals are

Algorithm 3.10: PPS-M2M

Input:
K: the number of subproblems;
K unit direction vectors: $v^1,...,v^K$;
S: the size of each subpopulation;
Q: a set of individual solutions and their F-values;
T_c: the control generation for $\varepsilon(k)$;
T_{max}: the maximum generation.

Output: a set of non-dominated and feasible solutions.

Initialization:
Decompose a population into K sub-populations ($M_1, ..., M_K$), each sub-population contains S individuals according to *AllocationSubPop*(Q, S, K);
Set $r_k = 1.0$, $PushStage = true$;
while $gen \leq T_{max}$ **do**
  for $k \leftarrow 1$ **to** K **do**
    foreach $x \in M_k$ **do**
      Randomly choose y from M_k;
      Apply genetic operators on x and y to generate a new solution z;
      $R := R \cup \{z\}$;
    end
    $Q := R \cup (\cup_{k=1}^{K} M_k)$;
    Use Q to set $M_1, ..., M_K$;
  end
  if $gen >= l$ **then**
    Calculate r_k; ;
  if $gen < T_c$ **then**
    if $r_k \leq \epsilon$ *and* $PushStage == true$ **then**
      $PushStage = false$;
      $\varepsilon(gen) = \varepsilon(0) = maxViolation$;
    else
      Update $\varepsilon(gen)$;
    end
  else
    $\varepsilon(k) = 0$;
  end
  for $k \leftarrow 1$ **to** K **do**
    if $|M_k| \leq S$ **then**
      randomly select $S - |M_k|$ solutions from Q and add them to M_k.
    else
      if $PushStage == true$ **then**
        result = *PushSubproblems*(M_k, S);
      else
        result = *PullSubproblems*(M_k, gen, T_{max});
      end
    end
  end
  $gen = gen + 1$;
end
Output the non-dominated and feasible solutions.

created in lines 6–10, while lines 12–13 handle the allocation of solution sets Q to the corresponding subpopulations. In Line 15, the maximum change rate of r_k is computed. Lines 16–25 provide a detailed description of how $\varepsilon(k)$ is dynamically updated. Lines 26–36 provide a detailed description of the entire process of dynamically updating the subpopulations. In the final step, as described in line 39, the algorithm identifies and extracts a set of solutions that are both non-dominated and feasible. It is worth noting that Algorithm 3.10 can be found in Ref. [258].

To provide a clear demonstration, Figure 3.9 presents a visual representation of the search process employed by the proposed PPS-M2M. At the beginning of the process, five directional vectors, denoted as $v^1, \ldots, v^5$, are strategically distributed evenly across the objective space. This distribution results in the formation of five unique and mutually exclusive subregions, labeled as $M_1, \ldots, M_5$. Figure 3.9a–c provides a detailed illustration of the process during the push search stage, where the population explores the search space without considering any constraints and gradually converges toward the unconstrained PF. As shown in these sub-figures, the population initially explores the search space in an unconstrained NSGA-II search manner. As the search progresses, the individuals gradually concentrate near the unconstrained PF, which represents the optimal solution set. It is worth noting that the existence of infeasible regions causes some solutions found on the unconstrained PF to be feasible, while others on the unconstrained PF become infeasible. In the subsequent pull search stage, these infeasible solutions are gradually guided toward the constrained PFs using the constrained NSGA-II search

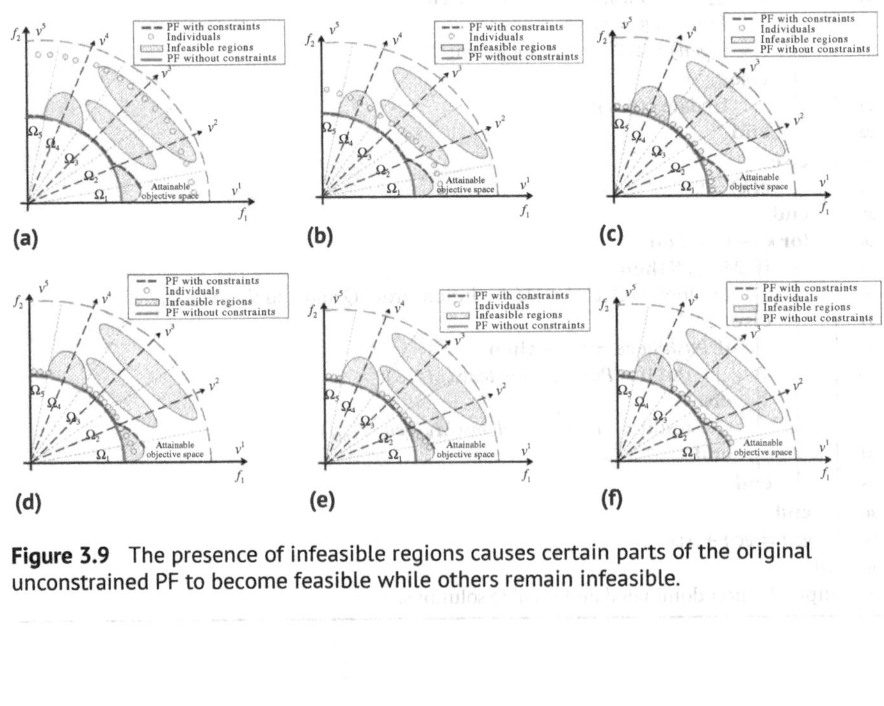

Figure 3.9 The presence of infeasible regions causes certain parts of the original unconstrained PF to become feasible while others remain infeasible.

approach. As shown in Figure 3.9d–f, this adjustment enables the population to better adapt to the constraints, optimizing the distribution of solutions and ultimately achieving convergence to a feasible and non-dominant set of solutions.

3.3.4 Key Differences Between Two PPS-based Algorithms

The key differences between two PPS-based algorithms lie in their decomposition strategies, search processes, and constraint-handling methods. Their primary differences are outlined below:

1. Decomposition Strategy: The PPS-M2M approach utilizes the M2M strategy, effectively breaking down a complex CMOP into several smaller and more manageable CMOPs. In contrast, PPS-MOEA/D employs a traditional decomposition approach that transforms a CMOP into multiple single-objective optimization subproblems.
2. Selection Mechanisms: The two algorithms use different strategies for selecting solutions. The search strategy of PPS-M2M combines a non-dominated sorting technique with the IEpsilon constraint-handling method, while PPS-MOEA/D relies on a combination of decomposition and the improved epsilon method to choose offspring solutions.
3. Applicability: PPS-M2M and PPS-MOEA/D are tailored for CMOPs with different characteristics. PPS-M2M performs better on CMOPs featuring imbalanced objectives and strict diversity constraints, whereas PPS-MOEA/D excels at solving CMOPs with extensive infeasible regions.

3.3.5 Experiment Results

In this section, the performance of the proposed PPS-M2M is compared against nine other algorithms: CM2M [259], MOEA/D-Epsilon [254], MOEA/D-SR [256], MOEA/D-CDP [256], C-MOEA/D [255], NSGA-II-CDP [45], MODE-ECHM [260], CM2M2 [261], and MODE-SaE [262]. These algorithms have been recognized for their effectiveness in addressing CMOPs. To evaluate the comparative performance of these algorithms, we employ the LIR-CMOP1-14 benchmark problems [257]. These problems are specifically selected due to their relatively small feasible regions, which provide a challenging and effective means to assess the constraint-handling capabilities of PPS-M2M.

In order to emphasize the differences between PPS-M2M and PPS-MOEA/D [14], a novel set of CMOPs, referred to as constraint imbalanced multi-objective optimization problems (CIMOPs), has been deliberately crafted for this purpose. These CIMOPs are inspired by the MOPs proposed in [51] and the DAS-CMOPs introduced in [263]. The objective functions of the CIMOPs remain consistent with those of MOPs [51], exhibiting an imbalanced nature. Meanwhile, the constraint functions incorporate diversity-hardness characteristics, a concept derived from DAS-CMOPs [263]. By combining imbalanced objectives with diversity-challenging constraints, the proposed CIMOPs serve as effective benchmarks for evaluating the performance and robustness of the PPS-M2M algorithm. It is worth noting that the specific details of the experimental setup and the associated parameter configurations are thoroughly explained in the Ref. [258].

3.3.5.1 Discussion of Experiments

Table 3.3 summarizes the mean and STD of IGD values for 10 algorithms evaluated across the LIR-CMOP1-14 test problems, each run independently 30 times. Similarly, Table 3.4 reports the mean and STD of HV values under the same conditions. The results clearly demonstrate that the proposed PPS-M2M algorithm outperforms the other algorithms in most test cases. Notably, PPS-M2M achieves significantly better performance on test problems LIR-CMOP1, LIR-CMOP5 to LIR-CMOP8, and LIR-CMOP10 to LIR-CMOP14, where it exceeds the competing methods by one to two orders of magnitude. To further validate this observation, we performed a Friedman aligned test (refer to Ref. [258] for details). The statistical results confirm that PPS-M2M is significantly superior to the other algorithms, reinforcing its strong performance and reliability.

To provide a more comprehensive analysis of the effectiveness of PPS-M2M, we present a detailed examination of the non-dominated solutions obtained by each algorithm when applied to three carefully selected representative problems: LIR-CMOP1, LIR-CMOP7, and LIR-CMOP11. These solutions correspond to the median HV values achieved by the respective algorithms. Furthermore, Figure 3.10 provides a visual representation of the feasible and infeasible regions associated with the three benchmark problems. These problems have been specifically chosen to illustrate three different levels of problem complexity, as detailed in [263].

For LIR-CMOP1, the feasible region is extremely small (Figure 3.10a), indicating that algorithms must overcome complex constraints to identify feasible solutions. In this problem, the feasible solution space is very limited, with infeasible regions occupying most of the objective space. This significantly increases the challenge for algorithms to converge to the true PF. Figure 3.11 provides a detailed comparison of the non-dominated solutions obtained by the 10 algorithms on this problem. From these results, in Figure 3.11a, b, e, h, i, and j, it is evident that six algorithms successfully converge to the true PF while maintaining a certain degree of solution diversity. Specifically, CM2M's convergence is inferior to PPS-M2M, as shown in Figure 3.11b. A likely reason for this is that CM2M mistakenly classifies some subregions as infeasible, preventing solutions in these regions from effectively converging to the true PF, which ultimately impacts

Table 3.3 Mean and STD of IGD performance of 10 comparison algorithms on the LIR-CMOP1-14. For simplicity, abbreviations are used: "Epsilon" represents MOEA/D-Epsilon, "CDP" refers to MOEA/D-CDP, "SR" corresponds to MOEA/D-SR, while "SaE" and "ECHM" stand for MODE-SaE and MODE-ECHM, respectively.

Test instance		PPS-M2M	CM2M [259]	CM2M2 [261]	Epsilon [254]	CDP [256]	SR [256]	C-MOEA/D [255]	NSGA-II-CDP [45]	SaE [262]	ECHM [260]
LIRCMOP1	mean	**1.330E−02**	3.106E−02	1.356E−02	5.74E−02	1.11E−01	1.81E−02	1.26E−01	3.23E−01	2.06E−02	5.429E−01
	std	4.407E−03	9.888E−03	2.993E−03	2.89E−02	5.04E−02	1.66E−02	7.03E−02	7.33E−02	1.15E−02	2.193E−01
LIRCMOP2	mean	1.485E−02	2.742E−02	**1.167E−02**	5.39E−02	1.43E−01	9.63E−03	1.40E−01	3.03E−01	1.51E−02	4.995E−01
	std	5.576E−03	8.596E−03	3.021E−03	2.13E−02	5.55E−02	7.23E−03	5.44E−02	7.24E−02	5.041E−03	2.062E−01
LIRCMOP3	mean	1.987E−02	4.383E−02	1.395E−02	8.81E−02	2.61E−01	1.78E−01	2.80E−01	4.08E−01	**7.44E−03**	7.866E−01
	std	7.371E−03	1.445E−02	5.867E−03	4.36E−02	4.33E−02	7.20E−02	4.21E−02	1.15E−01	1.91E−03	2.603E−01
LIRCMOP4	mean	2.504E−02	4.187E−02	1.196E−02	6.51E−02	2.53E−01	1.95E−01	2.59E−01	3.85E−01	**9.06E−03**	8.266E−01
	std	9.030E−03	2.371E−02	5.604E−03	3.01E−02	4.34E−02	6.40E−02	3.51E−02	1.35E−01	2.89E−03	2.754E−01
LIRCMOP5	mean	**3.922E−03**	2.941E−01	1.181E+00	1.15E+00	1.05E+00	1.04E+00	1.10E+00	5.53E−01	1.07E−01	2.087E+01
	std	1.515E−03	5.013E−01	1.826E−02	1.98E−01	3.63E−01	3.66E−01	2.99E−01	6.88E−01	2.00E−03	2.557E+00
LIRCMOP6	mean	**5.077E−03**	5.356E−01	9.915E−01	1.27E+00	1.09E+00	9.43E−01	1.31E+00	5.74E−01	4.95E−01	2.051E+01
	std	1.117E−02	5.509E−01	5.510E−01	2.95E−01	5.20E−01	5.90E−01	2.08E−01	4.21E−01	1.07E−01	2.994E+00
LIRCMOP7	mean	**3.711E−03**	5.237E−01	7.077E−02	1.51E+00	1.46E+00	1.08E+00	1.56E+00	2.38E−01	7.99E−02	2.824E+00
	std	1.993E−03	7.677E−01	5.172E−02	5.09E−01	5.58E−01	7.58E−01	4.24E−01	4.06E−01	4.07E−02	8.474E−01
LIRCMOP8	mean	**2.949E−03**	7.924E−01	6.582E−02	1.62E+00	1.38E+00	1.01E+00	1.58E+00	6.02E−01	3.31E−01	2.910E+00
	std	9.666E−05	7.455E−01	5.937E−02	3.05E−01	6.15E−01	7.24E−01	3.71E−01	7.39E−01	9.38E−01	8.183E−01

(Continued)

Table 3.3 (Continued)

Test instance		PPS-M2M	CM2M [259]	CM2M2 [261]	Epsilon [254]	CDP [256]	SR [256]	C-MOEA/D [255]	NSGA-II-CDP [45]	SaE [262]	ECHM [260]
LIRCMOP9	mean	**3.704E−01**	4.599E−01	4.197E−01	4.90E−01	4.81E−01	4.85E−01	4.81E−01	6.44E−01	5.04E−01	1.007E+00
	std	2.529E−02	6.426E−02	8.113E−02	4.22E−02	5.24E−02	4.78E−02	5.24E−02	1.60E−02	4.26E−06	1.685E−01
LIRCMOP10	mean	**1.572E−02**	2.291E−01	6.108E−01	2.13E−01	2.16E−01	1.92E−01	2.13E−01	5.97E−01	4.10E−01	1.258E+00
	std	4.506E−02	8.123E−02	1.375E−01	5.32E−02	6.81E−02	6.81E−02	4.63E−02	3.20E−02	2.86E−05	1.029E−01
LIRCMOP11	mean	**2.770E−02**	4.400E−01	5.135E−01	3.47E−01	3.42E−01	3.16E−01	3.81E−01	4.87E−01	5.33E−01	1.054E+00
	std	4.656E−02	1.038E−01	1.131E−01	9.28E−02	9.22E−02	7.49E−02	8.95E−02	1.05E−02	8.59E−02	1.196E−01
LIRCMOP12	mean	**9.310E−02**	1.488E−01	2.549E−01	2.52E−01	2.69E−01	2.06E−01	2.50E+00	5.80E−01	2.52E−01	1.028E+00
	std	9.310E−02	5.757E−02	1.592E−02	8.98E−02	9.06E−02	5.61E−02	9.63E−02	1.17E−01	7.79E−02	1.303E−01
LIRCMOP13	mean	**1.874E−01**	1.57E+00	1.263E+00	1.20E+00	1.21E+00	8.86E−01	1.18E+00	1.39E+01	8.97E−01	8.530E+00
	std	2.690E−02	1.05E−02	2.254E−01	3.06E−01	3.17E−01	5.76E−01	3.78E−01	2.26E+00	1.34E−01	7.869E−01
LIRCMOP14	mean	**1.746E−01**	2.39E+00	1.220E+00	1.02E+00	1.11E+00	1.03E+00	1.25E+00	1.36E+01	9.45E−01	8.568E+00
	std	2.840E−02	1.63E−02	2.192E−01	4.86E−01	3.98E−01	4.70E−01	5.30E−02	2.17E+00	1.75E−01	0.65187
Friedman test		**1.4286**	4.7857	4.6429	5.8571	6.1786	4.0714	7	7.4286	3.8214	9.7857

Table 3.4 Mean and STD of HV performance of 10 comparison algorithms on the LIR-CMOP1-14. For simplicity, abbreviations are used: "Epsilon" represents MOEA/D-Epsilon, "CDP" refers to MOEA/D-CDP, "SR" corresponds to MOEA/D-SR, while "SaE" and "ECHM" stand for MODE-SaE and MODE-ECHM, respectively.

Test instance		PPS-M2M	CM2M [259]	CM2M2 [261]	Epsilon [254]	CDP [256]	SR [256]	C-MOEA/D [255]	NSGA-II-CDP [45]	SaE [262]	ECHM [260]
LIRCMOP1	mean	**1.009E+00**	9.897E−01	1.008E+00	9.590E−01	7.540E−01	9.960E−01	7.410E−01	5.160E−01	2.292E−01	3.718E−01
	std	4.590E−03	1.322E−02	3.563E−03	3.280E−02	8.950E−02	2.910E−02	1.220E−01	5.570E−02	9.270E−03	1.871E−01
LIRCMOP2	mean	1.337E+00	1.321E+00	**1.345E+00**	1.280E+00	1.060E+00	1.340E+00	1.070E+00	8.240E−01	3.587E−01	5.971E−01
	std	7.093E−03	1.124E−02	4.409E−03	2.880E−02	1.080E−01	1.470E−02	9.100E−02	1.150E−01	2.839E−03	2.620E−01
LIRCMOP3	mean	**8.650E−01**	8.407E−01	8.645E−01	7.980E−01	4.860E−01	5.910E−01	4.710E−01	4.080E−01	2.054E−01	1.538E−01
	std	6.010E−03	1.787E−02	5.795E−03	3.930E−02	4.310E−02	1.070E−01	4.090E−02	2.880E−02	9.720E−04	1.793E−01
LIRCMOP4	mean	**1.095E+00**	1.051E+00	1.088E+00	1.020E+00	7.350E−01	8.150E−01	7.310E−01	6.170E−01	3.133E−01	1.707E−01
	std	8.869E−03	3.870E−02	3.633E−03	4.190E−02	5.440E−02	8.700E−02	5.160E−02	1.060E−01	2.900E−03	2.774E−01
LIRCMOP5	mean	**1.455E+00**	1.084E+00	1.513E−01	4.300E−02	1.630E−01	1.820E−01	9.720E−02	9.390E−01	2.603E−01	0.000E+00
	std	6.935E−03	6.097E−01	4.184E−01	2.350E−01	4.430E−01	4.390E−01	3.700E−01	3.210E−01	2.540E−01	0.000E+00
LIRCMOP6	mean	**1.118E+00**	5.337E−01	1.455E−01	5.400E−02	1.880E−01	3.020E−01	2.330E−02	4.130E−01	5.462E−02	0.000E+00
	std	3.888E−02	3.987E−01	3.034E−01	2.210E−01	3.870E−01	4.620E−01	1.280E−01	1.890E−01	1.090E−02	0.000E+00
LIRCMOP7	mean	**3.005E+00**	2.021E+00	2.776E+00	3.030E−01	3.740E−01	9.880E−01	2.040E−01	2.400E+00	2.587E−01	0.000E+00
	std	2.434E−02	1.349E+00	1.706E−01	9.070E−01	9.580E−01	1.270E+00	7.520E−01	6.520E−01	1.020E−02	0.000E+00
LIRCMOP8	mean	**3.016E+00**	1.498E+00	2.769E+00	1.060E−01	5.170E−01	1.100E+00	1.660E−01	1.900E+00	1.044E−01	0.000E+00
	std	2.494E−03	1.277E+00	2.149E−01	5.490E−01	1.050E+00	1.200E+00	6.110E−01	7.560E−01	1.480E−01	0.000E+00

(Continued)

Table 3.4 (Continued)

Test instance		PPS-M2M	CM2M [259]	CM2M2 [261]	Epsilon [254]	CDP [256]	SR [256]	C-MOEA/D [255]	NSGA-II-CDP [45]	SaE [262]	ECHM [260]
LIRCMOP9	mean	**3.152E+00**	2.826E+00	2.941E+00	2.740E+00	2.770E+00	2.750E+00	2.770E+00	2.060E+00	3.585E−01	1.110E+00
	std	8.752E−02	2.114E−01	1.698E−01	1.480E−01	1.840E−01	1.640E−01	1.840E−01	1.080E−02	9.970E−05	2.421E−01
LIRCMOP10	mean	**3.216E+00**	2.845E+00	1.847E+00	2.890E+00	2.880E+00	2.930E+00	2.890E+00	2.040E+00	4.913E−01	3.834E−01
	std	7.555E−02	1.720E−01	4.934E−01	1.020E−01	1.360E−01	1.350E−01	9.770E−02	4.450E−02	1.070E−04	2.594E−01
LIRCMOP11	mean	**4.343E+00**	3.115E+00	2.738E+00	3.340E+00	3.350E+00	3.380E+00	3.240E+00	3.110E+00	3.083E−01	1.280E+00
	std	1.401E−01	3.003E−01	5.532E−01	2.570E−01	2.570E−01	2.900E−01	2.550E−01	1.540E−02	6.890E−02	2.850E−01
LIRCMOP12	mean	**5.409E+00**	5.221E+00	4.900E+00	4.880E+00	4.830E+00	5.030E+00	4.890E+00	3.280E+00	5.077E−01	1.622E+00
	std	1.717E−01	1.878E−01	1.115E−01	3.170E−01	3.280E−01	1.750E−01	3.450E−01	3.610E−01	2.250E−02	3.114E−01
LIRCMOP13	mean	**4.874E+00**	4.566E−01	3.119E+00	4.550E−01	4.630E−01	1.890E+00	6.290E−01	0.000E+00	9.476E−02	0.000E+00
	std	1.068E−01	2.100E−02	2.746E+00	1.300E+00	1.420E+00	2.570E+00	1.710E+00	0.000E+00	6.010E−02	0.000E+00
LIRCMOP14	mean	**5.449E+00**	4.142E−01	2.313E+00	1.330E+00	8.810E−01	1.270E+00	1.800E−01	0.000E+00	3.178E−02	0.000E+00
	std	1.237E−01	5.110E−03	2.835E+00	2.450E+00	1.970E+00	2.290E+00	2.600E−01	0.000E+00	4.360E−02	0.000E+00
Friedman test		**1.1786**	3.9286	3.5	5.75	5.5357	3.8929	6.3571	6.6429	8.6429	9.5714

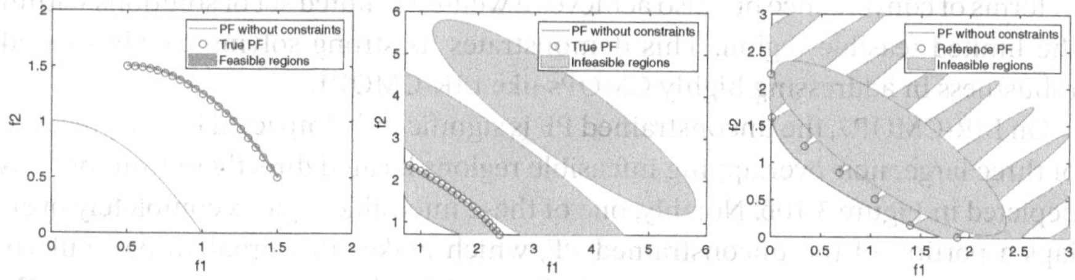

(a) LIR-CMOP1 (b) LIR-CMOP7 (c) LIR-CMOP11

Figure 3.10 The objective spaces of LIR-CMOP1, LIR-CMOP7, and LIR-CMOP11 feature distinct patterns in their feasible and infeasible regions. The unique distribution and shape of these regions highlight the varying levels of constraint complexity within each objective space.

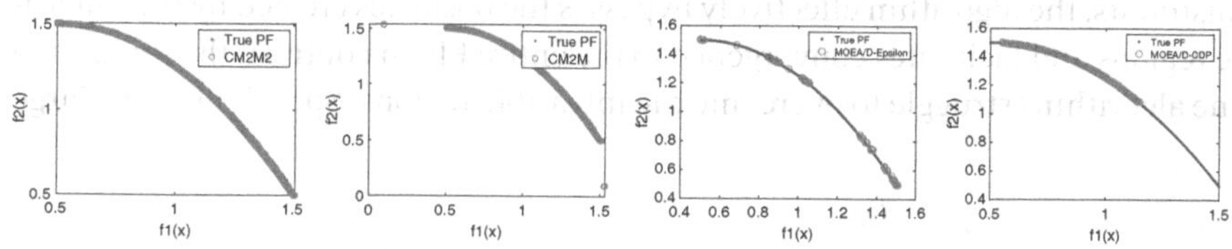

(a) PPS-M2M (b) CM2M (c) MOEA/D-Epsilon (d) MOEA/D-CDP

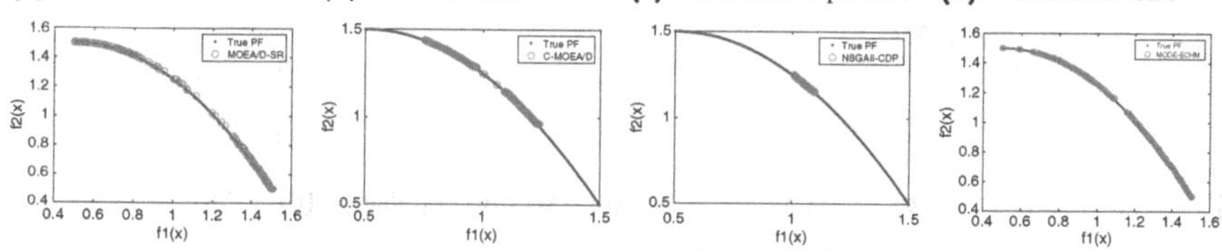

(e) MOEA/D-SR (f) C-MOEA/D (g) NSGA-II-CDP (h) MODE-ECHM

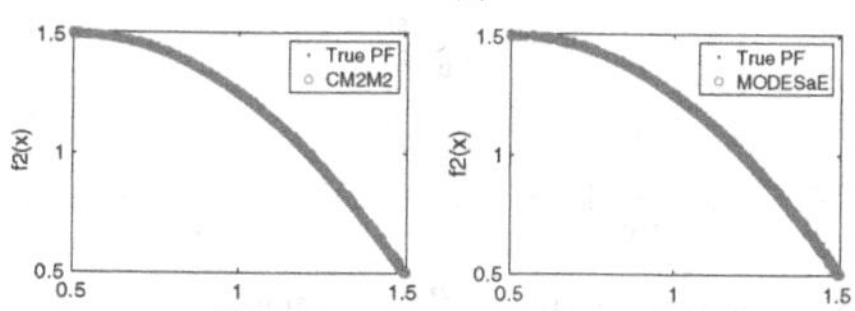

(i) CM2M2 (j) MODE-SaE

Figure 3.11 The feasible and non-dominated solution sets obtained by the 10 algorithms when solving LIR-CMOP1 under the same experiment conditions.

its overall convergence performance. For the other four algorithms, the solution diversity is relatively poor, as shown in Figure 3.11c, d, f, and g. Although these algorithms find certain non-dominated solutions, their distributions are overly concentrated, failing to sufficiently cover the true PF and revealing a lack of solution diversity. In summary, PPS-M2M not only outperforms the other algorithms in terms of convergence but also achieves a well-distributed set of solutions within the limited feasible region. This demonstrates its strong solving capability and robustness in addressing highly CMOPs like LIR-CMOP1.

On LIR-CMOP7, the unconstrained PF is significantly impacted by the presence of three large, non-overlapping infeasible regions located directly in front of it, as depicted in Figure 3.10b. Notably, one of these infeasible regions completely overlaps a portion of the unconstrained PF, which makes this problem particularly challenging in terms of convergence. In Figure 3.12, the performance of PPS-M2M and the other CMOPs on LIR-CMOP7 is illustrated. From these results, it is evident that PPS-M2M is the only algorithm capable of successfully crossing the infeasible regions and fully reaching the true PF, as shown in Figure 3.12a. This superior performance is primarily attributed to the design of PPS-M2M, where constraints are intentionally ignored during the push stage. By temporarily disregarding the constraints, the algorithm effectively bypasses the obstacles created by the infeasible regions and facilitates convergence to the entire PF. In contrast, the remaining nine algorithms struggle to overcome the infeasible regions. Specifically, two large

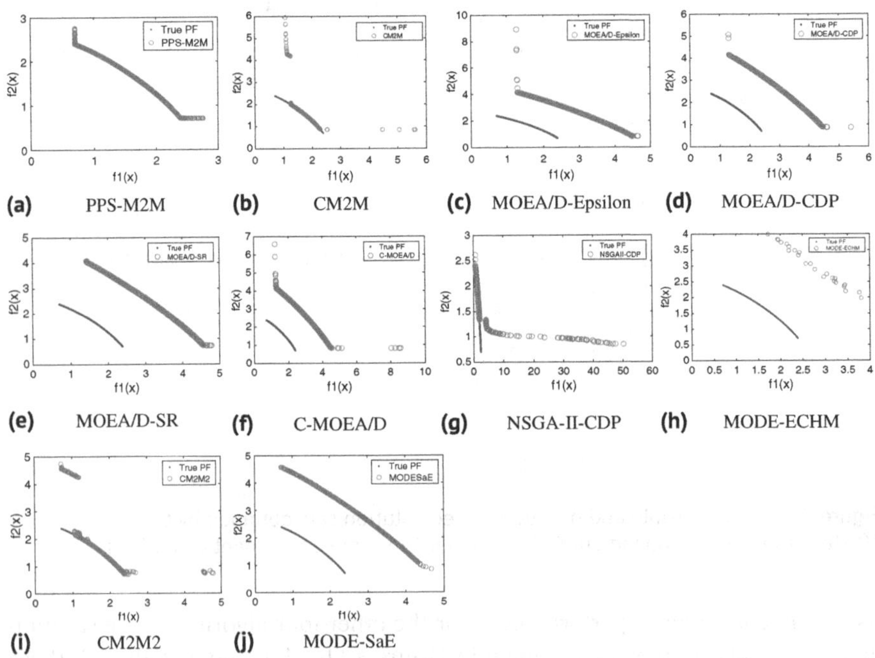

Figure 3.12 The feasible and non-dominated solution sets obtained by the 10 algorithms when solving LIR-CMOP7 under the same experiment conditions.

infeasible regions located directly in front of the PF act as significant barriers, preventing the populations of these algorithms from progressing toward the true PF. As a result, their solutions fail to achieve full coverage of the PF, demonstrating limited convergence ability in the presence of such challenging constraints.

As illustrated in Figure 3.10c, the PF of LIR-CMOP11 is discrete, which introduces significant challenges in maintaining solution diversity. This diversity-hard problem contains seven Pareto optimal solutions, where two solutions lie on the unconstrained PF, while the remaining five are located on the constrained PF. The discrete nature of the PF requires algorithms to accurately identify and cover all these optimal solutions despite the presence of complex constraints. The results obtained by the 10 tested algorithms on LIR-CMOP11 are presented in Figure 3.13. Notably, only PPS-M2M and CM2M successfully identify all seven Pareto optimal solutions, as shown in Figures 3.13a, b. However, PPS-M2M demonstrates superior convergence compared to CM2M. One possible explanation for this advantage is the design of PPS-M2M: during the final 10% of the maximum generation, all subpopulations are merged into a single unified population, which evolves using an enhanced epsilon-based constraint-handling mechanism. This strategy allows

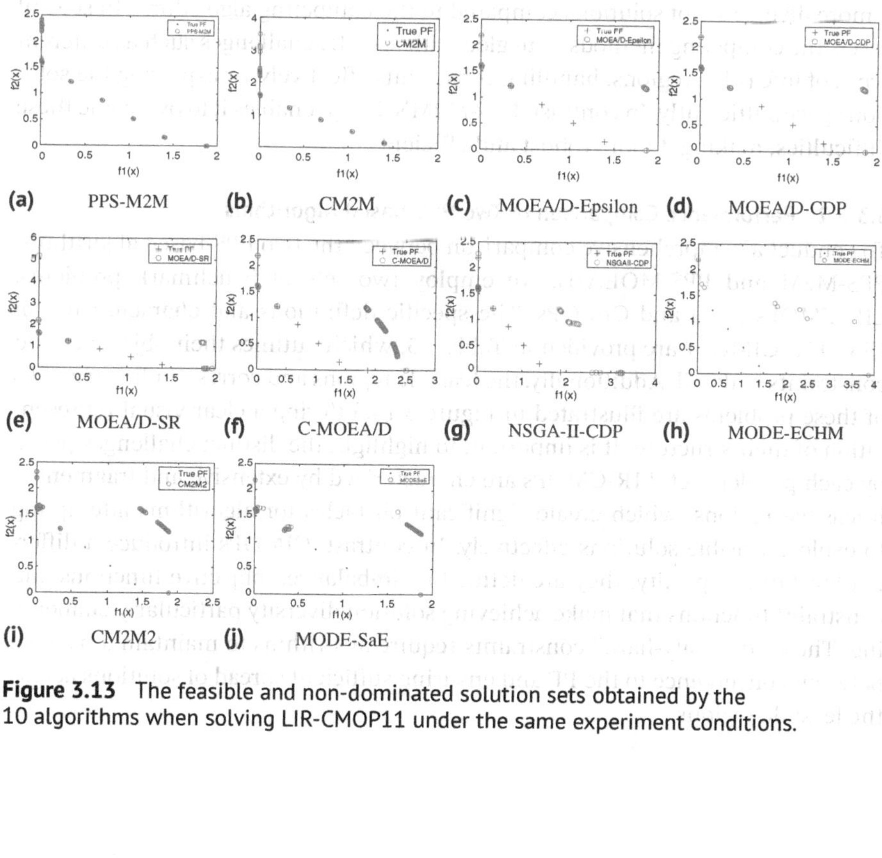

Figure 3.13 The feasible and non-dominated solution sets obtained by the 10 algorithms when solving LIR-CMOP11 under the same experiment conditions.

PPS-M2M to refine its solutions effectively and reach all optimal points. In contrast, the other eight algorithms fail to find the complete set of Pareto optimal solutions, as shown in Figure 3.13c, j. Their inability to overcome this challenge is primarily due to two factors. First, the infeasible regions obstruct the search trajectories of their populations, preventing them from reaching parts of the true PF. Second, these algorithms struggle with parameter settings for their constraint-handling mechanisms. The constraint landscape in LIR-CMOP11 is complex and poorly explored during the search process, making it difficult to adaptively tune these parameters to achieve optimal performance. In summary, PPS-M2M not only identifies all Pareto optimal solutions but also outperforms the other algorithms in terms of convergence and solution diversity, showcasing its robustness and effectiveness in addressing discrete, constraint-intensive problems like LIR-CMOP11.

From the observations and analysis presented above, it is evident that the proposed PPS-M2M algorithm demonstrates superior performance compared to the other nine CMOEAs. This advantage is particularly noticeable across the majority of the test instances evaluated in this study. The results consistently show that PPS-M2M not only achieves better convergence to the true PF but also maintains a more diverse set of solutions compared to the competing algorithms. In several cases, the competing methods struggled with specific challenges such as different types of infeasible regions, handling constraints effectively, or exploring the solution space sufficiently. In contrast, PPS-M2M's design enables it to overcome these difficulties, making it more robust and efficient.

3.3.5.2 Performance Comparison of Two PPS-based Algorithms

To conduct a comprehensive comparison between the two PPS-based algorithms, PPS-M2M and PPS-MOEA/D, we employ two sets of benchmark problems: LIR-CMOPs [257] and CIMOPs. The specific definitions and characteristics of CIMOP1–CIMOP7 are provided in Table 3.5, which outlines their objectives and constraints in detail. Additionally, the feasible regions and corresponding true PFs of these problems are illustrated in Figure 3.14, offering a clear visual representation of their structure. It is important to highlight the distinct challenges posed by each problem set. LIR-CMOPs are characterized by extensive and fragmented infeasible regions, which create significant obstacles for algorithms attempting to explore feasible solutions effectively. In contrast, CIMOPs introduce a different level of complexity: they are defined by imbalanced objective functions and constraint functions that make achieving solution diversity particularly challenging. These "diversity-hard" constraints require algorithms to maintain a balance between convergence to the PF and ensuring sufficient spread of solutions across the feasible regions.

Table 3.5 The mathematical functions of CIMOP1-7.

Problem	Objectives	Constraints
CIMOP1	$\begin{cases} \min & f_1(\mathbf{x}) = (1+g(\mathbf{x}))x_1 \\ \min & f_2(\mathbf{x}) = (1+g(\mathbf{x}))(1-\sqrt{x_1}) \\ \text{where} & g(\mathbf{x}) = 2\sin(\pi x_1)\sum_{i=2}^{n}(-0.9t_i^2 + \lvert t_i^{0.6}\rvert) \\ & t_i = x_i - \sin(0.5\pi x_1) \\ & n = 10, \mathbf{x} \in [0,1]^n \end{cases}$	$\begin{cases} c_1(\mathbf{x}) = \sin(a\pi x_1) - b \geq 0 \\ a = 20, b = 0 \end{cases}$
CIMOP2	$\begin{cases} \min & f_1(\mathbf{x}) = (1+g(\mathbf{x}))x_1 \\ \min & f_2(\mathbf{x}) = (1+g(\mathbf{x}))(1-x_1^2) \\ \text{where} & g(\mathbf{x}) = 10\sin(\pi x_1)\sum_{i=2}^{n}(\frac{\lvert t_i\rvert}{1+e^{5\lvert t_i\rvert}}) \\ & t_i = x_i - \sin(0.5\pi x_1) \\ & n = 10, \mathbf{x} \in [0,1]^n \end{cases}$	They are the same as those of CIMOP1
CIMOP3	$\begin{cases} \min & f_1(\mathbf{x}) = (1+g(\mathbf{x}))\cos(\frac{\pi x_1}{2}) \\ \min & f_2(\mathbf{x}) = (1+g(\mathbf{x}))\sin(\frac{\pi x_1}{2}) \\ \text{where} & g(\mathbf{x}) = 10\sin(\frac{\pi x_1}{2})\sum_{i=2}^{n}(\frac{\lvert t_i\rvert}{1+e^{5\lvert t_i\rvert}}) \\ & t_i = x_i - \sin(0.5\pi x_1) \\ & n = 10, \mathbf{x} \in [0,1]^n \end{cases}$	They are the same as those of CIMOP1
CIMOP4	$\begin{cases} \min & f_1(\mathbf{x}) = (1+g(\mathbf{x}))x_1 \\ \min & f_2(\mathbf{x}) = (1+g(\mathbf{x}))(1-x_1^{0.5}\cos^2(2\pi x_1)) \\ \text{where} & g(\mathbf{x}) = 1 + 10\sin(\pi x_1)\sum_{i=2}^{n}(\frac{\lvert t_i\rvert}{1+e^{5\lvert t_i\rvert}}) \\ & t_i = x_i - \sin(0.5\pi x_1) \\ & n = 10, \mathbf{x} \in [0,1]^n \end{cases}$	They are the same as those of CIMOP1
CIMOP5	$\begin{cases} \min & f_1(\mathbf{x}) = (1+g(\mathbf{x}))x_1 \\ \min & f_2(\mathbf{x}) = (1+g(\mathbf{x}))(1-\sqrt{x_1}) \\ \text{where} & g(\mathbf{x}) = 2\lvert\cos(\pi x_1)\rvert\sum_{i=2}^{n}(-0.9t_i^2 + \lvert t_i^{0.6}\rvert) \\ & t_i = x_i - \sin(0.5\pi x_1) \\ & n = 10, \mathbf{x} \in [0,1]^n \end{cases}$	They are the same as those of CIMOP1
CIMOP6	$\begin{cases} \min & f_1(\mathbf{x}) = (1+g(\mathbf{x}))x_1x_2 \\ \min & f_2(\mathbf{x}) = (1+g(\mathbf{x}))x_1(1-x_2) \\ \min & f_3(\mathbf{x}) = (1+g(\mathbf{x}))(1-x_1) \\ \text{where} & g(\mathbf{x}) = 2\sin(\pi x_1)\sum_{i=3}^{n}(-0.9t_i^2 + \lvert t_i^{0.6}\rvert) \\ & t_i = x_i - x_1x_2 \\ & n = 10, \mathbf{x} \in [0,1]^n \end{cases}$	$\begin{cases} c_1(\mathbf{x}) = \sin(a\pi x_1) - b \geq 0 \\ c_2(\mathbf{x}) = \cos(a\pi x_2) - b \geq 0 \\ a = 20, b = 0 \end{cases}$
CIMOP7	$\begin{cases} \min & f_1(\mathbf{x}) = (1+g(\mathbf{x}))\cos(\frac{x_1\pi}{2})\cos(\frac{x_2\pi}{2}) \\ \min & f_2(\mathbf{x}) = (1+g(\mathbf{x}))\cos(\frac{x_1\pi}{2})\sin(\frac{x_2\pi}{2}) \\ \min & f_3(\mathbf{x}) = (1+g(\mathbf{x}))\sin(\frac{x_2\pi}{2}) \\ \text{where} & g(\mathbf{x}) = 2\sin(\pi x_1)\sum_{i=3}^{n}(-0.9t_i^2 + \lvert t_i^{0.6}\rvert) \\ & t_i = x_i - x_1x_2 \\ & n = 10, \mathbf{x} \in [0,1]^n \end{cases}$	They are the same as those of CIMOP6

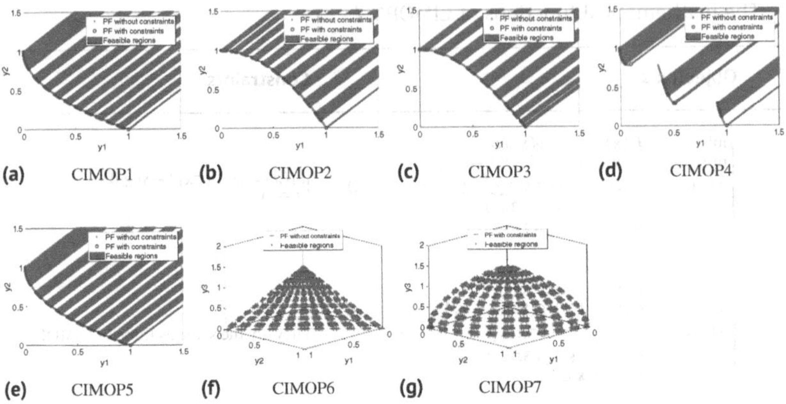

Figure 3.14 The feasible regions, the unconstrained PFs, and the true PFs under constraints for problems CIMOP1 through CIMOP7.

To begin the comparison, the performance of the proposed PPS-M2M and PPS-MOEA/D algorithms is evaluated on the LIR-CMOP1-14 test suite. The results, including IGD and HV values, are presented in Table 3.6. Additionally, a statistical analysis using the Friedman aligned test, as detailed in Ref. [258], is conducted to ensure the robustness and reliability of the comparisons. The data clearly reveal that PPS-MOEA/D achieves better performance compared to PPS-M2M across the LIR-CMOP1-14 problems. Specifically, PPS-MOEA/D consistently produces lower IGD values, indicating superior convergence to the true PF, while also achieving higher HV values, reflecting its ability to maintain a well-distributed set of solutions with good diversity. This trend suggests that PPS-MOEA/D is more effective at handling the complex constraints and infeasible regions commonly observed in the LIR-CMOP suite.

The performance of four algorithms (PPS-M2M, PPS-MOEA/D, CM2M, and CM2M2) on CIMOP1-7 is evaluated using two key performance metrics: IGD and HV. The results are presented in Tables 3.7 and 3.8. Notably, CM2M and CM2M2 are included in the comparison because they both adopt the M2M decomposition approach. The purpose of including these algorithms is to demonstrate that the superior performance of PPS-M2M is not solely attributed to the M2M approach but also to its effective constraint-handling mechanism embedded within the PPS framework. To further validate the results statistically, a Friedman aligned test is conducted, as described in Ref. [258]. The statistical analysis confirms that PPS-M2M achieves the highest ranking among the four compared CMOEAs. This indicates that PPS-M2M consistently outperforms the other algorithms across CIMOP1-7, highlighting the significant contribution of the constraint-handling strategy in the PPS framework.

Table 3.6 IGD and HV results of PPS-M2M and PPS-MOEA/D on LIR-CMOP1-14.

Test Instances		IGD		HV	
		PPS-M2M	PPS-MOEA/D [14]	PPS-M2M	PPS-MOEA/D [14]
LIRCMOP1	mean	1.330E−02	**6.413E−03**	1.009E+00	**1.016E+00**
	std	4.407E−03	1.938E−03	4.590E−03	1.580E−03
LIRCMOP2	mean	1.485E−02	**4.673E−03**	1.337E+00	**1.349E+00**
	std	5.576E−03	7.844E−04	7.093E−03	1.010E−03
LIRCMOP3	mean	1.987E−02	**8.545E−03**	8.650E−01	**8.703E−01**
	std	7.371E−03	5.184E−03	6.010E−03	2.650E−03
LIRCMOP4	mean	2.504E−02	**4.677E−03**	**1.095E+00**	1.093E+00
	std	9.030E−03	1.116E−03	8.869E−03	2.467E−03
LIRCMOP5	mean	3.922E−03	**1.837E−03**	1.455E+00	**1.462E+00**
	std	1.515E−03	9.263E−05	6.935E−03	2.919E−04
LIRCMOP6	mean	5.077E−03	**2.490E−03**	1.118E+00	**1.129E+00**
	std	1.117E−02	3.399E−04	3.888E−02	1.771E−04
LIRCMOP7	mean	3.711E−03	**2.797E−03**	3.005E+00	**3.015E+00**
	std	1.993E−03	9.854E−05	2.434E−02	2.663E−03
LIRCMOP8	mean	2.949E−03	**2.778E−03**	3.016E+00	**3.017E+00**
	std	9.666E−05	7.558E−05	2.494E−03	1.139E−03
LIRCMOP9	mean	3.704E−01	**9.940E−02**	3.152E+00	**3.570E+00**
	std	2.529E−02	1.519E−01	8.752E−02	2.242E−01
LIRCMOP10	mean	1.572E−02	**2.108E−03**	3.216E+00	**3.241E+00**
	std	4.506E−02	7.754E−05	7.555E−02	3.077E−04
LIRCMOP11	mean	2.770E−02	**2.832E−03**	4.343E+00	**4.390E+00**
	std	4.656E−02	1.359E−03	1.401E−01	2.217E−04
LIRCMOP12	mean	9.310E−02	**2.704E−02**	5.409E+00	**5.614E+00**
	std	5.548E−02	5.002E−02	1.717E−01	1.525E−01
LIRCMOP13	mean	1.874E−01	**6.455E−02**	4.874E+00	**5.710E+00**
	std	2.690E−02	2.177E−03	1.068E−01	1.275E−02
LIRCMOP14	mean	1.746E−01	**6.419E−02**	5.449E+00	**6.193E+00**
	std	2.840E−02	1.690E−03	1.237E−01	1.310E−02
Friedman test		2	1	1.9286	1.0714

Table 3.7 IGD results of PPS-M2M and the other three CMOEAs on CIMOP1-7.

Test instance		PPS-M2M	PPS-MOEA/D [14]	CM2M [259]	CM2M2 [261]
CIMOP1	mean	**2.05E−02**	2.81E−01	7.96E−01	5.22E−02
	std	1.13E−03	8.44E−02	6.36E−03	1.79E−02
CIMOP2	mean	**4.42E−02**	2.95E−01	6.38E−01	1.81E−01
	std	1.56E−02	1.29E−02	2.01E−02	3.18E−02
CIMOP3	mean	**1.19E−01**	3.51E−01	5.58E−01	2.80E−01
	std	1.13E−02	3.06E−02	1.54E−02	3.66E−02
CIMOP4	mean	**4.72E−03**	2.01E−01	6.47E−01	1.09E−02
	std	1.00E−03	2.92E−02	1.49E−02	3.44E−02
CIMOP5	mean	**4.45E−02**	3.76E−01	6.60E−02	8.26E−02
	std	5.28E−03	4.59E−02	1.25E−02	2.23E−02
CIMOP6	mean	**9.17E−02**	2.13E−01	6.98E−01	2.34E−01
	std	8.67E−03	4.46E−02	2.42E−02	2.46E−02
CIMOP7	mean	**1.64E−01**	3.65E−01	8.00E−01	3.67E−01
	std	1.87E−02	1.64E−02	5.11E−02	4.51E−02
Friedman test		1	2.8571	3.7143	2.4286

Table 3.8 HV results of PPS-M2M and the other three CMOEAs on CIMOP1-7.

Test instance		PPS-M2M	PPS-MOEA/D [14]	CM2M [259]	CM2M2 [261]
CIMOP1	mean	**1.065E+00**	6.623E−01	3.833E−01	1.032E+00
	std	1.359E−03	1.388E−01	5.250E−02	2.243E−02
CIMOP2	mean	**7.251E−01**	4.620E−01	7.109E−01	5.237E−01
	std	2.415E−02	1.478E−02	9.110E−03	2.600E−02
CIMOP3	mean	5.834E−01	2.470E−01	**6.630E−01**	3.384E−01
	std	2.507E−02	3.612E−02	2.260E−02	4.312E−01
CIMOP4	mean	**9.288E−01**	5.999E−01	7.856E−01	8.019E−01
	std	2.686E−03	1.747E−02	1.080E−03	7.203E−02
CIMOP5	mean	**1.028E+00**	6.550E−01	8.355E−01	9.617E−01
	std	7.116E−03	2.281E−02	6.810E−03	3.263E−02

Test instance		PPS-M2M	PPS-MOEA/D [14]	CM2M [259]	CM2M2 [261]
CIMOP6	mean	**1.437E+00**	1.326E+00	8.748E−01	9.505E−01
	std	1.227E−02	5.988E−02	1.440E−02	4.994E−02
CIMOP7	mean	**1.041E+00**	9.348E−01	8.319E−01	7.827E−01
	std	3.658E−02	6.182E−03	1.850E−02	3.010E−02
Friedman test		1.1429	3.2857	2.8571	2.7143

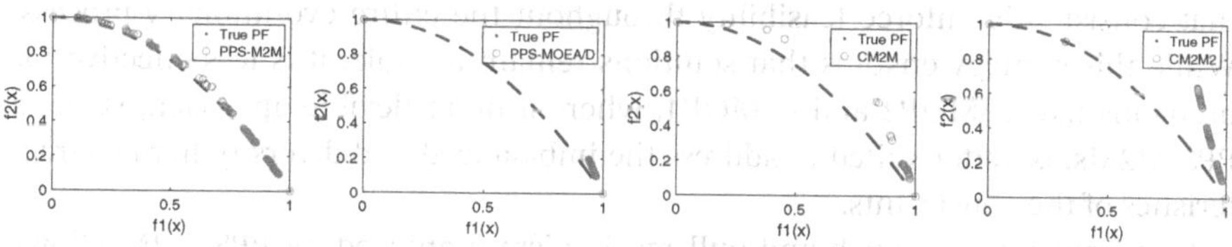

(a) PPS-M2M **(b)** PPS-MOEA/D **(c)** CM2M **(d)** CM2M2

Figure 3.15 The non-dominated solution sets obtained by the four algorithms when solving CIMOP2 under the same conditions.

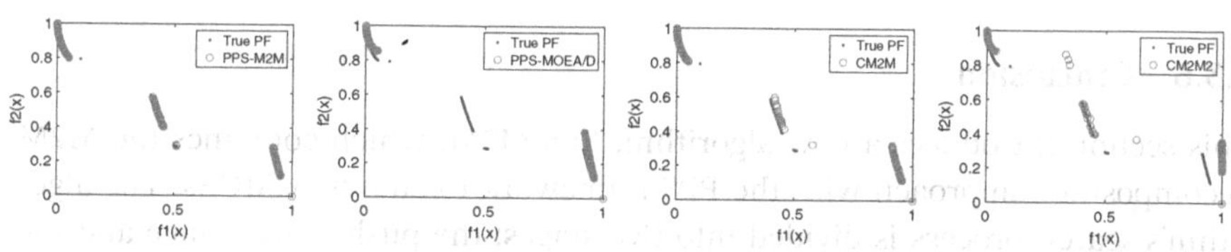

(a) PPS-M2M **(b)** PPS-MOEA/D **(c)** CM2M **(d)** CM2M2

Figure 3.16 The non-dominated solution sets obtained by the four algorithms when solving CIMOP4 under the same conditions.

To demonstrate the advantages of the proposed PPS-M2M in addressing CIMOPs, we use CIMOP2 and CIMOP4 as representative examples and plot the non-dominated solutions obtained by each algorithm corresponding to their median HV values. The results are depicted in Figures 3.15 and 3.16. From these figures, it is clear that PPS-M2M is able to identify the majority of Pareto optimal solutions, whereas the other three algorithms struggle to capture only a limited portion of the true PFs. This discrepancy is likely due to the diversity-hard nature of the constraint functions in CIMOP2 and CIMOP4. Specifically, the objective functions in these two problems are imbalanced, which poses a significant challenge for decomposition-based algorithms like MOEA/D, as previously discussed in [51]. MOEA/D's decomposition approach tends to focus on specific regions, failing to fully explore the true PFs. Additionally, the convergence performance of CM2M and CM2M2 is notably inferior to that of PPS-M2M. This is evident from Figures 3.15c, d and 3.16c, d, where many solutions produced by CM2M and CM2M2 fail to converge to the true PFs. A possible explanation for this poor performance lies in the relationship between the constrained and unconstrained PFs of CIMOP2 and CIMOP4: the constrained PFs are subsets of their corresponding unconstrained PFs. PPS-M2M handles this type of problem more effectively because it initially converges to the unconstrained PF without considering constraints. Through its unique push-and-pull mechanism, PPS-M2M then gradually refines solutions to satisfy the constraints, leading to better coverage and convergence. In contrast, CM2M and CM2M2 adopt constraint-handling mechanisms that consistently enforce feasibility throughout the entire evolutionary process. While this strategy ensures that solutions remain feasible, it is less effective for problems like CIMOP2 and CIMOP4, where a more flexible approach, such as PPS-M2Ms, is better suited to address the imbalanced and diversity-hard characteristics of the constraints.

In conclusion, the push-and-pull mechanism employed by PPS-M2M allows it to efficiently navigate complex constraint landscapes, outperforming CM2M, CM2M2, and PPS-MOEA/D in solving CIMOP2 and CIMOP4. This further highlights PPS-M2M's robustness and ability to address diversity and convergence challenges in CMOPs.

3.3.6 Conclusion

This section introduces a new algorithm, PPS-M2M, which combines the M2M decomposition approach with the PPS framework to address CMOPs. The algorithm's search process is divided into two stages: the push search stage and the pull search stage. In the push search stage, the M2M decomposition method is used to divide a CMOP into multiple simpler CMOPs, which are collaboratively solved without considering constraints. This approach allows subpopulations to traverse infeasible regions effortlessly while estimating constrained landscape information to guide the pull search stage. When the maximum rate of change between ideal and nadir points reaches a predefined threshold, the search switches to the pull search stage. In this stage, infeasible solutions obtained during the push stage are pulled toward feasible and non-dominated regions using an improved epsilon constraint-handling approach. In the final phase, all subpopulations are merged into a single population for further optimization. Experimental results demonstrate that PPS-M2M significantly outperforms nine other CMOEA algorithms on most LIR-CMOP1-14 problems. Compared to PPS-MOEA/D, PPS-M2M shows better performance on CIMOPs, while PPS-MOEA/D performs better on LIR-CMOPs. Overall, PPS-M2M is more suitable for addressing CMOPs with imbalanced objectives and diversity-challenging constraints, whereas PPS-MOEA/D is better suited for CMOPs with large infeasible regions.

3.4 SA-PPS

In a good deal of real-world engineering optimization problems, such as those in aerospace, autonomous driving, and medical device design, performing a large number of tests using simulation software or physical experiments is usually necessary. However, this approach is usually accompanied by significant computational costs and/or time expenses. Particularly in cases involving high-dimensional, multi-objective, and complex constraints, the difficulty of solving such optimization problems increases substantially. These problems are referred to as expensive constrained multi-objective optimization problems (ECMOPs). To address this challenge, this section proposes a novel evolutionary algorithm based on the PPS framework. By integrating Bayesian active learning, this algorithm is designed to efficiently solve ECMOPs and is named surrogate-assisted PPS (SA-PPS) [76]. The proposed algorithm organizes the optimization process into two distinct phases: the push search stage and the pull search stage. During the first stage, SA-PPS aims to guide the population to explore promising regions of the solution space thoroughly, thereby rapidly improving convergence. In the second stage, SA-PPS further aims on approximating the constrained PF while maintaining population diversity, avoiding premature convergence to local optima. In order to improve both the efficiency and ACC of the algorithm, the pull search stage incorporates a batch data selection strategy that leverages Bayesian active learning. This strategy leverages the Bayesian model's prediction of uncertainty in the solution space, prioritizing the sampling of regions of interest. As a result, the surrogate model can more accurately capture the local characteristics of the objective and constraint functions. This approach not only significantly reduces the number of samples required for model training but also markedly improves the overall optimization performance.

3.4.1 Proposed Method

The proposed SA-PPS is structured around three primary components, as illustrated in Algorithm 3.11: the push search stage, which focuses on unconstrained exploration; the pull search stage, which incorporates constraints for refined optimization; and a data selection mechanism that employs Bayesian active learning to intelligently prioritize training data. These components complement each other, forming an efficient and precise multi-objective optimization framework. Their core functions are outlined:

1. Push Search Stage: The SA-PPS algorithm integrates surrogate models with the NSGA-II evolutionary framework, utilizing their combined strengths to effectively direct the population's evolution toward optimal solutions. At this

stage, the algorithm does not consider the problem's constraints but focuses on exploring the unconstrained solution space. Candidate individuals are selected for expensive evaluations based on two key metrics: HV contribution and objective uncertainty. This approach allows the population to efficiently navigate through infeasible regions, facilitating a rapid progression toward the unconstrained PF.

2. Pull Search Stage: SA-PPS begins to incorporate constraints by constructing surrogate models for each objective and constraint, further refining the structure of the objective space. During this stage, reference vectors are used to divide the population into several subregions. This design of subregion division and tailored strategies effectively guides the population toward convergence on the constrained PF while ensuring population diversity.
3. Batch Data Selection Based on Bayesian Active Learning (BatchBALDS): To further enhance the efficiency and ACC of the pull search stage, SA-PPS incorporates BatchBALDS. With the accumulation of sample data, the time required to train the surrogate model increases significantly, frequently emerging as a major bottleneck in terms of efficiency, particularly in practical engineering applications where computational resources are limited. BatchBALDS actively selects the most representative training data, significantly accelerating the training process of the surrogate model while ensuring its predictive performance.

The detailed process of the proposed SA-PPS is shown in Figure 3.17. The algorithm begins by initializing the population and performing an expensive evaluation to obtain the true objective function and constraint function values of the initial population. The algorithm then evaluates the current search state to determine the appropriate stage of operation. If the search is in the push stage, surrogate models are created exclusively for the optimization objectives, and candidate solutions are identified based on the push stage search criteria. On the other hand, if the search is in the pull stage, the algorithm applies the BatchBALDS strategy to strategically select training data, builds surrogate models for both the objectives and constraints, and filters candidate solutions using the pull stage search criteria for further refinement.

The switching mechanism between the push search stage and the pull search stage is determined by the HV value change rate γ_{HV}. The transition from the push search stage to the pull search stage occurs when the value of γ_{HV} falls below the predefined threshold δ. The core purpose of this switching mechanism is to dynamically adjust the search mode of the algorithm, balancing global exploration and local optimization. The specific definition of the HV value change rate is as follows:

$$\gamma_{HV} = \frac{HV_k - HV_{k-l}}{\max(HV_{k-l}, \Delta)} \tag{3.6}$$

Algorithm 3.11: SA-PPS.

Input: The max number of true evaluations FE_{max}, the number of selected candidates N_{sel}, the surrogate evaluated population size N_{surr}, the number of generations for surrogate assisted search t_{max}

Output: The archive of expensive evaluated solutions P_{arch}

$P_{arch} \leftarrow$ generate N individuals by Latin hypercube sampling;
stage $\leftarrow 0$;
while $FE < FE_{max}$ **do**
    if *stage* $== 0$ **then**
        $P_{cand} \leftarrow$ PushSearch($P_{arch}, N_{sel}, N_{surr}, t_{max}$);
    else
        $P_{cand} \leftarrow$ PullSearch($P_{arch}, N_{sel}, N_{surr}, t_{max}$);
    end
    Evaluate each solution in P_{cand} with a real function;
    $P_{arch} \leftarrow P_{arch} \cup P_{cand}$;
    update FE;
    if *stage* $== 0$ *and* ($\gamma_{HV} < \delta$ *or* $FE > FE_{max}/2$) **then**
        stage $\leftarrow 1$;
    end
end

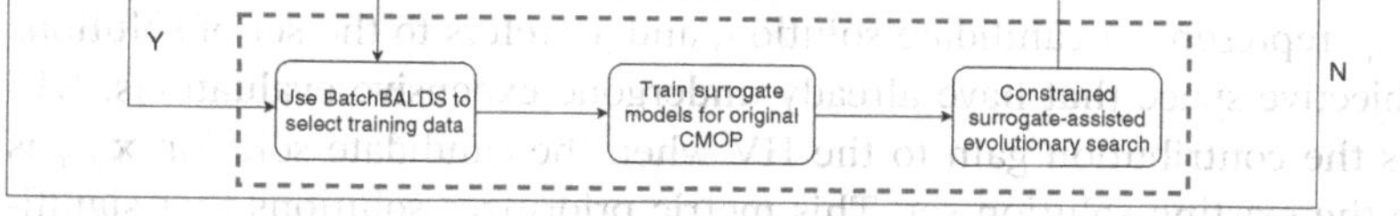

Figure 3.17 Flowchart of the proposed SA-PPS.

where HV_k indicates the HV value corresponding to the population of the current k-th generation, while HV_{k-l} refers to the HV value of the population from l generations prior. The term Δ, a small constant fixed at $1e^{-6}$, is introduced to ensure numerical stability by avoiding a potential division by zero. This metric is used to quantify the convergence status of the population during each generation of evolution. As the HV change rate diminishes and approaches an exceedingly small value, it serves as an indicator that the population is nearing convergence, with minimal progress being made in generating new non-dominated solutions. So, continuing the push search is unlikely to yield significant progress and may instead waste valuable computational resources. Therefore, the proposed SA-PPS algorithm transitions to the pull search stage at this phase, incorporating constraint conditions to perform more precise local searches in the solution space, thereby improving search efficiency and optimization performance.

It is worth noting that, to prevent excessive resource consumption in the push search stage, if the number of expensive evaluations reaches 50% of the maximum allowable evaluations and the algorithm still does not meet the transition condition, it will force a transition from the push stage to the pull stage.

3.4.2 Push Search Stage

In the push search stage, a separate surrogate model is constructed for each objective function, resulting in a total of M surrogate models to approximate the distribution of the true objective functions. At this stage, the algorithm employs the classical NSGA-II as the optimizer to guide the evolution and search of the population. Unlike the pull stage, the goal of the push stage is to perform a global exploration of the solution space, focusing on rapidly approaching the unconstrained PF.

When selecting candidate solutions for expensive evaluations, the algorithm combines two key metrics: hypervolume improvement (HVI) and individual uncertainty. This combination of metrics aims to balance the exploration capability of solutions with the reliability of the model's predictions. The formula for calculating *HVI* is as follows:

$$HVI(\mathbf{x}_{\text{surr}}) = HV(\hat{\boldsymbol{f}}(\mathbf{x}_{\text{surr}}) \cup \boldsymbol{y}_e) - HV(\boldsymbol{y}_e) \tag{3.7}$$

where $\mathbf{x}_{\text{surr}}$ represents a candidate solution, and $\boldsymbol{y}_e$ refers to the set of solutions in the objective space that have already undergone expensive evaluations. *HVI* measures the contribution gain to the HV when the candidate solution $\mathbf{x}_{\text{surr}}$ is added to the existing solution set. This metric prioritizes solutions that significantly enhance the diversity of the solution set and expand the PF, thereby accelerating the population's convergence toward the unconstrained PF.

However, in certain cases, such as when the number of non-dominated solutions in $\boldsymbol{y}_e$ is smaller than the specified selection number N_{sel}, many candidate solutions may have an *HVI* value close to zero. In such situations, relying solely on the *HVI* metric makes it difficult to effectively differentiate the quality of candidate solutions. To address this, we introduce an uncertainty-based evaluation method, which selects candidate solutions for expensive evaluations based on their average uncertainty. The formula for calculating the average uncertainty is as follows:

$$\hat{\sigma}_{\text{mean}}(\mathbf{x}) = \frac{1}{m}\sum_{i=1}^{m}\hat{\sigma}_i(\mathbf{x}) \tag{3.8}$$

where $\hat{\sigma}_i(\mathbf{x})$ is calculated using the trained Gaussian process regression (GPR) model and represents the uncertainty of the individual $\mathbf{x}$ for the i-th objective. The average uncertainty $\hat{\sigma}_{\text{mean}}(\mathbf{x})$ reflects the overall predictive uncertainty of the candidate solution $\mathbf{x}$ across all objectives.

Algorithm 3.12 presents the detailed steps of the push search process. In this algorithm, all individuals that have undergone expensive evaluations are used as the basis to construct a training dataset, which is then utilized to train surrogate models for each objective function separately (as shown in line 1). These surrogate models approximate the true objective functions, significantly reducing the number of expensive evaluations required and improving the efficiency of the search process. Next, the algorithm employs the classical NSGA-II as the optimizer to guide the evolution of the population based on the surrogate evaluated objective values (as shown in line 2). During this stage, NSGA-II leverages the objective function values and the dominance relationships among individuals to drive the population toward more promising regions, thereby expanding the coverage of the solution space. After ω generations of evolution, the algorithm selects the individual with the highest HV indicator value from the surrogate evaluated population and adds it to the candidate solution set (as shown in lines 8–11). This process ensures that the selected solution makes a significant contribution to the HV of the current solution set, accelerating the population's progression toward the unconstrained PF. However, in some cases, if the maximum HV indicator value in the population is 0, it indicates that the candidate solutions in the current population no longer contribute to the HV. In such situations, relying solely on the HV indicator metric fails to effectively distinguish between candidate solutions. To address this issue, the algorithm incorporates an uncertainty-based selection strategy, where the individual with the highest average uncertainty $\hat{\sigma}_{\text{mean}}$ is selected and added to the candidate solution set (as shown in lines 12–14).

3.4.3 Pull Search Stage

In the stage, the primary objective of the algorithm is to efficiently locate and explore the constraint PF. To achieve this, the study proposes an adaptive infill criterion based on subregion division. This method not only precisely guides the population toward the constraint PF but also effectively maintains population diversity. Specifically, for the solutions in a given subregion Δ_i, which belong to

Algorithm 3.12: PushSearch.

Input: The archive of expensive evaluated solutions P_{arch}, the number of selected candidates N_{sel}, the surrogate evaluated population size N_{surr}, the number of generations for surrogate assisted search $t_{\max}$

Output: Candidate solutions for expensive evaluation P_{arch}

$S_{\text{models}} \leftarrow$ Train GPR models separately for each objective using the individuals in P_{arch};
$P_{\text{surr}} \leftarrow$ NSGA-II($P_{\text{arch}}, S_{\text{models}}, t_{\max}$);
$P_{\text{cand}} = \emptyset$;
foreach $\mathbf{x} \in P_{\text{surr}}$ **do**
  $\hat{\sigma}_{\text{mean}}(\mathbf{x}) \leftarrow \frac{1}{m} \sum \hat{\sigma}_i(\mathbf{x})$;
end
for $i \leftarrow 1 : N_{\text{sel}}$ **do**
  foreach $\mathbf{x} \in P_{surr}$ **do**
    $HVI_{\mathbf{x}} \leftarrow HV(\mathbf{x} \cup P_{\text{cand}} \cup P_{\text{arch}}) - HV(P_{\text{cand}} \cup P_{\text{arch}})$;
  end
  $\mathbf{x} = \underset{\mathbf{x} \in P_{\text{surr}} \setminus P_{\text{cand}}}{\text{argmax}} \; HVI_{\mathbf{x}}$;
  if $HVI_{\mathbf{x}} = 0$ **then**
    $\mathbf{x} \leftarrow \underset{\mathbf{x} \in P_{\text{surr}} \setminus P_{\text{cand}}}{\text{argmax}} \; \hat{\sigma}_{\text{mean}}(\mathbf{x})$;
    $P_{\text{cand}} \leftarrow P_{\text{cand}} \cup \mathbf{x}$;
  else
    $P_{\text{cand}} \leftarrow P_{\text{cand}} \cup \mathbf{x}$;
  end
end

the expensive evaluated population or the surrogate evaluated population, their distributions can be classified into three distinct scenarios:

1. At least one feasible solution exists in the subregion, indicating that the region contains solutions satisfying the constraints.
2. Only infeasible solutions exist in the subregion, indicating that the solutions in this region do not meet the constraint conditions.
3. No solutions exist in the subregion, indicating that the region has not been explored or lacks candidate solutions.

Based on the three possible scenarios, nine distinct cases can be identified, as shown in Figure 3.18. In this figure, the letters A, B, and C denote the three possible distribution scenarios observed within the expensive evaluated population,

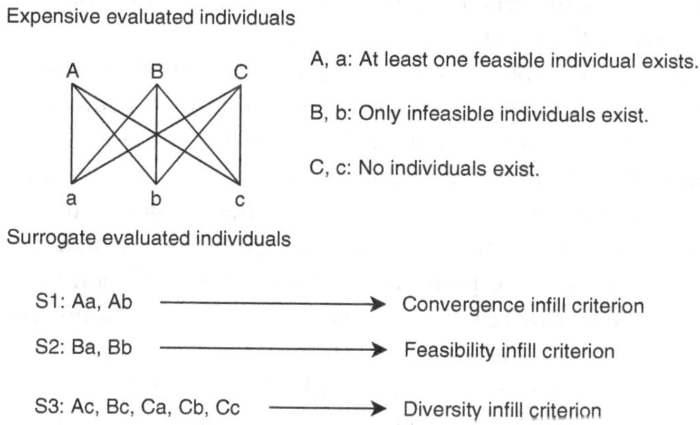

Figure 3.18 An adaptive infill criterion based on subregion division.

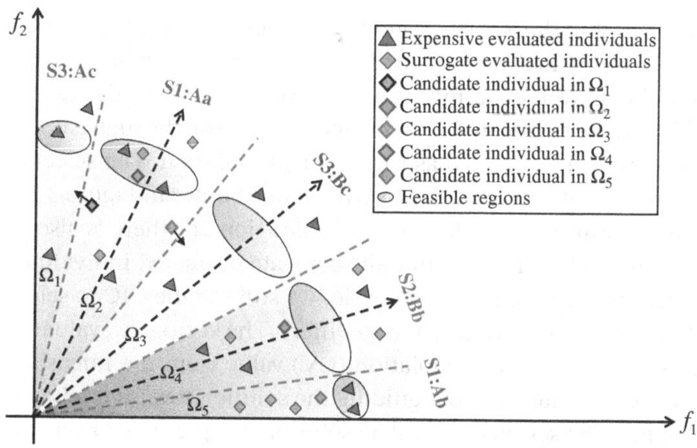

Figure 3.19 Illustrations for selecting candidate individuals in the objective space.

whereas a, b, and c refer to the corresponding three distribution scenarios identified within the surrogate evaluated population.

The specific correspondence between Figures 3.18 and 3.19 can be described as follows: Figure 3.19 provides a visual representation of the procedure used to select candidate individuals within the objective space, offering a clear depiction of how the selection process is structured and carried out. In this illustration, the objective space is partitioned into five distinct subregions, denoted as Ω_1, Ω_2, Ω_3, Ω_4, and Ω_5. Each of these subregions aligns with one of the five scenarios depicted in Figure 3.18, specifically corresponding to S3:Ac, S1:Aa, S3:Bc, S2:Bb, and S1:Ab.

This correspondence clarifies the relationship between the different subregions and the distribution states of the evaluated populations.

In the cases of S3:Ac and S3:Bc, subregions Ω_1 (or Ω_3) contain expensive evaluated individuals but lack surrogate evaluated individuals. To address this situation, the algorithm employs the diversity infill criterion (DIC) to select candidate solutions. Under this criterion, the diversity values of the surrogate evaluated individuals within the neighborhood are computed. These values are derived by combining two components: the Chebyshev decomposition value, which measures the deviation from a reference point, and the constraint violation value, which quantifies the degree to which a solution fails to satisfy the constraints. This approach effectively measures the distribution of surrogate evaluated individuals in the solution space. The surrogate evaluated individual with the smallest diversity value is ultimately selected as the candidate solution, ensuring a uniform distribution of the population in the objective space. In Figure 3.19, the candidate solutions selected from Ω_1 and Ω_3 are marked with diamonds outlined in blue and yellow, respectively.

For the case of S1:Aa, within the feasible region of subregion Ω_2, there is at least one expensive evaluated individual and one surrogate evaluated individual. In this scenario, we employ the convergence infill criterion (CIC) to select the candidate solution. Specifically, the surrogate evaluated individual with the highest HVI value from the feasible region is selected as the candidate solution. The selected candidate solution is represented by a diamond with a blue border in Figure 3.19. For the case of S1:Ab, within the feasible region of subregion Ω_5, there is also at least one expensive evaluated individual, but all surrogate evaluated individuals are located in the infeasible region. In this scenario, we still use the CIC to select the candidate solution, but the selection criterion differs. The surrogate evaluated individual with the smallest constraint violation (CV) value from the infeasible region is chosen as the candidate solution, guiding the solution toward gradually satisfying the constraints. The selected candidate solution is represented by a diamond with a purple border in Figure 3.19.

In the case of S2:Bb, all individuals within subregion Ω_4, whether expensive evaluated individuals or surrogate evaluated individuals, are located in the infeasible region. To address this situation, the algorithm employs the feasibility infill criterion (FIC) to select candidate solutions. Specifically, the surrogate evaluated individual with the highest feasibility value is selected from the infeasible region as the next candidate solution. This selection strategy effectively guides the population toward the feasible region, gradually reducing constraint violations. The selected candidate solution is highlighted in Figure 3.19 with a red-bordered diamond, clearly indicating its position and status in the solution space.

The detailed descriptions of the infill strategies focusing on convergence, feasibility, and diversity are as follows:

1. **CIC**

To ensure that individuals within a subregion consider both distribution and convergence during the search process, CIC combines HVI with CDP, forming a candidate solution selection criterion known as HVI-CDP. This criterion balances exploration and exploitation, optimizing the diversity and quality of the solution set. According to HVI-CDP, solution $\mathbf{x}_a$ is deemed superior to solution $\mathbf{x}_b$ when one or more of the following conditions are met:

   a. Both $\mathbf{x}_a$ and $\mathbf{x}_b$ are infeasible, but the constraint violation value $CV(\mathbf{x}_a)$ of $\mathbf{x}_a$ is smaller than that of $\mathbf{x}_b$ ($CV(\mathbf{x}_b)$).
   b. $\mathbf{x}_a$ is a feasible solution, while $\mathbf{x}_b$ is infeasible.
   c. Both $\mathbf{x}_a$ and $\mathbf{x}_b$ are feasible, and the HVI value of $\mathbf{x}_a$ is greater than that of $\mathbf{x}_b$.

2. **FIC**

The objective of FIC is to prioritize individuals with the highest feasibility as candidate solutions, ensuring that the population gradually approaches the feasible domain. To quantify the feasibility of surrogate evaluated individuals, the algorithm introduces a feasibility indicator I_{fea}, where a smaller value of I_{fea} indicates higher feasibility of the surrogate evaluated individual. Specifically, the definition of I_{fea} varies depending on the feasibility status of the individual:

   a. If $\mathbf{x}$ is an infeasible solution, I_{fea} represents the total constraint violation value and is calculated as follows:

$$I_{\text{fea}} = \sum_{i=1}^{p} \max(\hat{g}_i(\mathbf{x}), 0) + \sum_{j=1}^{q} \max(\hat{h}_j(\mathbf{x}) - \delta, 0) \tag{3.9}$$

   Here, $\hat{g}_i(\mathbf{x})$ represents the predicted value of the inequality constraint, $\hat{h}_j(\mathbf{x})$ represents the predicted value of the equality constraint, and δ is a small tolerance constant to account for the precision of equality constraints.

   b. If $\mathbf{x}$ is a feasible solution, I_{fea} represents the sum of the constraint function values and is calculated as follows:

$$I_{\text{fea}} = \sum_{i=1}^{p} \min(\hat{g}_i(\mathbf{x}), 0) + \sum_{j=1}^{q} \min(\hat{h}_j(\mathbf{x}) - \delta, 0) \tag{3.10}$$

   From the definitions above, it can be observed that when $\mathbf{x}$ is predicted to be an infeasible solution, its I_{fea} value equals the sum of all constraint violation values. When $\mathbf{x}$ is predicted to be a feasible solution, its I_{fea} value equals the sum of the constraint function values. A smaller I_{fea} value indicates that the solution $\mathbf{x}$ lies deeper within the feasible domain, implying stronger feasibility. By leveraging this indicator, FIC effectively guides the population toward the feasible domain.

3. DIC

The objective of DIC is to encourage the population to explore different subregions as broadly as possible, avoiding convergence to local optima and thereby improving the comprehensiveness of the search and the diversity of solutions. When a subregion lacks expensive evaluated individuals or surrogate evaluated individuals, the algorithm designs a corresponding subproblem for that subregion as part of the infill criterion. The objective function of this subproblem is defined as follows:

$$f_{\mathrm{sub}}(\mathbf{x}|\lambda) = g_{\mathrm{tch}}(\mathbf{x}|\lambda) + CV(\mathbf{x}) \tag{3.11}$$

Here, $g_{\mathrm{tch}}(\mathbf{x}|\lambda)$ represents the Chebyshev decomposition function, which measures the deviation of a solution from a reference point, and $CV(\mathbf{x})$ represents the constraint violation value, which evaluates the feasibility of the solution. Through this mechanism, DIC encourages the population to cover unexplored regions, further enhancing the global exploration capability of the solution space.

Algorithm 3.13 presents the detailed pseudocode for the pull search stage. In lines 1–2, the BatchBALDS method is first used to select training data, and based on this data, surrogate models for the objective functions and constraints are trained separately, providing a reliable modeling foundation for subsequent optimization. In line 3, the algorithm utilizes the CCMO optimizer to optimize the surrogate models, replacing the real function evaluations, which significantly reduces computational costs while maintaining optimization efficiency. In lines 5–6, the algorithm associates each individual in the surrogate evaluated population P_{surr} and the expensive evaluated archive set A with a weight vector λ_i, laying the groundwork for the subsequent division of the objective space into regions. In line 7, the objective space is divided into N_{sel} non-overlapping subregions, denoted as Δ_i, to ensure a fine-grained optimization process. Specifically, the strategy for dividing the regions is based on the number of candidate solutions, N_{sel}. First, N_{sel} weight vectors are evenly generated within the objective space. Then, for each point in the objective space, the distances from this point to all the weight vectors are calculated, and the point is assigned to the subregion associated with the nearest weight vector. Consequently, the objective space is effectively divided into N_{sel} non-overlapping subregions. In lines 10–11, the surrogate evaluated population is divided into N_{sel} subpopulations, and the state of the solutions within each subregion Δ_i is evaluated to provide a basis for selecting candidate solutions. Subsequently, an appropriate candidate selection criterion is applied based on the state of each subregion: in lines 12–13, if the state of Δ_i is $S1$, the CIC criterion is used to select a candidate solution; in lines 14–15, if the state is $S2$, the FIC criterion is adopted; and in lines 16–17, if the state is $S3$, the DIC criterion is applied.

Algorithm 3.13: PullSearch.

Input: The archive of expensive evaluated solutions P_{arch}, the number of selected candidates N_{sel}, the surrogate evaluated population size N_{surr}, the number of generations for surrogate assisted search t_{max}

Output: Candidate solutions for expensive evaluation P_{arch}

$D \leftarrow$ BatchBALDS(P_{arch}, N_{D1}, N_{D2});

$S_{models} \leftarrow$ Train GPR models separately for each objective and constraint using the individuals in D;

$P_{surr} \leftarrow$ CCMO(P_{arch}, S_{models}, t_{max});

Set uniform reference vectors $\{\lambda_1, \ldots, \lambda_{N_{sel}}\}$;

Assign each individual $\mathbf{x} \in P_{surr}$ to its closest reference vector;

Assign each individual $\mathbf{x} \in P_{arch}$ to its closest reference vector;

Divide the objective space into N_{sel} subregions Δ_i, for $i \in \{1, \ldots, N_{sel}\}$;

$P_{cand} = \emptyset$;

for $i \leftarrow 1 : N_{sel}$ **do**

  $P_{surr}^{\Delta_i} \leftarrow P_{surr}$ in Δ_i;

  Determine the state S_{Δ_i} of subregion Δ_i;

  if S_{Δ_i} *is S1* **then**

    $\mathbf{x} \leftarrow$ HVICDP($P_{surr}^{\Delta_i}$, 1);

  else if S_{Δ_i} *is S2* **then**

    $\mathbf{x} \leftarrow \max_{\mathbf{x} \in P_{surr}^{\Delta_i}} I_{fea}(\mathbf{x})$;

  else if S_{Δ_i} *is S3* **then**

    $\mathbf{x} \leftarrow \min_{\mathbf{x} \in P_{surr}} g_{tch}(\mathbf{x}|\lambda_i) + CV(\mathbf{x})$;

  end

  $P_{cand} \leftarrow P_{cand} \cup \mathbf{x}$;

end

3.4.4 BatchBALDS

As the size of the training dataset increases, the training time for the surrogate model grows significantly, potentially leading to substantial computational overhead. Moreover, training data located far from the PF can interfere with the model's learning process, preventing it from focusing on critical regions and thereby reducing its prediction ACC near the constraint PF. To address these issues, we propose BatchBALDS. This method optimizes the selection of training data in two stages: in the first stage, a set of samples, $D1$, is selected to enhance the prediction ACC of the surrogate model near the constraint PF region; in the second stage, an additional set of samples, $D2$, is selected to reduce the global prediction error of the surrogate model across the entire objective space.

To generate the sample set $D1$, we select data from both the infeasible and feasible regions near the constraint PF. For sample selection in the infeasible region, we draw inspiration from the BiCo algorithm [264]. Specifically, the constraint violation values are treated as an additional optimization objective, and non-dominated sorting is performed. Through this sorting process, a set of non-dominated solutions is obtained, from which the individuals classified as infeasible solutions for the original problem are defined as the population U, representing the samples located on the infeasible side of the constraint PF. Subsequently, the constraints of the original optimization problem are ignored, and another round of non-dominated sorting is performed on population U. In this new sorting structure, the individuals in the earlier non-dominated layers are farther from the constraint PF, while those in the later layers are closer to the constraint PF. Therefore, to filter samples, we begin with the last non-dominated layer and progressively select individuals, using crowding distance to choose individuals within the same non-dominated layer. This selection process essentially operates as the reverse of the selection step in NSGA-II. In the feasible region, we directly apply NSGA-II to select samples, ensuring that the chosen data accurately represents the distribution of solutions within the feasible region. For the construction of dataset $D2$, we employ a batch data selection method based on Bayesian active learning. By selecting a set of data points with the highest uncertainty, our goal is to minimize the surrogate model's prediction error on a global scale. To achieve this, we introduce a greedy approximation strategy, where at each iteration, the individual with the highest uncertainty is selected and added to dataset $D2$. This process continues until the number of selected samples reaches a predefined threshold.

Algorithm 3.14 provides a detailed explanation of the BatchBALDS method. First, in lines 1–4, samples are selected from individuals near the constraint PF, which are then used to construct the initial training dataset $D1$, ensuring that the dataset effectively covers critical regions. Next, as shown in lines 8 and 20, the GPR model is trained for each objective function and constraint, establishing the predictive framework for the surrogate model. Following this, to further optimize the training dataset $D2$, the algorithm employs a greedy iterative strategy, prioritizing the selection of individuals with the highest uncertainty and adding them to $D2$. This step corresponds to the operations in lines 10–17 and 21–22. Finally, the algorithm removes duplicate individuals from $D2$ to ensure data uniqueness and merges $D1$ with $D2$ to form the final training dataset, which is then used for the current round of GPR model training.

In the process of sample selection, it is necessary to compute the uncertainty $\sigma(\mathbf{x}|D2)$. According to the fundamental assumptions of Gaussian processes, in the absence of noise, the joint distribution of the test point $\mathbf{x}^*$ and the training set $D1$

Algorithm 3.14: BatchBALDS.

Input: The archive of expensive evaluated solutions P_{arch}, the number of selected candidates in the first stage N_{D1}, the number of selected candidates in the second stage N_{D2}

Output: The data for training surrogate models D_{train}

1 $U \leftarrow$ Select infeasible solutions from P_{arch} and construct a new solution set using CV as an additional objective;
2 $D1_1 \leftarrow$ Select $N_{D1}/2$ individuals from U using the NSGA-II selection operator in reverse order;
3 $D1_2 \leftarrow$ Select $N_{D1}/2$ individuals from P_{arch} using the NSGA-II selection operator;
4 $D1 \leftarrow D1_1 \cup D1_2$;
5 $D_{\text{pool}} \leftarrow P_{\text{arch}} \backslash D1$;
6 $D2 \leftarrow \emptyset$;
7 **for** $i \leftarrow 1 : m$ **do**
8   Use $D1$ to train the GPR model for the i-th objective;
9     $D_{\text{sel}} \leftarrow \emptyset$;
10   **for** $j \leftarrow N_{D2}$ **do**
11   **foreach** $\mathbf{x} \in D_{\text{pool}} \backslash D_{\text{sel}}$ **do**
12             $S_{\mathbf{x}} \leftarrow \sigma(\mathbf{x}|D_{\text{sel}})$;
13   **end**
14         $\mathbf{x}_j = \underset{\mathbf{x} \in D_{\text{pool}} \backslash D_{\text{sel}}}{\text{argmax}} \; S_{\mathbf{x}}$;
15         $D_{\text{sel}} \leftarrow D_{\text{sel}} \cup \mathbf{x}_j$;
16   **end**
17     $D_2 \leftarrow D_2 \cup D_{\text{sel}}$;
18 **end**
19 **for** $i \leftarrow 1 : q$ **do**
20   Use $D1$ to train the GPR model for the i-th constraint;
21     $D_{\text{sel}} \leftarrow$ Use the same method as shown in lines 10-16 to obtain the training data;
22     $D_2 \leftarrow D_2 \cup D_{\text{sel}}$;
23 **end**
24 Remove duplicate individuals from $D2$;
25 $D_{\text{train}} \leftarrow D1 \cup D2$;

(containing N_{D1} samples) as well as the dataset $D2$ (containing l samples) can be modeled as a Gaussian distribution, expressed as follows:

$$\begin{bmatrix} \mathbf{y}_{D1 \cup D2} \\ y^* \end{bmatrix} \sim N(\begin{bmatrix} \mu(X_{D1 \cup D2}) \\ \mu(\mathbf{x}^*) \end{bmatrix}, \begin{bmatrix} K(X_{D1 \cup D2}, X_{D1 \cup D2}) & K(\mathbf{x}^*, X_{D1 \cup D2})^T \\ K(\mathbf{x}^*, X_{D1 \cup D2}) & K(\mathbf{x}^*, \mathbf{x}^*) \end{bmatrix}) \tag{3.12}$$

where $K(X_{D1 \cup D2}, X_{D1 \cup D2})$ denotes the covariance matrix between the samples in $D1$ and $D2$, and $K(\mathbf{x}, X_{D1 \cup D2})$ represents the covariance vector between the test point $\mathbf{x}^*$ and the training data $X_{D1 \cup D2}$. The specific definitions of these covariance matrices and vectors are as follows:

$$K(X_{D1 \cup D2}) = \begin{bmatrix} k(x_{D1}^1, x_{D1}^1) & \dots & k(x_{D1}^1, x_{D2}^l) \\ \vdots & \ddots & \vdots \\ k(x_{D2}^l, x_{D2}^1) & \dots & k(x_{D2}^l, x_{D2}^l) \end{bmatrix} \tag{3.13}$$

$$K(\mathbf{x}^*, X_{D1 \cup D2}) = \begin{bmatrix} k(\mathbf{x}^*, x_{D1}^1) & \dots & k(\mathbf{x}^*, x_{D2}^i) \end{bmatrix} \tag{3.14}$$

Since the conditional distribution of a Gaussian distribution is still Gaussian, the predictive variance $\sigma(\mathbf{x}^*)$ corresponding to the test point $\mathbf{x}^*$ can be derived mathematically as follows [265]:

$$\sigma(\mathbf{x}^*) = K(\mathbf{x}^*, X_{D1 \cup D2}) K(X_{D1 \cup D2}, X_{D1 \cup D2})^{-1} (\mathbf{y}_{D1 \cup D2} - \mu(X_{D1 \cup D2})) \tag{3.15}$$

3.4.5 Experimental Study

To comprehensively evaluate the performance of SA-PPS, we select nine representative algorithms for comparative analysis, including MGSAEA [266], KTS [267], M1-2 [268], M2-2 [268], CCMO [68], MultiObjectiveEGO [269], SA-CRVEA-AS [270], PPS-MOEA/D [14], and PAC-MOO [271]. For the comparison experiments, we use two widely adopted test problem sets: MW [272] and LIRCMOP [257]. The number of decision variables is uniformly set to 10 to ensure fairness and comparability in the experiments.

For all comparing algorithms, the parameter configurations are set according to [76]. In addition, three commonly used metrics are adopted: IGD [78], HV [273], and the success rate (SR) of finding feasible solutions. Among them, the IGD and HV metrics mainly measure the convergence and diversity of the algorithms, while SR assesses the proportion of successful runs in which feasible solutions are found out of 30 independent runs. Additionally, the Wilcoxon rank-sum test with a significance level of 0.05 is applied to perform statistical analysis of the results, enabling us to determine whether the performance differences between algorithms are significant. It is important to note that, for any algorithm failing to find feasible solutions in a given test case, the IGD value is assigned a large constant (set to 1e20 in this study), and the HV value is set to 0. This approach ensures consistency and reliability in the statistical tests.

3.4.5.1 Experimental Results on MWs

Table 3.9 summarizes the Wilcoxon rank sum significance test results for SA-PPS and nine other ECMOEA algorithms on the MW1-MW14 test problems. From the IGD and HV metrics, it is evident that SA-PPS achieves the best performance on six test problems, demonstrating exceptional capability in balancing population convergence and diversity. Following SA-PPS, KTS, MGSAEA, MultiObjectiveEGO, and SA-CRVEA-AS perform best on four, two, one, and one test problems, respectively. This indicates that while other algorithms show strengths in specific cases, SA-PPS consistently exhibits more comprehensive and stable performance across both IGD and HV metrics. In terms of SR, SA-PPS achieves a 100% SR on 12 test problems within the MW test set, significantly outperforming the comparison algorithms. Specifically, CCMO, PPS-MOEA/D, MultiObjectiveEGO, M1-2, M2-2, MGSAEA, KTS, PAC-MOO, and SA-CRVEA-AS achieve 100% SRs on 6, 1, 8, 6, 6, 10, 9, 2, and 10 test problems, respectively. Although MGSAEA and SA-CRVEA-AS also demonstrate high SRs on some test problems, SA-PPS shows a clear advantage across a wider range of problems. This indicates that SA-PPS is more effective at identifying feasible solutions, particularly when tackling problems with higher constraint complexity, and exhibits superior applicability and robustness.

Figure 3.20 illustrates the feasible non-dominated solution sets with median HV values obtained by 10 algorithms after 30 independent runs on the MW13 test problem. The constrained PF of MW13 lies in three discontinuous and narrow feasible regions, posing significant challenges to population-based search algorithms. However, as shown in the figure, SA-PPS demonstrates remarkable advantages in terms of both convergence and diversity. These advantages primarily stem from two key factors. First, during the push stage, SA-PPS conducts its search without considering constraints, effectively bypassing the limitations imposed by the narrow feasible regions. This significantly enhances the algorithm's exploratory capabilities in the early stages of optimization. Second, SA-PPS incorporates an adaptive region partitioning strategy, which dynamically adjusts the distribution

Table 3.9 Wilcoxon ranksum significance test results (win/loss/tie) of SA-PPS and the other 9 ECMOEAs on MW1-MW14 and LIRCMOP1-LIRCMOP14.

Test problem set	Metrics	Algorithms									
		PPS-MOEA/D	CCMO	Multi Objective EGO	M1–2	M2–2	MGSAEA	KTS	PAC-MOO	SA-CRVEA-AS	SA-PPS
MW1–14	IGD	0/14/0	0/13/1	1/12/1	2/11/1	2/11/1	2/9/3	4/6/4	0/14/0	3/6/5	—
	HV	0/14/0	0/13/1	1/12/1	0/13/1	0/13/1	1/10/3	4/6/4	0/14/0	0/10/4	—
LIRCMOP1–14	IGD	0/14/0	0/14/0	0/14/0	0/14/0	1/12/1	3/11/0	6/6/2	0/14/0	1/5/8	—
	HV	0/14/0	0/14/0	0/14/0	0/14/0	0/13/1	3/11/0	6/6/2	0/14/0	1/10/3	—

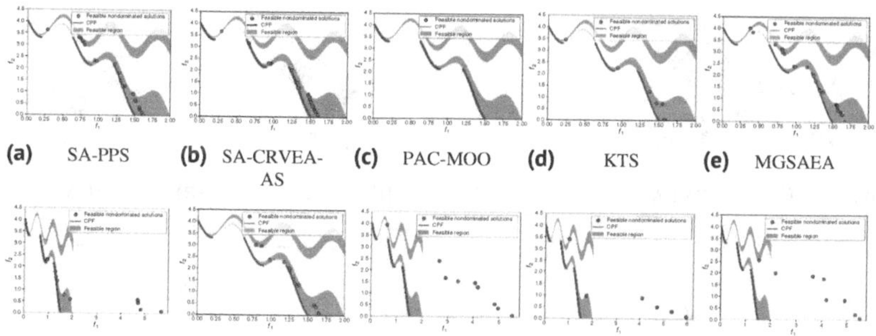

(a) SA-PPS (b) SA-CRVEA-AS (c) PAC-MOO (d) KTS (e) MGSAEA

(f) M1-2 (g) M2-2 (h) Multi objective EGO (i) CCMO (j) PPS-MOEA/D

Figure 3.20 The feasible non-dominated solution sets with median HV values obtained by 10 algorithms after 30 independent runs on the MW13 test problem.

of solutions and further strengthens the diversity and global search ability of the population. In comparison, while MGSAEA and KTS manage to identify solutions near the three PF regions, their convergence performance still requires substantial improvement. M1-1 and M2-2, on the other hand, can only approach one of the PFs, and their diversity remains inferior to that of SA-PPS. Furthermore, MultiObjectiveEGO, CCMO, and PPS-MOEA/D exhibit significantly poorer performance in both diversity and convergence compared to SA-PPS. It is worth noting that SA-CRVEA-AS achieves results comparable to those of SA-PPS on MW13, indicating its competitiveness in handling such complex problems. However, PAC-MOO fails to find any feasible solutions in this test, further highlighting the stability and robustness of SA-PPS when tackling highly constrained optimization problems.

Figure 3.21 illustrates the feasible non-dominated solution sets corresponding to the median HV values obtained by SA-PPS and the other nine comparison algorithms after 30 independent runs on the MW14 test problem. The most notable characteristic of the MW14 problem is that its PF consists of multiple disconnected surfaces, which poses significant challenges for any algorithm in terms of global exploration capability, local exploitation efficiency, and solution distribution. As shown in the figure, SA-PPS outperforms the other nine comparison algorithms significantly. On MW14, SA-PPS not only successfully identifies all the disconnected PF surfaces but also achieves a uniform distribution of solutions within each PF region. This indicates that SA-PPS strikes an effective balance between global exploration and local exploitation when addressing complex, disconnected PFs. The key to this exceptional performance lies in its push search stage, which adopts a global exploration strategy free from constraint limitations, allowing the algorithm to search the solution space more comprehensively.

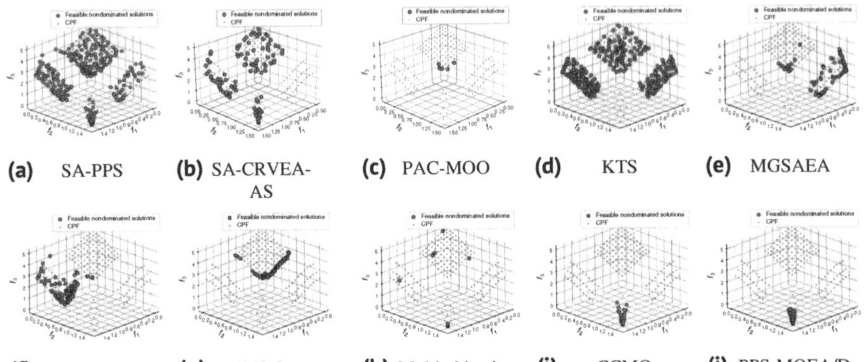

(a) SA-PPS **(b)** SA-CRVEA-AS **(c)** PAC-MOO **(d)** KTS **(e)** MGSAEA

(f) M1-2 **(g)** M2-2 **(h)** Multi objective EGO **(i)** CCMO **(j)** PPS-MOEA/D

Figure 3.21 The feasible non-dominated solution sets with median HV values obtained by 10 algorithms after 30 independent runs on the MW14 test problem.

Additionally, the region partitioning strategy employed during the pull search stage enhances the diversity and global distribution of solutions. In contrast, none of the other nine comparison algorithms are able to locate all the PF surfaces in MW14. While some algorithms can identify certain PF regions, their exploration of distant or narrow PF surfaces is notably inadequate, resulting in a significant decline in solution diversity and coverage. For instance, certain algorithms exhibit strong performance in regions close to the initial population but fail to adequately explore more distant PF surfaces. Furthermore, some algorithms struggle with sufficient population expansion in such complex, disconnected problems, highlighting their limitations. In summary, the superior performance of SA-PPS not only demonstrates its outstanding capability in addressing complex and disconnected PF problems but also underscores its highly coordinated global exploration and local exploitation capabilities.

3.4.5.2 Experimental Results on LIRCMOPs

We conducted an in-depth analysis of the performance of the SA-PPS algorithm and nine other comparison algorithms on the LIRCMOP test suite. A notable characteristic of the LIRCMOP test problems is that some problems have extremely narrow feasible solution regions, posing significant challenges to the search capabilities of algorithms. Table 3.9 summarizes the Wilcoxon rank-sum test results of SA-PPS and the nine ECMOEA algorithms on the LIRCMOP1-LIRCMOP14 problems. The results clearly indicate that SA-PPS significantly outperforms CCMO, PPS-MOEA/D, M1-2, MultiObjectiveEGO, and PAC-MOO across all test problems, demonstrating strong performance advantages. Further analysis shows that compared to M2-2, MGSAEA, KTS, and SA-CRVEA-AS, SA-PPS achieves significantly better performance on 12, 11, 6, and 4 problems, respectively. This performance gap may stem from the considerable distance between constrained PFs and unconstrained PFs, which requires substantial computational resources for the population to transition from unconstrained PFs to constrained PFs. However, the infill criterion introduced by SA-PPS effectively accelerates this transition process, significantly improving the algorithm's efficiency and capability in finding feasible solutions under complex constraints. Specifically, SA-PPS achieved the best performance on LIRCMOP3-4, LIRCMOP7, and LIRCMOP12-14, showcasing its adaptability and robustness in solving complex problems. On the other hand, KTS demonstrated excellent performance on six problems, including LIRCMOP1-2, LIRCMOP5, and LIRCMOP9-11. Its success can largely be attributed to its adaptive search mode, which effectively reduces computational overhead when searching on unconstrained PFs.

Figures 3.22 and 3.23 illustrate the feasible non-dominated solution sets with median HV values obtained by SA-PPS and the other nine comparison algorithms after 30 independent runs on the LIRCMOP12 and LIRCMOP14 test problems, respectively. These results clearly highlight the significant advantages of SA-PPS in addressing complex CMOPs. First, as shown in Figure 3.22, SA-PPS demonstrates superior convergence performance on the LIRCMOP12 problem compared to the other algorithms. This remarkable performance can be attributed to the synergistic effects of its PPS stages. During the push search stage, SA-PPS temporarily disregards constraints, enabling the algorithm to explore the entire solution space more broadly. This characteristic effectively reduces the interference

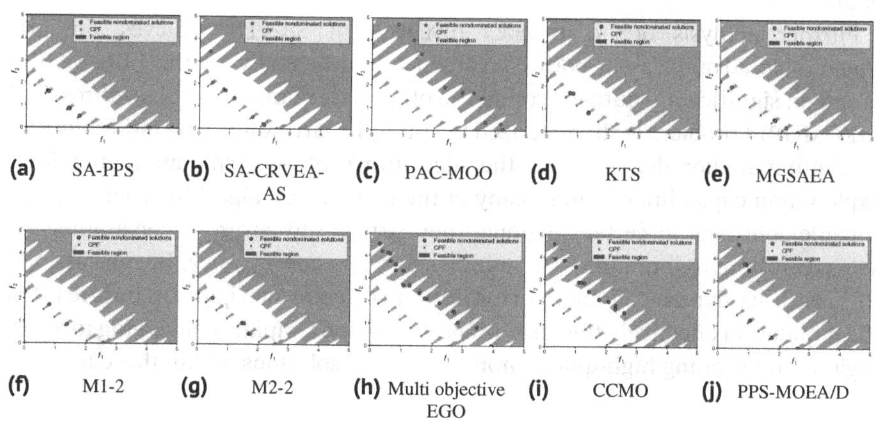

(a) SA-PPS (b) SA-CRVEA-AS (c) PAC-MOO (d) KTS (e) MGSAEA

(f) M1-2 (g) M2-2 (h) Multi objective EGO (i) CCMO (j) PPS-MOEA/D

Figure 3.22 The feasible non-dominated solution sets with median HV values obtained by 10 algorithms after 30 independent runs on the LIRCMOP12 test problem.

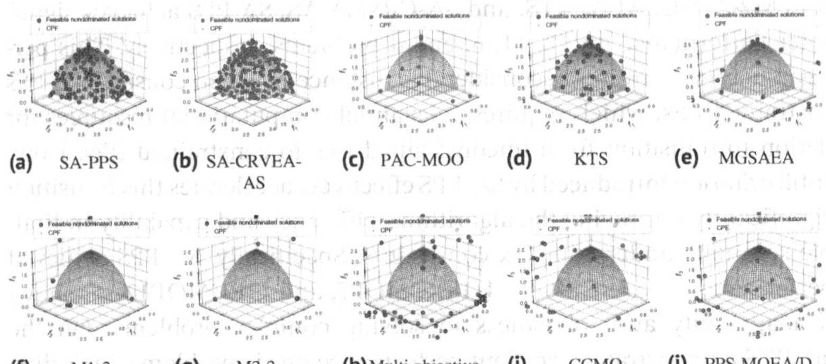

(a) SA-PPS (b) SA-CRVEA-AS (c) PAC-MOO (d) KTS (e) MGSAEA

(f) M1-2 (g) M2-2 (h) Multi objective EGO (i) CCMO (j) PPS-MOEA/D

Figure 3.23 The feasible non-dominated solution sets with median HV values obtained by 10 algorithms after 30 independent runs on the LIRCMOP14 test problem.

caused by infeasible regions in the evolutionary path, preventing the population from getting trapped in local feasible areas. Meanwhile, during the pull search stage, SA-PPS employs an infill criterion based on subregion partitioning to optimize the selection of candidate solutions. This method allows for precise adjustments to the diversity of solutions in different regions, ensuring well-distributed and comprehensive solutions near the complex and discontinuous constrained PF. Consequently, SA-PPS exhibits outstanding performance in searching for non-dominated solutions in LIRCMOP12, successfully identifying all disconnected constrained PF regions while maintaining high uniformity and diversity of the population.

Further analysis of Figure 3.23 reveals that SA-PPS achieves similarly remarkable advantages in both convergence and diversity on the LIRCMOP14 problem, significantly outperforming the other nine comparison algorithms. The LIRCMOP14 problem is characterized by multiple narrow and discontinuous PFs, presenting higher demands on the algorithm's global exploration and local exploitation capabilities. While many of the comparison algorithms achieve reasonable solutions in certain regions, they struggle to cover all constrained PF areas, resulting in limitations in both diversity and convergence. In contrast, SA-PPS, with its dynamically adaptive solution selection mechanism, ensures that the population effectively traverses infeasible regions, rapidly converging to all PF regions and locating high-quality non-dominated solutions within those regions.

3.4.6 Conclusion

This paper introduces a novel SA-PPS designed specifically for solving complex ECMOPs. By incorporating a PPS framework and a batch data selection strategy based on Bayesian active learning, SA-PPS demonstrates remarkable advantages in terms of convergence, diversity, and computational efficiency. Specifically, during the push stage, the algorithm rapidly approximates the unconstrained PFs by temporarily ignoring constraints, laying a solid foundation for subsequent searches. In the pull stage, a new infill criterion effectively guides the population toward the constrained PFs. Additionally, the batch data selection strategy improves the efficiency of expensive evaluations by eliminating redundant data and selecting high-value solutions as training samples. This ensures the ACC of the surrogate model near the constrained PFs while maintaining overall predictive reliability.

Experimental results show that SA-PPS significantly outperforms state-of-the-art algorithms, including CCMO, PPS-MOEA/D, MGSAEA, and KTS, on multiple test problems. Its advantages are particularly evident in solving problems with narrow feasible regions and disconnected PFs. The superior performance of SA-PPS can be attributed to the synergy between the PPS stages, along with its adaptive region partitioning and Bayesian active learning mechanisms. However, while SA-PPS achieves outstanding results in most cases, the ACC of the surrogate model may decline when dealing with ECMOPs involving high-dimensional decision variables. Therefore, future research should focus on enhancing the algorithm's capability to handle high-dimensional decision spaces and improving the robustness of the surrogate models under such conditions.

3.5 Genetic U-Net

The conventional approach to designing neural network architectures heavily depends on human expertise combined with a repetitive trial-and-error process. This method often proves inefficient, consuming significant amounts of time and resources while still running the risk of failing to discover the optimal network structure. To overcome these limitations, this book introduces a novel framework for automating the architecture design process through evolution-based neural architecture search [179]. Specifically tailored for vision systems in intelligent robots, this method leverages evolutionary algorithms to explore and optimize network architectures autonomously. By removing the reliance on manual intervention, the proposed approach streamlines the design process, improving efficiency and enhancing the likelihood of achieving superior network performance.

This section begins by introducing the search space for neural network architectures, providing a foundation for understanding the range of potential solutions. Next, it details the method used to encode a given network architecture into a binary string representation, which serves as a compact and manageable format for optimization. The final part of this section focuses on the functioning of the genetic algorithm (GA). The GA operates by iteratively improving candidate solutions through a set of evolutionary operations, namely crossover, which combines information from two parent solutions; mutation, which introduces variations to explore new possibilities; and selection, which prioritizes the most competitive architectures for progression. These operations enable the GA to effectively navigate the search space and identify high-performing network architectures.

3.5.1 Search Space and Encoding Mechanism

3.5.1.1 Search Space

As depicted in Figure 3.24, the U-Net architecture comprises two main components: an encoder, denoted as E, and a decoder, referred to as D. These components are constructed using multiple blocks, represented as $e_i(i = 0, 1, 2, 3)$ for the encoder and $d_j(j = 0, 1, 2)$ for the decoder. The U-Net is structured into separate levels, labeled $S_k(k = 0, 1, 2, 3)$, where the feature dimensions remain uniform across each level as the network processes data from top to bottom. To enhance the flow of information, skip connections are implemented at all stages except the final one. These connections transmit features with semantic content, obtained from the encoder, directly to their corresponding decoder levels. This mechanism reinforces the connection between the encoder and decoder while mitigating issues caused by the vanishing gradient during model training [274, 275]. In the decoder, features obtained through skip connections must be fused with those produced by the up-sampling process. The two widely used methods for feature combination are concatenation and element-wise addition. In its original form, U-Net uses concatenation, which combines output feature representations by stacking them together. However, as demonstrated by certain U-Net variants [276], element-wise addition has proven to be an effective alternative.

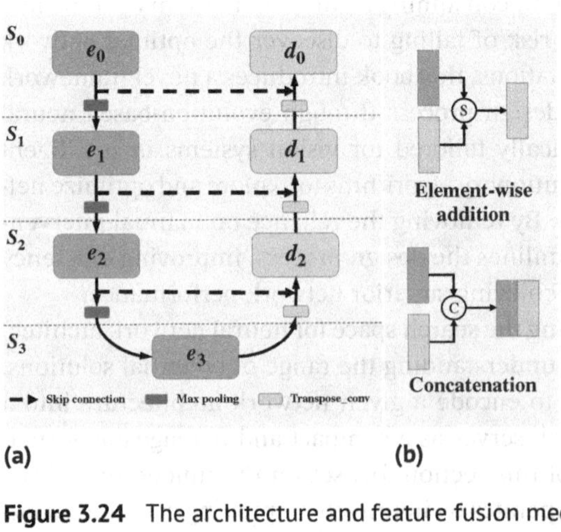

(a) (b)

Figure 3.24 The architecture and feature fusion mechanisms of U-Net.

The key distinction between these methods is that concatenation produces larger feature maps, which increases the computational burden, whereas element-wise addition reduces this complexity by performing a direct sum operation. To balance performance and computational efficiency in this study, we opted for element-wise addition as the feature fusion method, as it reduces computational overhead while maintaining effective information integration. Figure 3.24b highlights the differences between these two fusion strategies.

The proven success of U-Net and its various extensions highlights the effectiveness of the U-shaped encoder–decoder architecture in solving complex tasks. Inspired by this, we incorporated the U-shaped design as the backbone for our proposed model. The architecture, as shown in Figure 3.24a, consists of seven building blocks organized across four distinct stages. This modular U-shaped structure allows flexibility in the design process. By fine-tuning the internal configurations of these blocks (such as the number of convolutional layers, filter sizes, or activation functions), it becomes possible to achieve well-optimized architectures tailored to specific tasks. The hierarchical arrangement of the stages further enables the model to extract and integrate multi-scale features effectively, ensuring both spatial and semantic information is well preserved throughout the network.

3.5.1.2 Components and Their Encoding Mechanism

In the original U-Net architecture, the internal structure of each block consists of two fundamental layers: a 3×3 convolutional layer followed by a ReLU activation function. This simple yet effective design serves as the foundation of U-Net's performance. However, many U-Net variants [277–280] have demonstrated that modifying the internal configuration of these blocks can significantly enhance the model's capability. For instance, several variants have replaced the basic block structure with more advanced designs, such as the ResNet block [281], the DenseNet block [282], and the InceptionNet block [283]. These modifications allow the network to capture richer features, improve gradient flow, and enhance learning efficiency. For example, ResNet blocks introduce skip connections to alleviate the vanishing gradient problem, while DenseNet blocks establish dense connections that promote feature reuse across layers. Similarly, InceptionNet blocks combine multi-scale convolutions to extract features at various receptive fields. Such improvements emphasize the critical role that the internal structures of blocks play in determining the overall performance of U-Net and its extensions.

The internal structures of the building blocks are derived from Genetic CNN [284], providing a flexible design that can accommodate various types of network architectures. This adaptable structure allows the blocks to represent a wide range of architectures, including popular and well-established ones such as ResNet [281], DenseNet [282], and VGG [285]. Each building block can be viewed as a directed acyclic graph, where nodes and edges collectively determine the flow of computation within the block. In this representation, each node corresponds to an operation unit (e.g., a convolution or activation function) or a sequence of operations, while the edges signify the flow of data between nodes. A directed edge connecting two nodes indicates that the output feature map from the preceding node is passed as input to the succeeding node. An essential property of this design is that all feature maps within a block maintain the same dimensions, which ensures compatibility and eliminates the risk of size mismatches during computation. In scenarios where a node receives input from multiple edges, the corresponding feature maps are combined through element-wise summation, effectively integrating information from different sources.

Figure 3.25 provides four examples that demonstrate how connections are established between nodes within a block. To encode these connections, a binary string representation is utilized. The string encodes the presence or absence of connections between nodes, where each bit corresponds to a specific connection. Assuming the maximum number of intermediate nodes in a block is K, the total number of possible internode connections is calculated as $1 + 2 + 3 + \ldots + (K - 1)$, which simplifies to $\frac{K(K-1)}{2}$ bits. Each bit in the binary string represents a potential connection between a pair of nodes. For example, the first bit encodes the connection between $node_1$, $node_2$, the next two bits represent the connections between $node_1$, $node_3$ and between $node_2$, $node_3$, and so on. A bit value of 1 indicates that the two nodes are connected, while a 0 signifies no connection. To ensure valid encoding, certain structural rules must be followed. The default input node (depicted in

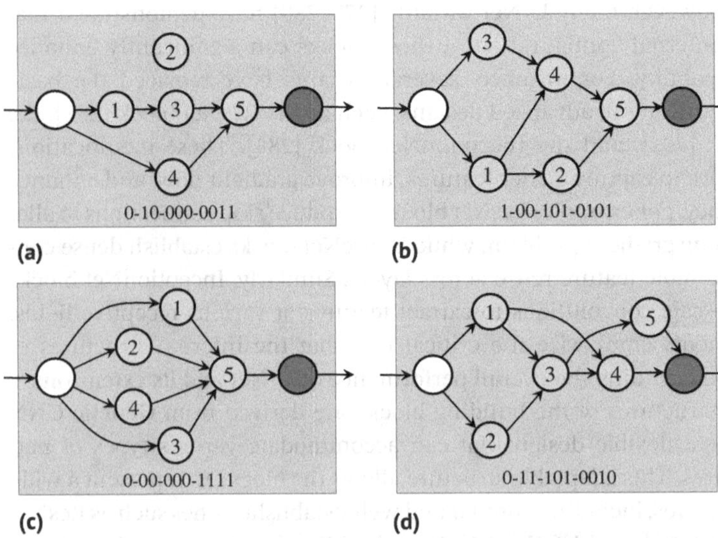

Figure 3.25 Examples of binary encoding for inter-node connections.

white in Figure 3.25) connects only to intermediate nodes that serve as successors but have no predecessors. Conversely, the default output node (marked in gray) is linked exclusively to intermediate nodes that have predecessors but no successors. If an intermediate node neither receives input nor produces output (such as $node_2$ in Figure 3.25a), it is removed from the final network architecture. As a result, the total number of valid nodes within a block can vary dynamically, depending on the encoding.

In Genetic CNN [284], all nodes are predefined to follow a fixed sequence of operations, specifically 3×3 convolution, BN [286], and the ReLU activation function. However, in this work, we expand the design space by offering 16 distinct operation sequences, as outlined in Table 3.10. These operation sequences serve as candidates for nodes, allowing both the optimal structures and the best-suited operations within blocks to be discovered simultaneously through the search process. Each operation sequence is uniquely identified by an ID and consists of several commonly used basic computational units in CNNs, such as 3×3 convolution, 5×5 convolution, ReLU [287], Mish [288], and instance normalization [289]. The Mish activation function, a more recent proposal, shares some similarities with ReLU but possesses additional desirable properties, including continuous differentiability and non-monotonic behavior. These attributes have demonstrated superior performance across various tasks, which is why both ReLU and Mish are incorporated into the search space. The key differences among the 16 operation sequences stem from variations in critical factors, such as convolutional kernel sizes, activation functions, the positioning of activations, and the

Table 3.10 The operation sequences for the nodes.

ID	Operation sequence	ID	Operation sequence
0	3×3 conv → ReLU	8	ReLU → 3×3 conv
1	3×3 conv → Mish	9	Mish → 3×3 conv
2	3×3 conv → IN → ReLU	10	IN → ReLU → 3×3 conv
3	3×3 conv → IN → Mish	11	IN → Mish → 3×3 conv
4	5×5 conv → ReLU	12	ReLU → 5×5 conv
5	5×5 conv → Mish	13	Mish → 5×5 conv
6	5×5 conv → IN → ReLU	14	IN → ReLU → 5×5 conv
7	5×5 conv → IN → Mish	15	IN → Mish → 5×5 conv

IN is instance normalization, ReLU represents Rectified Linear Unit, and Mish indicates a self-regularized non-monotonic neural activation function.

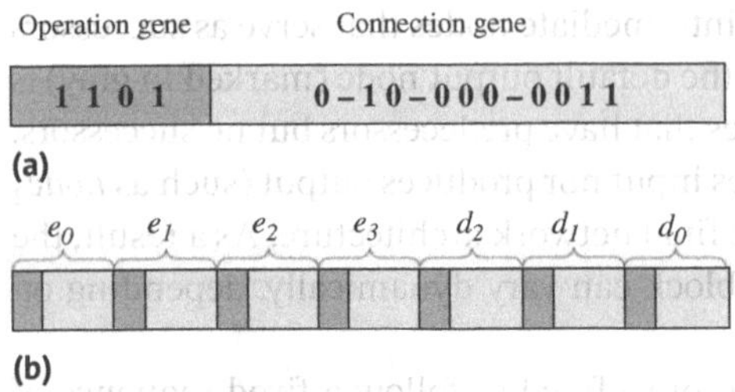

Figure 3.26 Block gene and genotype representation of a neural network architecture.

normalization methods. To efficiently encode these options, a four-bit binary encoding scheme is adopted, where each combination of bits corresponds to one of the 16 operation sequences presented in Table 3.10. To simplify the design, it is assumed that all nodes within a single block share the same operation sequence. As a result, each block gene comprises two components: an operation gene, encoded with four bits to represent the operation sequence, and a connection gene that encodes node connections, as illustrated in Figure 3.26a. In total, the architecture's genotype consists of seven block genes, which collectively define the structure and operations of the network, as shown in Figure 3.26b. This approach allows each block to adopt a unique combination of operations and connections, enabling the search algorithm to explore highly diverse architectures.

3.5.1.3 Neural Architecture Parameter Number

The architectural parameter number is determined by multiple aspects, including the stage number in the U-shaped backbone, the intermediate node number within each block, and the channel number allocated for convolution operations at each node. To balance architectural flexibility with manageable parameter sizes, we initially define the backbone structure for the search space, thereby limiting the total number of parameters. For additional control, we cap the intermediate nodes in each block at a maximum of 5 and restrict the channels used for convolution to 20. These choices help limit the complexity of the architecture. These choices simplify the architecture, ensuring that most network designs within this search space contain fewer than 0.4 million parameters. In contrast, these conventional models often exceed several million parameters, with some reaching tens of millions. By imposing this constraint on the architectural parameters, the search space is more focused, allowing the optimization process to identify better-performing architectures that achieve a higher parameterization efficiency.

3.5.2 Evolutionary Algorithm

Figure 3.27 provides an intuitive illustration of how Genetic U-Net leverages evolutionary algorithms to achieve the automatic design and optimization of neural

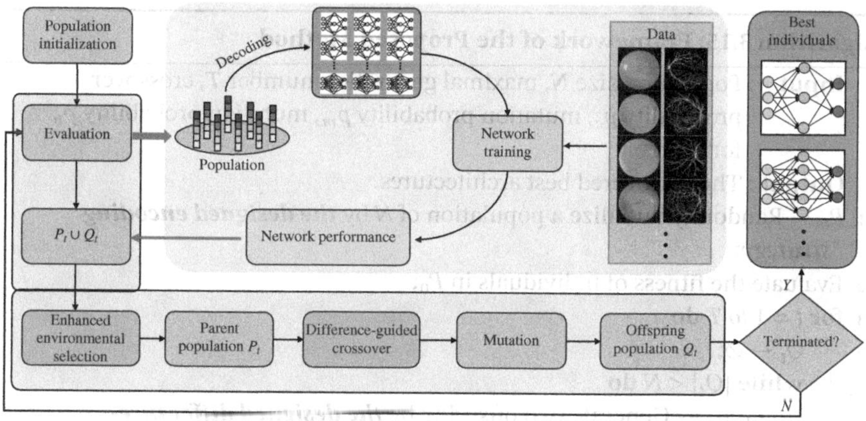

Figure 3.27 Evolutionary design and optimization of neural network architectures with Genetic U-Net. [18] / Springer Nature / CC BY 4.0.

network architectures. In this complex iterative evolutionary process, each individual represents a unique neural network architecture, with its fitness determined by the performance of the corresponding architecture on specific tasks or application scenarios. Specifically, the fitness evaluation process is data-driven, relying on rigorous performance assessments of the architectures to ensure that the evolutionary trajectory consistently moves toward optimal performance. The detailed pseudocode for this process is clearly outlined in Algorithm 3.15. The entire procedure begins with the random generation of N initial individuals, which form the starting population and represent the initial solution space for neural network architecture design. Once the initialization is complete, the algorithm iteratively evolves the population over T generations. In each generation, the population undergoes three key evolutionary operations: crossover, mutation, and selection. After generating new individuals in each generation, the algorithm trains the architectures represented by these individuals from scratch and evaluates their performance on the provided dataset. This training-from-scratch approach ensures that the evaluation results accurately reflect the architectures' capabilities in real-world applications, without being influenced by initial conditions or data distribution biases. This evaluation process provides a precise basis for calculating fitness, ensuring the reliability of the evolutionary process.

3.5.2.1 Crossover Operation

Crossover is a critical evolutionary operation whose primary goal is to facilitate the exchange of information between individuals, thereby generating offspring with superior traits and enhancing the overall performance of the algorithm.

Algorithm 3.15: Framework of the Proposed Method

Input : Population size N, maximal generation number T, crossover probability p_c, mutation probability p_m, mutation probability p_b for each bit.

Output : The discovered best architectures.

$P_0 \leftarrow$ Randomly initialize a population of N by ***the designed encoding strategy***;
Evaluate the fitness of individuals in P_0;
for $t = 1$ *to* T **do**
  $Q_t \leftarrow \emptyset$;
  while $|Q_t| < N$ **do**
    $o_1, o_2 \leftarrow$ Generate two offspring by ***the designed difference-guided crossover*** operation with the probability p_c from P_{t-1};
    $o_1, o_2 \leftarrow$ Apply the mutation with the mutation probability p_m and a flipping probability of p_b to o_1 and o_2;
    $Q_t \leftarrow Q_t \cup o_1 \cup o_2$;
  end
  Evaluate the fitness of individuals in Q_t;
  $P_t \leftarrow$ Select N individuals from $P_{t-1} \cup Q_t$ using ***the proposed environmental selection***;
end
return the individuals with the best fitness in P_t.

Numerous studies [290, 291] have shown that introducing multiple crossover points during the crossover operation can significantly improve the algorithm's search capability and efficiency in solving complex problems. Therefore, this chapter adopts a multi-point crossover strategy, allowing parent individuals to exchange multiple segments of genetic material, which generates a diverse set of offspring and enhances the population's exploration capacity. However, relying solely on traditional multi-point crossover strategies is not sufficient to maximize search efficiency and improve the quality of the generated offspring. To further optimize the crossover operation, we propose a difference-guided crossover method. The core idea of this method is to prioritize the selection of two parent individuals with significant structural differences for crossover, thereby increasing the diversity and potential adaptability of the offspring. This strategy aims to prevent premature convergence of the population (i.e., early stagnation at local optima) and to improve the algorithm's global optimization capability by exploring a broader solution space. The detailed steps of this method are outlined in Algorithm 3.16.

Algorithm 3.16: Difference-guided Crossover Operation.

Input : The population P_t, the probability for crossover operation p_c, the difference threshold of crossover operation μ.
Output : Two offspring o_1, o_2.

$o_1, o_2 \leftarrow \varnothing$;
for $j = 1$ *to* 10 **do**
  $p_1, p_2 \leftarrow$ Select two individuals from P_t by binary tournament selection;
  diff $\leftarrow$ Compute the difference between p_1 and p_2;
  if *diff* $> \mu$ **then**
    break;
  end
end
$r \leftarrow$ Randomly generate a number from (0, 1);
if $r < p_c$ **then**
  len $\leftarrow$ Compute the length of p_1 and p_2;
  ints $\leftarrow$ Randomly choose ten different integers from [0, *len*] and sort them;
  $(i_0, i_1), (i_2, i_3), (i_4, i_5), (i_6, i_7), (i_8, i_9) \leftarrow$ Divide *ints* into five pairs in order;
  for $k = 1$ *to* 5 **do**
    $p_1, p_2 \leftarrow$ Exchange $p_1[i_{2k-2} : i_{2k-1}]$ and $p_2[i_{2k-2} : i_{2k-1}]$;
  end
  $o_1, o_2 \leftarrow p_1, p_2$;
end
else
  $o_1, o_2 \leftarrow p_1, p_2$;
end
return o_1, o_2.

In Algorithm 3.16, the process begins by using binary tournament selection [292] to randomly select two individuals, p_1 and p_2, from the current population. For the initially selected individuals p_1 and p_2, the Hamming distance [293] between them is calculated. To standardize the process, the Hamming distance is normalized to the range [0, 1], and the normalized value is denoted as *diff*, which quantifies the structural similarity or difference between p_1 and p_2. The computed difference value *diff* is then compared with a predefined threshold u: (1) if *diff* $> u$, it is considered that p_1 and p_2 have sufficient structural differences and can be designated as parent individuals for the crossover operation. This selection mechanism ensures that the parent individuals have adequate diversity in their structures, increasing the likelihood of the offspring exploring the global optimum; (2) if $diff \leq u$, the parent individuals need to be reselected. In this case, binary tournament selection is performed again to select a new pair of individuals, p_1 and p_2, in an attempt to find individuals with greater differences. To avoid an infinite loop during the reselection process, a maximum number of reselection attempts is set. If, after 10 attempts, no pair of individuals with a difference greater than the threshold u is found, the last selected pair of p_1 and p_2 will be directly designated as the parent individuals. Once the parent individuals p_1 and p_2 are finalized, they undergo a crossover operation with a crossover probability of p_c.

3.5.2.2 Mutation Operation

The mutation operation is a crucial component of evolutionary algorithms, primarily aimed at introducing randomness to enhance population diversity. This diversity, in turn, strengthens the algorithm's ability to explore the solution space and effectively avoids getting trapped in local optima. Mutation injects new genetic information into the population, broadening the search space and providing greater robustness and flexibility during the optimization process. In the method proposed in this chapter, the mutation operation is primarily applied to the offspring generated through crossover. Specifically, each offspring has a probability of p_m to undergo mutation, and when mutation occurs, each bit (gene) is flipped independently with a probability of p_b. This design ensures that the mutation operation introduces moderate randomness while avoiding excessively drastic structural changes. Typically, p_b is set to a relatively small value (e.g., 0.05) to ensure that the overall characteristics of an individual are not significantly altered by mutation. This control mechanism not only preserves the stability of the population but also introduces sufficient diversity at the individual level.

3.5.2.3 Environmental Selection

In general, the most commonly used selection strategies in GA are tournament selection and roulette selection. The core objective of these methods is to probabilistically select higher-quality individuals based on their fitness, allowing them to enter the next generation. However, these strategies have a significant potential drawback: in some cases, they may fail to retain the best individuals in the current population [294]. This omission can directly lead to a decline in algorithm performance and negatively impact the final optimization results. This issue is especially critical in complex optimization problems, where the inability to preserve the best individuals may prevent the algorithm from quickly converging toward the global optimum, significantly reducing overall efficiency. On the other hand, explicitly selecting the top N fittest individuals to form the next generation can effectively avoid the loss of the best individuals. However, this approach introduces another serious issue: a dramatic reduction in population diversity.

The lack of diversity makes the population more prone to premature convergence, meaning that the algorithm can get trapped in local optima [295, 296], thereby severely limiting its global search capability [297]. This problem is particularly pronounced in complex or multi-modal solution spaces, where maintaining population diversity is crucial to avoiding local optima and ensuring robust global exploration. Therefore, relying solely on elitist selection strategies is also not a viable solution. To balance convergence and population diversity in the selection mechanism, we propose a hybrid selection method that combines elitist and non-elitist individuals. Specifically, this method retains a portion of the best-performing individuals (elitist individuals) to ensure that critical high-fitness genes are not lost during evolution, while also introducing a certain proportion of non-elitist individuals to enhance population diversity. The core idea of this hybrid mechanism is to simultaneously leverage the high-fitness traits of elitist individuals and the exploratory potential of non-elitist individuals, thereby accelerating the algorithm's convergence while preventing premature convergence to local optima.

Algorithm 3.17 provides a detailed description of the environmental selection process, which is a critical step in evolutionary algorithms. The purpose of environmental selection is to filter out high-quality individuals for the next generation population, thereby achieving a balance between convergence and diversity. The goal of this process is to select the fittest individuals from the current population P_t and the offspring population Q_t to form the next generation population P_{t+1}. Specifically, this process can be divided into two stages. First, the current population P_t and the offspring population Q_t are merged into a single set, $P_t \cup Q_t$. Within this set, all individuals are sorted based on their fitness values, and the top five fittest individuals are selected. These five elite individuals are directly retained and added to the next generation population P_{t+1}. At the same time, they are removed from $P_t \cup Q_t$. The purpose of this step is to ensure the preservation of high-quality genes within the population, preventing the loss of optimal individuals due to randomness or other operations during the evolutionary process, thereby enhancing the overall performance and stability of the algorithm. Next, from the remaining individuals in $P_t \cup Q_t$ (after removing the top five), binary tournament selection is applied to select $|P_t| - 5$ individuals, which are then added to the next generation population P_{t+1}. Binary tournament selection is an efficient and probabilistic method in which two individuals are randomly drawn from the population, and the one with the higher fitness value is selected. This method not only ensures the selection of high-quality individuals but also introduces a degree of randomness, thus maintaining population diversity and reducing the risk of premature convergence to local optima. Through these two steps, the size of the next generation population P_{t+1} is strictly controlled to match the size of the current population P_t. This design ensures the efficient use of computational resources

Algorithm 3.17: Environmental Selection.

Input : The parent population P_t, the offspring population Q_t.
Output : The population for the next generation P_{t+1}.
$P_{t+1} \leftarrow \emptyset$; $P_{best} \leftarrow$ Select the best five individuals from $P_t \cup Q_t$; $P_t \cup Q_t \leftarrow$ Remove P_{best} from $P_t \cup Q_t$; $P_{t+1} \leftarrow P_{t+1} \cup P_{best}$; **while** $|P_{t+1}| < |P_t|$ **do**
    $p \leftarrow$ Select an individual from P_t by binary tournament selection;
    $P_{t+1} \leftarrow P_{t+1} \cup p$;
end
return P_{t+1}.

during the evolutionary process while providing a stable population baseline for subsequent operations.

3.5.2.4 Fitness Evaluation

The fitness of an individual is determined by the $F1$ score of its corresponding neural network architecture on the validation dataset. It is worth mentioning that the validation dataset used in this study is not limited to retinal vessel segmentation tasks but also includes a variety of computer vision tasks such as object detection, depth estimation, image denoising, and image dehazing. This multi-task adaptability broadens the application scope of Genetic U-Net, allowing it to demonstrate its capabilities across a wide range of image processing fields. Algorithm 3.18 provides a detailed explanation of the process for evaluating the fitness of individuals within the population. First, each individual is mapped to its corresponding neural network architecture through an inverse binary encoding process. This step ensures that every individual in the population has a clear and unique structural definition, laying a solid foundation for subsequent training and optimization. During the network initialization phase, He initialization [298] is applied to the architecture's weights, ensuring that the network has a well-distributed parameter initialization. This helps prevent gradient vanishing or exploding problems and enhances the stability and efficiency of the training process. Once initialized, the network proceeds to the training stage. This study employs the Lookahead optimization method [299] based on the Adam optimizer [300] to optimize the network architecture. This combined strategy not only accelerates convergence but also provides greater stability in weight updates, making it particularly effective in multi-task environments for improving network performance. During the first 80 epochs of training, the network focuses on learning the core features of the data. From the 80th epoch onward, the model is evaluated on the validation dataset at the end of each epoch, continuing this process until the 130th epoch. Throughout the training and evaluation process, the $F1$ score is used as the primary metric to measure model performance. After each epoch, the $F1$ score of the network on the

Algorithm 3.18: Evaluate Fitness.

Input : The population P_t for fitness evaluation, training data D_{train}, validation data D_{valid}.
Output : The population P_t with fitness.

```
foreach individual in P_t do
    arch ← Transform the individual to its corresponding architecture;
    Apply weight initialization to arch;
    epoch ← 0;
    F1 − scorebest ← 0;
    for epoch < 130 do
        Train arch on D_train by gradient descent for an epoch;
        if epoch > 80 then
            F1 − score ← Evaluate the trained arch on D_valid;
            if F1 − score > F1 − score_best then
                F1 − score_best ← F1 − score;
            end
        end
    end
    Set F1 − score_best as the fitness of individual;
end
return P_t.
```

validation dataset is recorded, and the highest F1 score achieved during the entire training process is selected as the final fitness score of the individual.

3.5.3 Experimental Setup

3.5.3.1 Loss Function

For the vessel segmentation task, vessel pixels make up less than 10% of all pixels, with the vast majority belonging to the non-vessel class. This significant class imbalance poses a challenge that must be addressed effectively to ensure accurate segmentation. To tackle this issue, focal loss [301], originally developed to mitigate sample imbalance in object detection tasks, is adopted as the loss function in this study. The mathematical formulation of the focal loss is:

$$Loss = -\sum_{n=1}^{m}(\alpha y_n(1-\hat{y}_n)^{\omega}\log \hat{y}_n + (1-\alpha)(1-y_n)\hat{y}_n^{\omega}\log(1-\hat{y}_n)) \tag{3.16}$$

where y and $\hat{y}$ represent the ground truth and model predictions, respectively. The variable n corresponds to the index of the current sample, while m denotes the total number of samples. Additionally, the parameter α controls the balance between positive (vessel) and negative (non-vessel) samples, and ω determines the weighting factor to adjust the emphasis between easy and hard samples.

3.5.3.2 Experimental Datasets

We use four publicly available datasets: DRIVE [302], CHASE_DB1 [303], STARE [304], and HRF [305]. A detailed description of these datasets is provided in Table 3.11. These datasets contain images captured by various devices, resulting in differences in image resolution and size. In contrast, the HRF dataset includes annotations from a single expert. The division of training and test sets across these datasets follows the method described in [306]. For the DRIVE dataset, we adopt the official split, which consists of 20 images for training and 20 for testing. In the STARE dataset, we use a leave-one-out strategy, where each image is alternately used as a test sample while the others serve as the training set. In the CHASE_DB1 dataset, a total of 28 images are available, of which the initial 20 images are designated for training purposes, whereas the last 8 images are utilized as the test set. The HRF dataset comprises a total of 45 images, which are divided into three distinct categories: healthy individuals, patients with diabetic retinopathy, and those with glaucomatous conditions. Within each category, there are 15 images in total. For training purposes, the first 5 images are selected, while the remaining 10 images are allocated to the test set to evaluate the model's performance. As a result, we generate these masks using the method described in [307]. This consistent approach to dataset preparation ensures a fair and comparable evaluation across all experiments.

3.5.3.3 Evaluation Metrics

In the context of this experimental study, retinal vessel segmentation represents a classic binary classification problem. The primary objective is to determine the class of each pixel within a fundus image, categorizing it as either part of a vessel or belonging to the non-vessel region. The model generates a probability map, where each pixel is assigned a value indicating the likelihood of it being part of the vessel

Table 3.11 Four publicly used datasets.

Dataset	Quantity	Resolution	Train-test split
DRIVE	40	565 × 584	20/20
STARE	20	700 × 605	Leave-one-out
CHASE_DB1	28	999 × 960	20/8
HRF	45	3504 × 2336	15/30

class. During the experiments, a probability threshold of 0.5 was applied to these values to determine the final classification, categorizing pixels as either vessels or non-vessels based on whether their assigned probability exceeded this threshold. Specifically, when a vessel pixel is correctly classified, it is considered a true positive (*TP*); if it is misclassified, it is considered a false positive (*FP*). Similarly, when a non-vessel pixel is correctly classified, it is regarded as a true negative (*TN*); if misclassified, it is a false negative (*FN*). To comprehensively evaluate the model's performance, we adopted the five metrics listed in Table 3.12. It is important to note that, unless otherwise specified, all performance metrics are calculated solely within the field of view (FOV). This restriction ensures that the evaluation is not affected by irrelevant areas outside the FOV, thereby providing a more accurate reflection of the model's effectiveness in retinal vessel segmentation tasks.

It is important to note that this experiment is divided into two stages. In the first stage, the DRIVE dataset is used for architecture search, where the last 5 images from the training set are used as the validation set, and the remaining 15 images are used for training. Additionally, in this stage, the performance of the architectures obtained through the search is tested on other datasets (STARE, CHASE_DB1, and HRF). The specific settings for the hyperparameters of Genetic U-Net and other related parameters can be found in Ref. [179]. In the second stage, the discovered architectures were trained from scratch to comprehensively assess their performance on retinal vessel segmentation across multiple datasets. The training settings were largely consistent with those in the first stage, with the primary difference being an extended number of epochs to ensure convergence.

3.5.4 Results and Analysis

In the study, the vessel segmentation task is divided into two main stages. In the first stage, GA is employed to perform an automated search for neural network architectures, aiming to discover efficient architectures that meet the task requirements. In the second stage, the best architectures identified in the first stage are

Table 3.12 Five used evaluation metrics.

Metrics	Description
ACC (*accuracy*)	$ACC = (TP + TN) / (TP + TN + FP + FN)$
SE (*sensitivity*)	$SE = TP / (TP + FN)$
SP (*specificity*)	$SP = TN / (TN + FP)$
F1-score (*F1*)	$F1 = (2 \times TP) / (2 \times TP + FP + FN)$
AUROC	Area under the ROC curve.

trained from scratch (i.e., with weights initialized anew) to comprehensively validate their actual performance in the retinal vessel segmentation task.

We apply the proposed method to design network architectures, which are then thoroughly tested and validated on four widely used public datasets: CHASE_DB1, HRF, STARE, and DRIVE, as shown in Tables 3.13–3.16. To evaluate the effectiveness of the method, we compare the performance of the generated network models with that of current mainstream methods, which are primarily based on convolutional neural networks.

In these four tables, the experimental results of existing methods, except for CS^2-Net [318] and CE-Net [278], are directly taken from the original papers. The performance results of CS^2-Net and CE-Net, however, are obtained by re-evaluating them under the same training settings as the architectures discovered in this study. It is worth noting that existing methods marked with a star calculate their performance metrics based on pixels within the FOV, while those without a star do

Table 3.13 Comparison with existing methods on DRIVE dataset.

Methods	Year	ACC	SE	SP	F1	AUROC	Params(M)
Vega et al. [308]	2015	0.9412	0.7444	0.9612	0.6884	N/A	N/A
Li et al. [306]	2015	0.9527	0.7569	0.9816	N/A	0.9738	N/A
Orlando et al. [309]	2016	N/A	0.7897	0.9684	0.7857	N/A	N/A
Fan and Mo [310]	2016	0.9612	0.7814	0.9788	N/A	N/A	N/A
Liskowski et al. [307]	2016	0.9535	0.7811	0.9807	N/A	0.979	48.00
Mo and Zhang [311]	2017	0.9521	0.7779	0.9780	N/A	0.9782	N/A
Yan et al.* [312]	2018	0.9542	0.7653	0.9818	N/A	0.9752	31.35
Alom et al. [313]	2019	0.9556	0.7792	0.9813	0.8171	0.9784	1.07
Jin et al. [314]	2019	0.9566	0.7963	0.9800	0.8237	0.9802	0.88
Bo Wang et al. [315]	2019	0.9567	0.7940	0.9816	0.8270	0.9772	N/A
Yicheng Wu et al. [316]	2019	0.9578	0.8038	0.9802	N/A	0.9821	1.70
Mou Lei et al. [317]	2019	0.9594	0.8126	0.9788	N/A	0.9796	56.03
CE-Net* [278]	2019	0.9545	0.8276	0.9735	0.8243	0.9794	15.28
CS^2-Net* [318]	2021	0.9553	0.8154	0.9757	0.8228	0.9784	8.91
Genetic U-Net W/ FOVs	-	0.9577	**0.8300**	0.9758	**0.8314**	**0.9823**	**0.27**
Genetic U-Net W/o FOVs	-	**0.9707**	**0.8300**	**0.9843**	**0.8314**	**0.9885**	**0.27**

Table 3.14 Comparison with existing methods on STARE dataset.

Methods	Year	ACC	SE	SP	FI	AUROC
Vega et al. [308]	2015	0.9483	0.7019	0.9671	0.6614	N/A
Li et al.* [306]	2015	0.9628	0.7726	0.9844	N/A	0.9879
Orlando et al.* [309]	2017	N/A	0.7680	0.9738	0.7644	N/A
Fan and Mo [310]	2016	0.9654	0.7834	0.9799	N/A	N/A
Liskowski et al.* [307]	2016	0.9566	0.7876	0.9754	N/A	0.9785
Mo and Zhang [311]	2018	0.9674	0.8147	0.9844	N/A	0.9885
Yan et al.* [312]	2018	0.9612	0.7581	0.9846	N/A	0.9801
Alom et al. [313]	2019	0.9712	0.8292	0.9862	0.8475	0.9914
Bo Wang et al. [315]	2019	0.9661	0.8074	0.9821	0.8037	0.9812
Yicheng Wu et al. [316]	2019	0.9661	0.8132	0.9814	N/A	0.9860
CE-Net* [278]	2019	0.9641	0.8093	0.9797	0.8054	0.9834
CS^2-Net* [318]	2021	0.9651	0.8329	0.9784	0.8141	0.9851
Genetic U-Net W/ FOVs	-	**0.9719**	**0.8658**	0.9846	**0.8630**	**0.9921**
Genetic U-Net W/o FOVs	-	**0.9792**	**0.8658**	**0.9886**	**0.8630**	**0.9942**

Table 3.15 Comparison with existing methods on the CHASE_DB1 dataset.

Methods	Year	ACC	SE	SP	FI	AUROC
Li et al.* [306]	2015	0.9581	0.7507	0.9793	N/A	0.9716
Fan and Mo [310]	2016	0.9573	0.7656	0.9704	N/A	N/A
Yan et al.* [312]	2018	0.9610	0.7633	0.9809	N/A	0.9781
Alom et al. [313]	2019	0.9634	0.7756	0.9820	0.7928	0.9815
Jin et al. [314]	2019	0.9641	0.7595	0.9878	0.8143	0.9832
CE-Net* [278]	2019	0.965	0.8406	0.9813	0.8363	0.9871
CS^2-Net* [318]	2021	0.9670	0.8396	0.9813	0.8420	0.9875
Genetic U-Net W/ FOVs	-	**0.9667**	**0.8463**	0.9845	**0.8223**	**0.9880**
Genetic U-Net W/o FOVs	-	**0.9769**	**0.8463**	**0.9857**	**0.8223**	**0.9914**

Table 3.16 Comparison with existing methods on the HRF dataset.

Methods	**Year**	**ACC**	**SE**	**SP**	**F1**	**AUROC**
Orlando et al.* [309]	2016	N/A	0.7794	0.9650	0.7341	N/A
Yan et al.* [312]	2018	0.9437	0.7881	0.9592	N/A	N/A
Jin et al. [314]	2019	0.9651	0.7464	**0.9874**	N/A	N/A
CE-Net* [278]	2019	0.9613	0.7805	0.9798	0.7895	0.9766
CS^2-Net* [318]	2021	0.9618	0.7890	0.9795	0.7935	0.9758
Genetic U-Net W/ FOVs	-	**0.9667**	**0.8220**	0.9818	**0.8179**	**0.9872**
Genetic U-Net W/o FOVs	-	**0.9715**	**0.8220**	0.9839	**0.8179**	**0.9891**

not explicitly clarify this in their respective papers. To comprehensively demonstrate the performance of the proposed method, this study reports the results of the discovered architectures under both conditions: within and beyond the FOV. Additionally, to ensure fairness in comparisons, the same data split as the other methods is used in this study.

In experiments across all datasets, the proposed method surpasses all comparison methods in the two comprehensive metrics, F1 and AUROC. This demonstrates that the discovered architecture achieves optimal overall performance without the need for any post-processing or special pre-processing techniques. Moreover, in SE (sensitivity), the proposed method also performs exceptionally well, significantly outperforming other comparison methods. This indicates that the discovered architecture is more effective in detecting vessel pixels. Notably, ACC, as a balance between SE and SP (specificity), is more susceptible to SP due to the significantly larger non-vessel region compared to the vessel region in fundus images. Despite this inherent challenge, the proposed method still achieves comparable or even superior performance in ACC compared to other methods, showcasing its excellent adaptability and balance.

Another significant advantage is that the discovered architecture has an extremely low number of parameters, only 0.27M, making the model more lightweight compared to other methods. This not only reduces the model's computational complexity and storage requirements but also makes it more suitable for real-world applications with limited resources. It is worth mentioning that this study only employed simple data augmentation strategies such as flipping and rotation. Compared to the complex and diverse data augmentation techniques used in many other methods, the preprocessing required by this approach is much simpler and more efficient.

Finally, the proposed method also demonstrates excellent transferability. The architecture discovered on the DRIVE dataset can be successfully applied to three other datasets (STARE, HRF, and CHASE_DB1) without any structural adjustments, achieving outstanding performance. This further validates the generalization ability of the architecture identified in this study and its robustness across different data distributions, showcasing its wide applicability in retinal vessel segmentation tasks.

3.5.5 Conclusion

This section introduces a neural network architecture design method called Genetic U-Net, based on GP, for retinal vessel segmentation tasks. By automating the search and optimization of network architectures, this method overcomes the limitations of traditional network design, which often relies on manual expertise and trial-and-error processes. The design philosophy is grounded in using GAs for evolution, employing operations such as crossover, mutation, and selection to iteratively optimize the topology and operational types of the network.

In the experiments, the search space of Genetic U-Net includes a wide variety of module variants (e.g., DenseNet, ResNet, and InceptionNet) and incorpo rates a multi-objective optimization strategy to balance structural exploration with performance enhancement. Fitness evaluation is centered on the F1 score of the network on the validation set, with zero-start training ensuring the reliability of the evaluation results. Experiments conducted on four public datasets (DRIVE, STARE, CHASE_DB1, and HRF) demonstrate that Genetic U-Net significantly improves segmentation performance, outperforming existing methods. Furthermore, the architecture has a parameter size of only 0.27M, drastically reducing computational complexity.

The advantages of this method lie in its automated design process, high efficiency, and robust performance. Validations on different datasets showcase Genetic U-Net's strong transferability and adaptability, making it suitable for resource-constrained real-world applications. In the future, its potential for expansion to more tasks can be explored, further advancing the application of automated network design.

3.6 Evolving Hybrid Bond Graph Using GP

GP, as an evolutionary computation method, is widely applied across various fields. Initially, the core goal of GP is to automatically evolve computer programs that meet specific requirements by simulating the processes of natural selection and genetic evolution. It provides a groundbreaking approach to program automation and design. However, as research progresses, scientists gradually discover that the potential of GP extends far beyond the generation of tree-structured programs. Its flexibility and scalability enable it to adapt to a broader range of complex systems, making it a powerful tool for solving interdisciplinary and intricate problems.

Specifically, the application of GP has gradually expanded from the evolution of programs for single-purpose tasks to multiple fields, such as neural network design, electronic circuit optimization [319], the generation of complex graph structures, and the modeling of biological morphology. This cross-disciplinary adaptability is attributed to the continuous innovation and expansion of GP methods by researchers. For instance, to meet the design requirements of different fields, various innovative encoding methods have been proposed, such as Cellular Encoding and Edge Encoding [320], along with their variations. These methods have significantly broadened the applicability of GP, enabling it to address increasingly complex and diverse engineering problems.

Inspired by Cellular Encoding, some researchers further combined bond graphs (BGs) with GP, resulting in the development of the bond graph and genetic programming (BGGP) method. By integrating evolutionary computation with BG technology, BGGP enhances the design capabilities of BG and has achieved significant results in several case studies. For example, BGGP has demonstrated outstanding performance in electrical filter design [109, 110], achieving comprehensive improvements from performance optimization to automated design.

In addition, BGGP has been successfully applied to the redesign of typewriters [37], where it not only optimized traditional mechanical structures but also proposed innovative design solutions. Meanwhile, the method has also been used for the design of mechanical vibration absorbers [321], where it generated devices with excellent vibration damping performance through simulated evolutionary processes, providing new solutions for improving the stability and reliability of mechanical systems.

More complex application scenarios include the co-design of controllers and physical components in vehicle suspension systems [111, 135]. In this case, BGGP significantly improved the dynamic performance and comfort of vehicles by optimizing both the controller design and the physical suspension structure simultaneously. Furthermore, BGGP has made important breakthroughs in the design of Micro-Electro-Mechanical Systems (MEMS) [101]. MEMS, as a complex multiphysics coupled system, requires the consideration of mechanical, electrical, and thermal factors in its design. Through automated optimization, BGGP successfully generated MEMS designs that meet specific functional requirements, further validating its adaptability and efficiency in solving cross-disciplinary problems.

The success of the BGGP method across such a wide range of fields can be largely attributed to its core mechanism. During the evolutionary process, BGs often contain junctions and components that can be further optimized. Specifically, at the beginning of the evolutionary process for an individual, BGGP first generates a BG from a generic embryo. Before establishing the state-space equations for the BG, researchers need to apply a BG simplification algorithm to streamline its structure. Once a simplified HBG is obtained, the number of switches within it is passed to the lookahead controller. Following this, the integration of the state-space equations is carried out. The system is then simulated under multiple different scenarios, with performance data recorded for each scenario. These performance metrics are aggregated and calculated to derive a fitness value, which is ultimately returned to the evolutionary algorithm to further optimize the structure of the BG.

This systematic evaluation process ensures that the designs generated by BGGP are highly optimized in terms of both performance and structure. Unlike traditional methods, BGGP operators can act not only on the nodes (components) of the BG but also on its edges (bonds). This flexibility allows BGGP to efficiently handle design tasks with complex topological structures. Figure 3.28 provides a visual summary of this evaluation process, illustrating the foundation of BGGP's successful application across different domains.

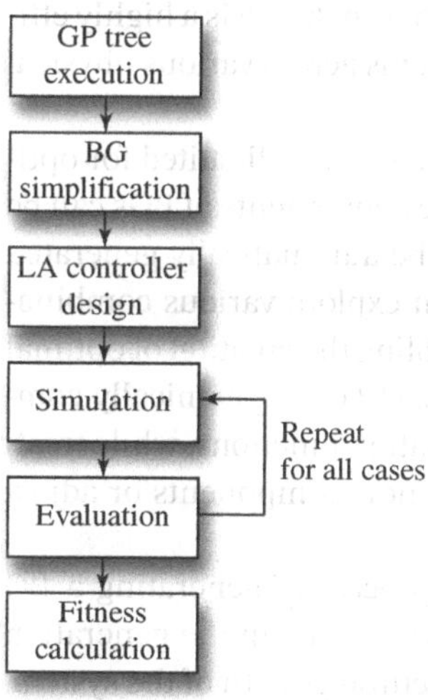

Figure 3.28 General process of BGGP evolutionary design. In the figure, LA controller is the abbreviation of lookahead controller. The LA controller can also be replaced with other controller models based on BG representations.

3.6.1 Bond Graph and GP

BGs are a graphical modeling method used to represent energy exchange in physical systems. Their core concept is modeling power transmission through bonds between components. In BGs, power is decomposed into two fundamental components, effort and flow, whose product represents power. This power-based modeling approach not only provides an intuitive representation of systems but also enables modeling across multiple physical domains within a single framework, offering a unified and efficient tool for the analysis and design of complex systems.

The flexibility of this modeling method stems from the high generality of the components in BGs. The components in BGs can describe energy generation (S_e, S_f), storage (C, I), transformation (GY, TF), and dissipation (R), while also allowing for the formation of complex energy flow networks through node connections. Specifically, 0-nodes are used to connect components sharing the same effort value, while 1-nodes are used to connect components sharing the same flow value. To accommodate the modeling needs of dynamic systems and discrete phenomena, BGs also introduce switch components (Sw) to support hybrid modeling. Switch components can simulate sources with zero-effort or zero-flow depending on their state, enabling accurate modeling of ideal switching behaviors. The modular nature of these components and connection methods makes BGs a highly efficient and flexible tool for modeling and analyzing systems across various physical domains.

This highly modular characteristic makes BGs particularly well-suited for optimization through GP. Within the framework of GP, the components of BGs can be treated as primitives, while the overall structure can be automatically generated and optimized using evolutionary algorithms. GP can explore various combinations and connections of components within BGs, enabling the creation of optimal designs that meet specific requirements. For example, GP can dynamically combine components for storage, transformation, or dissipation functions, while treating nodes and bonds as modification points to insert new components or adjust existing structures.

In greater detail, from the perspective of GP, the process of generating a BG begins with an "embryo." The concept of the embryo is crucial in the generation of BGs as it provides the initial framework for the structural growth of the system. An embryo typically consists of two parts: a non-modifiable subsystem and modifiable sites. The non-modifiable subsystem refers to the fixed, predetermined components of the system, which often represent known physical properties or design constraints that cannot be altered. For instance, this could be a specified input source, an output load, or other critical components that must remain unchanged during the design process. On the other hand, the modifiable sites represent areas of the system where further exploration and optimization are possible. These modifiable sites are carefully designed and defined within the embryo to ensure that the new structures of the system can naturally "grow" while adhering to the specified design objectives.

Depending on the design goals, the scale of the embryo can be categorized into two scenarios: small-scale embryos and large-scale embryos. For small-scale embryos, the embryo itself contains only a minimal initial structure, and the majority of the system's final design must "grow" from scratch through the evolutionary process. This scenario is often applicable to high-uncertainty design tasks where the designer has little to no clear understanding of the target design and relies heavily on the algorithm's exploration capabilities to generate a complete system structure. In contrast, large-scale embryos already include most of the system's design structure, requiring only optimization or redesign of certain subsystems. This approach is generally suited to scenarios where the design objectives are more clearly defined. In such cases, the embryo provides a relatively stable framework, allowing the algorithm to focus on local optimization. This not only saves computational resources but also ensures that the generated designs meet the specified functional requirements.

It is worth noting that previously generated solutions can be used as new embryos, enabling continuous evolutionary design. This approach is particularly useful for the gradual optimization or iterative development of complex systems. For example, an initial design can be generated using the algorithm to create a BG that meets the basic functional requirements. This design can then be used as a new embryo and reprocessed by the GP algorithm to optimize specific performance metrics, such as energy efficiency, response speed, or stability.

Figure 3.29 presents a concise and representative example illustrating the entire process of generating a BG using a GP tree. In this example, the system begins from an initial state, referred to as the "embryo" (as shown in Figure 3.29b), which defines the basic structural framework of the system and its potential growth points. The embryo consists of two key components: a fixed, non-modifiable structure and modifiable growth points. The non-modifiable structure is represented here by the input flow source S_f and the inertial component I_1, which together define the known and immutable physical properties of the system, such as the energy input source and the initial system inertia. Meanwhile, R_1, as a terminal resistor, represents the output load to which the system will connect. This is a common modeling approach for components in the electrical domain.

The embryo also defines the modification points where the system structure can grow. Among these, the 0-junction is marked as a "node modification point," while the bond associated with R_1 is marked as a "bond modification point." These modification points serve as growth "anchors," providing reference positions for the subsequent expansion and adjustment of the structure.

Starting from the embryo state, the primitives in the GP tree (as shown in Figure 3.29a) are sequentially applied to these modification points, gradually

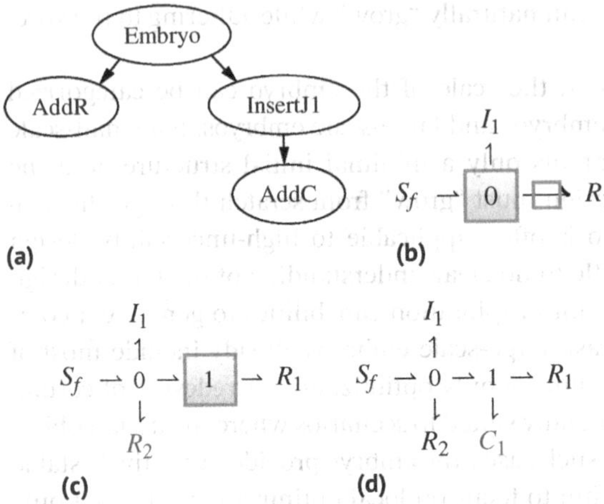

Figure 3.29 Entire process of generating a BG using a GP tree. (a) GP tree. (b) Initial embryo state. (c) An intermediate state. (d) Complete BG structure.

expanding the system structure according to the design rules. In this example, the AddR primitive is first applied to the 0-junction modification point, adding a new resistor component R_2. Next, the InsertJ1 primitive is applied to the bond modification point, inserting a new 1-junction, which acts as an entirely new modification point and provides further possibilities for subsequent expansions. Finally, the AddC primitive utilizes the newly inserted 1-junction modification point to add a capacitor component C_1.

By following the sequence of operations described in the GP tree (Figure 3.29a), the system's BG evolves step by step from the initial embryo state (Figure 3.29b) to an intermediate state (Figure 3.29c), and finally to the complete BG structure shown in Figure 3.29d. Throughout this process, the design objectives of GP are realized. The addition and removal of components, the insertion of junctions, and the expansion of modification points are all carefully executed to ensure that the final BG meets the functional requirements of the system while maintaining the potential for further optimization.

3.6.2 Basic Primitives

To successfully evolve a fully functional BG, it is essential to define a set of primitives that can alter the topological structure of the BG. These primitives play a central role in the evolutionary process by guiding the structural growth of the BG and defining the modification points along with their operational rules. Figure 3.30 lists the key primitives used in this approach, most of which are redesigned or extended based on the original BG primitives.

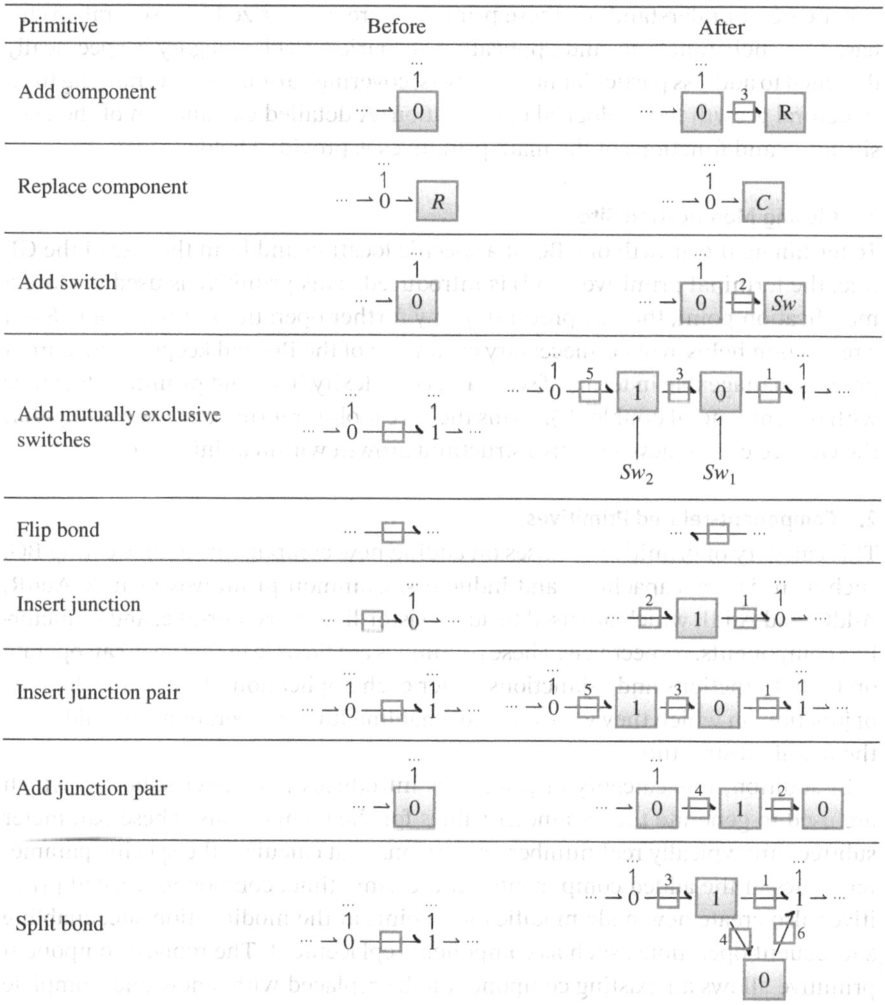

Figure 3.30 A set of primitives based on BG.

These primitives are carefully designed to operate on a single modification point as their basic principle. After each operation, the newly created modification point is returned as a parameter, allowing subsequent primitives to further modify it. This parameter-returning mechanism ensures that the BG structure can expand in a controlled manner, avoiding chaotic growth and topological disorder. In this way, the evolution process of the BG becomes more efficient and manageable, ultimately generating a structure that meets the design requirements while leaving room for further optimization.

For ease of understanding, these primitives are categorized into several groups based on their functions and application scenarios. Each category is specifically designed to address particular needs of BGs, covering various aspects ranging from structural growth to topological optimization. A detailed explanation of the classification and functions of the main primitives is provided below:

1. Closing Modification Site

To terminate the growth of a BG at a specific location and limit the size of the GP tree, the terminal primitive (End) is introduced. This primitive is used to close a modification point, thereby preventing any further operations at that point. Such a restriction helps avoid unnecessary expansion of the BG and keeps the evolution process manageable in terms of scale and complexity. The End primitive, together with the ephemeral double (E), forms the core tools for termination, ensuring that the GP tree can achieve effective structural growth within a finite size.

2. Component-related Primitives

This category of primitives focuses on adding new components to an existing BG, such as resistors, capacitors, and inductors. Common primitives include AddR, AddC, and AddI, which are used to add resistor-like, capacitor-like, and inductor-like components, respectively. These primitives are flexible in that they can operate on both 0-junctions and 1-junctions. After each application, they return the type of junction on which they were applied, enabling further operations to build upon the modified structure.

In addition, this category of primitives introduces parameter subtrees, which are used to generate the parameter values for the components. These parameter subtrees are typically real-number expressions that calculate the specific parameter values of the added components. At the same time, component-related primitives also create new node modification points at the modification site, enabling subsequent operations, such as component replacement. The replace component primitive allows an existing component to be replaced with a new one, complete with its own independent parameter subtree, thereby enabling more fine-grained structural optimization.

The Switch Addition Primitive is an important extension of the component-related primitives, enabling the creation of hybrid systems. By adding switch components, BGs can achieve multi-state dynamic behavior. However, to prevent excessive system complexity, the number of switches in HBGs should be strictly limited. Too many switches can lead to an exponential increase in control complexity. For example, when using a lookahead controller, the system's simulation time can grow rapidly with the number of switches. Therefore, limiting the number of switches not only reduces computational overhead but also ensures the controllability of the system.

3. **Junction-related Primitives**

In the initial design, junction-related primitives included two basic forms, InsertJ0 and InsertJ1, which are used to insert 0-junctions and 1-junctions, respectively. These primitives allow new junctions to be added to the BG without considering the specific type of the target junction. However, due to the properties of adjacent junctions of the same type, this can result in the consecutive insertion of redundant junctions. These redundant junctions are later merged into a single junction during the simplification process. Although this phenomenon does not affect the topology of the simplified BG, it can lead to unnecessary bloat in the GP tree, increasing the computational burden.

Figure 3.31 clearly illustrates a typical issue of ineffective junction insertion. In this example, multiple 0-junctions are consecutively inserted into the BG through operations in the GP tree. However, these junctions do not provide any meaningful contribution to the final structure. During the subsequent operation of adding resistor components, these redundant 0-junctions only lead to unnecessary topological complexity. In the following simplification steps, these redundant junctions are merged, and the multiple resistor components are reduced into a single equivalent resistor, R_{eq}. Ultimately, these redundant elements are eliminated, and the simplified BG retains only the essential structure.

To mitigate the bloating effect, a Junction Pair Insertion Primitive is introduced. Unlike the insertion of a single junction, this primitive simultaneously inserts a connected pair of 0-junction and 1-junction. The sequence of the junctions is specifically designed to ensure that nodes of the same type do not directly connect to each other, thereby reducing the likelihood of redundant junction insertions. While the junction pair insertion primitive significantly reduces bloating, it does not entirely eliminate the occurrence of ineffective topological subtrees. During the simplification phase, the size of the generated BG is typically reduced by approximately 60%, further enhancing the optimization of the BG.

4. **Loop Creation Primitive**

The loop creation primitive is a key structural generation tool in the evolution of BGs. Among these, the SplitBond primitive introduces loops into the system by dividing the power flow into two parallel paths. This effectively adds loops to the system, enhancing its topological complexity and enabling it to exhibit more diverse dynamic behaviors. For example, in certain multi-input, multi-output systems, the presence of loops can be critical for optimizing energy distribution, improving system stability, or achieving specific functionalities.

5. **Flip Bond Primitive**

The flip bond primitive functions similarly to the operation of reversing component polarity in circuit evolution. In circuit design, reversing polarity can change the direction of current or voltage to meet different functional requirements. Similarly, in BGs, flipping the direction of a bond can be used to precisely control the flow of power within the system. This capability allows for flexible adjustments of energy flow paths within loops, enabling more efficient energy management and optimization of system performance.

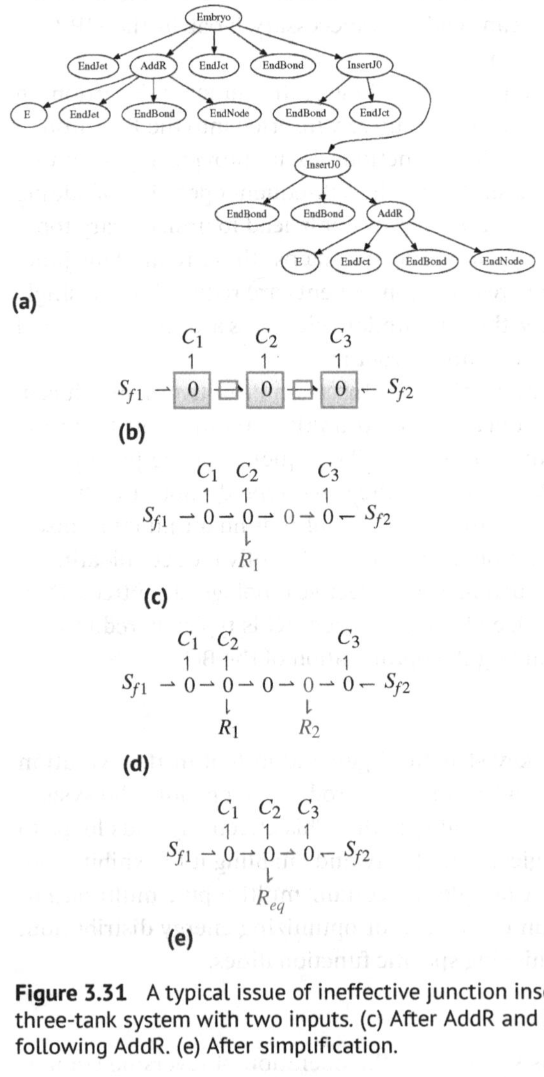

Figure 3.31 A typical issue of ineffective junction insertion. (a) GP tree. (b) Embryo of a three-tank system with two inputs. (c) After AddR and InsertJ10. (d) After InsertJ01 and following AddR. (e) After simplification.

3.6.3 Genetic Operators

The application of genetic operators in evolving HBGs is similar to the operators used in traditional GP. However, since the evolution of HBGs involves simultaneous optimization of both topological structures and parameter values, the process is significantly more complex than single-objective optimization. To better address the requirements of this task, these operators have been specifically improved. The improvements focus on the following two key aspects: (1) balancing topological exploration and parameter optimization: in traditional genetic operations, focusing solely on topology or parameter optimization can limit the search space or lead to premature convergence of the population to local optima. The improved operators, while preserving the ability to explore topological changes, further enhance the optimization of parameters. This ensures an effective balance between topological exploration and parameter optimization. (2) Enhancing population diversity to avoid premature convergence: to prevent excessive homogenization of the population and to encourage the generation of diverse solutions, a speciation operator has been introduced, thereby expanding the set of operators. The speciation operator manages the population by dividing it into subgroups based on different characteristics. This mechanism effectively maintains the diversity of the population, reducing the risk of premature convergence.

1. Crossover and Mutation:

In GP, crossover and mutation operators typically select nodes in the tree randomly for modification, meaning that, in theory, all nodes have an equal probability of being selected. However, since parameter nodes are defined by numerical subtrees, which coexist with structural nodes within the same tree, parameter nodes and structural nodes theoretically share the same probability of selection. In practice, however, numerical nodes usually outnumber structural modification nodes by a large margin, making parameter modification nodes far more likely to be selected during crossover or mutation.

This phenomenon introduces a significant issue: the slowing down of structural exploration. As crossover and mutation operations tend to favor parameter nodes for optimization, structural nodes have fewer opportunities to be modified, which can lead to premature structural convergence. This limits the diversity and potential of topological exploration, ultimately affecting the overall effectiveness of the system's evolution.

To address this issue, it is necessary to explicitly control the exploration process of both structures and parameters. Ref. [322] proposed a probability-based improvement method, which first determines whether to modify parameters or structures, and then applies crossover or mutation operations exclusively to the corresponding type of nodes. This method effectively avoids the problem of operators being overly biased toward a specific type of node by clearly dividing the weights of structure and parameter modifications.

In particular, since newly generated structures require some time to adjust their parameters and demonstrate their potential, a bias mechanism is introduced for parameter nodes. Specifically, this mechanism increases the probability of parameter modifications, giving new structures more opportunities to optimize their parameters. This allows them to fully showcase their performance before deciding whether to discard them. Typically, the probability ratio between structure and parameter modifications is set as follows: $p_{\text{struct}} = 0.15$, $p_{\text{param}} = 0.85$.

To further achieve balance during parameter evolution, Ref. [322] proposed an explicit node selection control method. In this approach, the root nodes of all numerical subtrees are first listed, and one of them is randomly selected with equal probability. Subsequently, crossover or mutation operations are applied only to the subtree rooted at the selected node. This process ensures that all parameters have an equal chance of being modified through mutation or crossover, thereby preventing excessive focus on certain overly large numerical subtrees. This mechanism not only maintains balance during parameter evolution but also effectively avoids overemphasis on specific node types during the evolutionary process.

Through the above improvements, crossover and mutation operations achieve a more balanced distribution between structural exploration and parameter optimization. This significantly enhances the efficiency of evolution and the diversity of solutions while effectively mitigating the issue of premature convergence caused by insufficient structural modifications.

2. Speciation Operator:
Premature convergence has long been a significant challenge in evolutionary computation. To achieve a proper balance between exploration and exploitation, researchers have proposed numerous mechanisms over the years. In GAs, several diversity maintenance techniques, such as crowding, deterministic crowding, and fitness sharing, have demonstrated excellent performance when dealing with problems involving a fixed number of parameters. These methods effectively prevent the population from becoming trapped in local optima by encouraging its distribution across multiple regions of the search space. However, when a problem requires not only parameter optimization but also the evolution of topological structures, the limitations of these traditional methods become apparent. For instance, solely relying on fitness sharing may fail to address the complexity introduced by topological changes, making it insufficient to fully tackle such problems.

When a problem involves only structural evolution, researchers have developed specialized strategies to maintain population diversity. One approach is to use multi-objective fitness, which enforces diversity by calculating the distance between GP trees. Another approach is implicit fitness sharing, where individuals are grouped based on their performance outcomes (i.e., phenotypes) rather than their genotypes. These methods effectively enhance diversity during structural optimization.

However, when a problem requires simultaneous optimization of both structure and parameters, the complexity of the search space increases significantly. In such cases, enhancing structural diversity alone is insufficient. This is because the performance of a specific structure can only be accurately assessed after its parameters have been thoroughly optimized. Without adequate exploration of parameters, even structurally promising designs may be mistakenly discarded due to insufficient parameter optimization. Therefore, under these circumstances, achieving a balance between structural diversity exploration and parameter optimization becomes essential

Moreover, the structural space formed by GP trees exhibits highly nonlinear characteristics. In GP, due to weak causality, modifying a single node often leads to significant changes in the phenotype's behavior. This property makes traditional methods based on GP tree distance metrics inefficient for assessing performance similarity, as they fail to accurately reflect the true fitness of individuals. This issue further complicates the challenges of maintaining diversity and evaluating fitness effectively.

To address the aforementioned issues, researchers have proposed several approaches, such as Structure Fitness Sharing (SFS) and Hierarchical Fair Competition (HFC), to improve the balance between structural exploration and parameter optimization. These techniques achieve enhanced diversity maintenance and search efficiency by managing population distribution in a more refined manner.

Specifically, in traditional fitness sharing methods, the population is distributed around multiple peaks in the search space, with the number of individuals allowed near each peak being proportional to the height of that peak. Inspired by this approach, SFS treats each GP tree structure as a "peak" in the structure space. It tracks changes in the BG structure (rather than parameter values) and marks a new "peak" only when a significant structural change occurs. This strategy effectively controls the amount of parameter optimization surrounding each new structure, preventing excessive focus on parameter optimization at the expense of structural exploration. As a result, it achieves a dynamic balance between structural exploration and parameter optimization.

In addition, HFC further enhances this balance. The core idea of HFC is to divide structures with under-optimized parameters into low-fitness subpopulations, allowing them to compete in a less restrictive environment. This provides room for immature solutions to develop while preventing premature convergence within a single subpopulation. However, despite its significant improvements in search efficiency, HFC does not completely resolve the issue of structural convergence. This is particularly evident in search spaces with broad local minima, where most solutions tend to become trapped, resulting in a decline in exploration capability.

To address this issue, this chapter proposes combining HFC with SFS based on the simplified BG structure to further enhance population diversity. Specifically, SFS performs a comprehensive comparison of BGs after executing the GP tree and applying the simplification algorithm. It classifies individuals into different species and dynamically adjusts their fitness values based on the attributes of each species. In SFS, the fitness adjustment factor for individual i is calculated based on the size of the species n_s^i to which it belongs, as follows:

$$\eta_{\mathrm{SFS}}^i = \left(\frac{n_s^i}{N}\right)^{-\alpha} \tag{3.17}$$

where N represents the maximum number of individuals allowed per species, and α is an adjustment parameter, typically set to 1.5. Through this fitness adjustment mechanism, SFS effectively balances the prioritization of structure exploration and parameter optimization within the population, preventing the neglect of structure exploration.

Additionally, this chapter draws inspiration from the neuro-evolution approach NEAT and incorporates an aging factor into the fitness calculation. The aging factor is designed to identify and eliminate strong individuals that are trapped in local optima, preventing them from monopolizing resources and hindering further exploration of the population. Specifically, the aging adjustment factor for individual i is calculated using the following formula:

$$\eta_{\mathrm{age}}^i = \begin{cases} 1 - \dfrac{a_s^i - A}{\beta}, & \text{if } a_s^i > A \\ 1, & \text{otherwise} \end{cases} \tag{3.18}$$

where a_s^i represents the age (in generations) of the species s to which individual i belongs, A is the age threshold at which fitness begins to be affected, and β is a parameter that regulates the aging speed. If $\eta_{\mathrm{age}}^i < 0$, it is set to $\eta_{\mathrm{age}}^i = 0$.

Finally, the fitness adjustment factor η^i for an individual is calculated as the product of the SFS adjustment factor and the aging factor:

$$\eta^i = \eta_{\mathrm{SFS}}^i \eta_{\mathrm{age}}^i \tag{3.19}$$

If $\eta^i > 1$, it is capped at $\eta^i = 1$. Finally, the fitness value Φ^i of individual i is adjusted based on the adjustment factor as follows:

$$\Phi^i = \eta^i \Phi^i \tag{3.20}$$

By incorporating SFS and the aging factor, this approach effectively mitigates the issue of premature convergence while enhancing the population's exploration capability in the search space. This enables a well-balanced trade-off between structural exploration and parameter optimization. As a result, it provides a more efficient and robust optimization tool for evolving HBGs, facilitating the discovery of more diverse and high-performance solutions.

3.6.4 Conclusion

This section provides a detailed explanation of how BGs are implemented in automated design, emphasizing the use of GP techniques to achieve everything from constructing basic modules to optimizing complex systems. Through GP's evolutionary mechanisms, the topological structures and parameter values of BGs can be automatically explored and optimized during the design process.

Specifically, this section first explains the fundamental building blocks of BGs and how these modules can be dynamically adjusted and expanded using GP operators, such as crossover and mutation. Genetic operations play a central role in the evolution of BGs, significantly improving the efficiency of automated design by balancing parameter optimization with structural exploration. To address common challenges in evolutionary computation, such as premature convergence, this section also introduces several advanced methods, including the speciation operator and its mechanisms, such as SFS and aging factors. These techniques effectively enhance population diversity and improve coverage in the search space.

Furthermore, this section lays the theoretical foundation for more complex real-world applications and provides support for the case studies presented in Chapter 4. Chapter 4 will showcase practical design examples based on the proposed methods, further demonstrating the potential and effectiveness of combining BGs with GP in system design. Collectively, this content highlights that utilizing GP for automated BG design is an efficient, flexible, and scalable approach, offering a novel pathway for modeling and optimizing complex systems.

4

Application of MODENA

4.1 Morphology Design Automation

MODENA for intelligent robot morphologies refers to the use of advanced optimization techniques to create the robot's structure, including both mechanical components and supporting infrastructures. Current studies in this field generally fall into two main categories: (1) fixing the topological structures and optimizing the geometric parameters [323–327]; (2) developing a library of parametric modules for robot morphologies [328–330], and then reconfiguring those parameterizable modules to achieve simultaneous optimization of topologies and geometric parameters. In the subsequent sections, two practical case studies will be used to demonstrate the effectiveness of MODENA in mechanical ontology design optimization. These case studies highlight real-world applications of MODENA, demonstrating its effectiveness and potential in robot design.

The first case focuses on the optimal design of a teaching manipulator, where the primary objective is to optimize its mechanical structure. This involves refining the arrangement of components to enhance overall performance and efficiency. The second case deals with the redesign of an electric typewriter driver mechanism, where the goal is to improve both its topology and the geometric parameters. This optimization is carried out using an advanced approach known as the improved BGGP method, which enables a more efficient search for optimal solutions by combining both topological and parameter optimization in a unified framework.

Through these case studies, the aim is to provide a comprehensive overview of how MODENA can be used to automate the design of intelligent robot ontologies and provide insights into the theoretical aspects and practical applications of this innovative design framework.

Modular Design Automation of Intelligent Robot Systems, First Edition. Zhun Fan et al.

4.1.1 Teaching Manipulator Design

1. Design of the Teaching Manipulator

It is a labor-intensive and time-consuming task for researchers to plan and specify manipulator trajectories using a programming system. This challenge becomes even more pronounced when frequent adjustments and redesigns of the trajectories are required to accommodate various applications in a dynamic production line. In this section, a teaching manipulator is developed using an evolutionary algorithm to address the aforementioned problem. As illustrated in Figure 4.1, the human operator is able to grasp the manipulator's end-effector and easily guide it along an appropriate trajectory. While the end-effector follows the specified path, the rotational movements of each joint are captured and stored to generate a motion sequence [13]. Using the recorded motion sequence, other manipulators can replicate the teaching manipulator's trajectory within the production line. However, if there are problems in the design of the teaching manipulator, it may seriously affect the operator's perception of motion during the teaching process. For instance, excessive or fluctuating resistance from the end-effector will prevent the operator from manipulating the manipulator correctly as required. Moreover, the teaching manipulator must be designed to ensure that its joints can maintain static balance when stopped at any position. Consequently, developing an optimal design that simultaneously considers all of these factors presents a significant challenge for human designers. To achieve the robot's optimal design, a robot design optimization process is carried out. In this section, a model of the teaching manipulator is created, and the corresponding optimization problem is formulated as a CMOP.

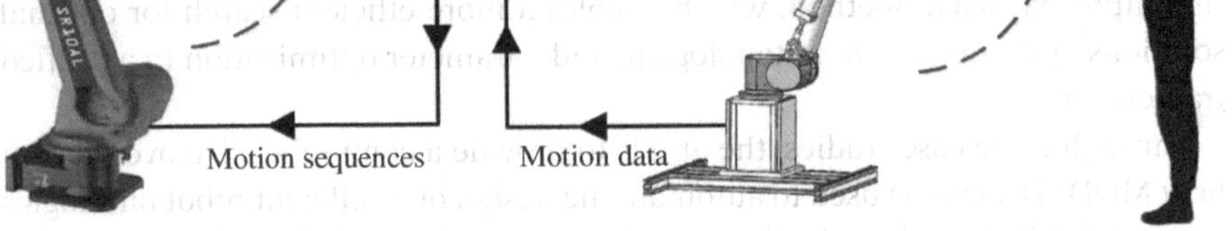

Figure 4.1 The human operator drives the teaching manipulator to execute the desired trajectory. Motion data from the manipulator is captured by the computer and subsequently transmitted as recorded motion sequences to the manipulators deployed on the production line.

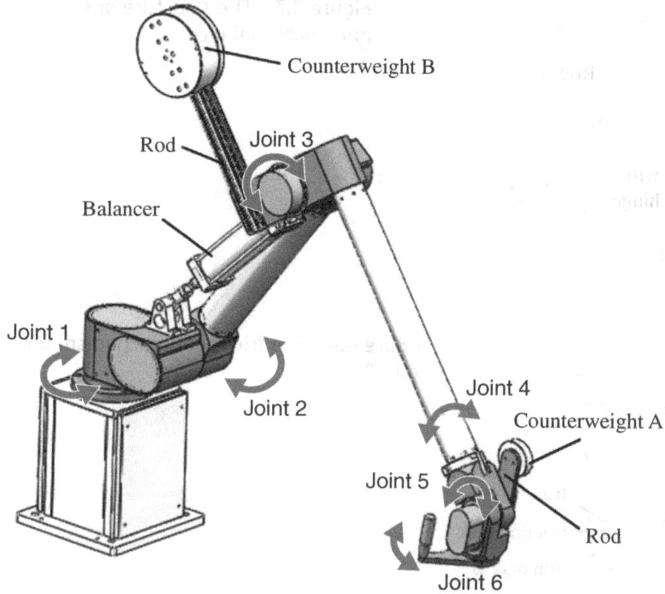

Figure 4.2 A teaching manipulator with 6-DOF.

Figure 4.2 illustrates the teaching manipulator developed for this study, which features six degrees of freedom (DOFs). The link lengths of this manipulator are set to be identical to those used in the production line manipulators. This design choice ensures compatibility, enabling the motion data generated by the teaching manipulator to be directly transferred and utilized as motion sequences for the production line robots. Unlike traditional manipulators, the teaching manipulator has no actuators, as it is driven by a human operator. To capture the rotation data of each joint, Joint 6 encoders are installed at their corresponding joints [13]. Some specific structures are incorporated to maintain joint balance in static conditions, thereby reducing the operating force required of the human operator. These include one pneumatic balancer installed at Joint 2; Joint 2 counterweights positioned at Joint 3 and Joint 5, and Joint 3 friction mechanisms integrated into Joint 1, Joint 2, and Joint 3. Figure 4.3 illustrates the mechanism of the pneumatic balancer, and Figure 4.4 shows the frictional components located in Joint 1.

Using the D-H (Denavit-Hartenberg) convention [331], the teaching manipulator's forward kinematics is established. As illustrated in Figure 4.5, the reference inertial frame is represented by the coordinate system $o_0x_0y_0z_0$, and based on this, other coordinate systems $o_ix_iy_iz_i$ are constructed, where i takes values from 1 to 6. Table 4.1 presents the D-H parameters. Notably, the coordinate systems of Joint 6 and the end-effector are assumed to be identical.

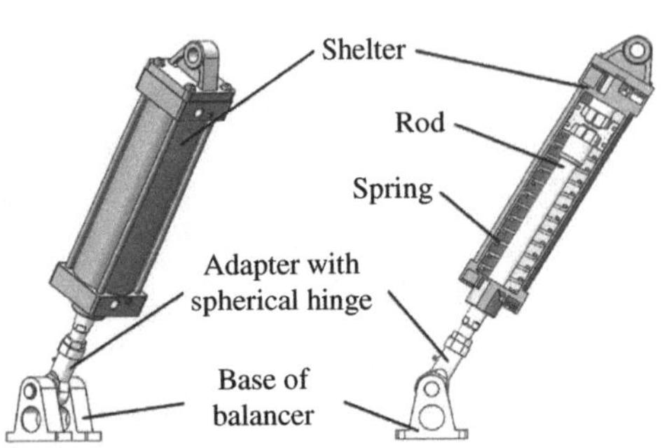

Figure 4.3 The structure of the pneumatic balancer.

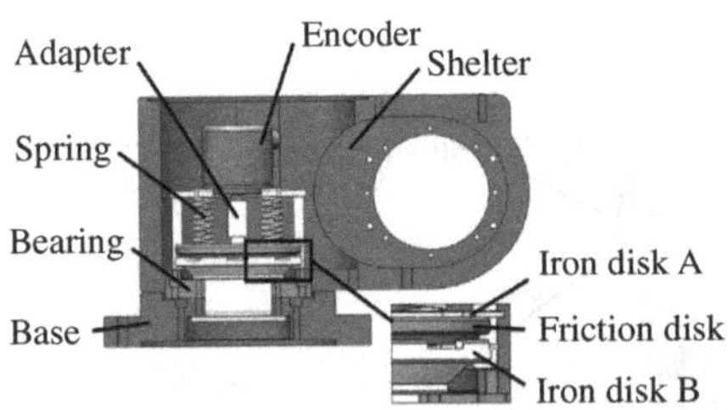

Figure 4.4 The frictional mechanism for Joint 1.

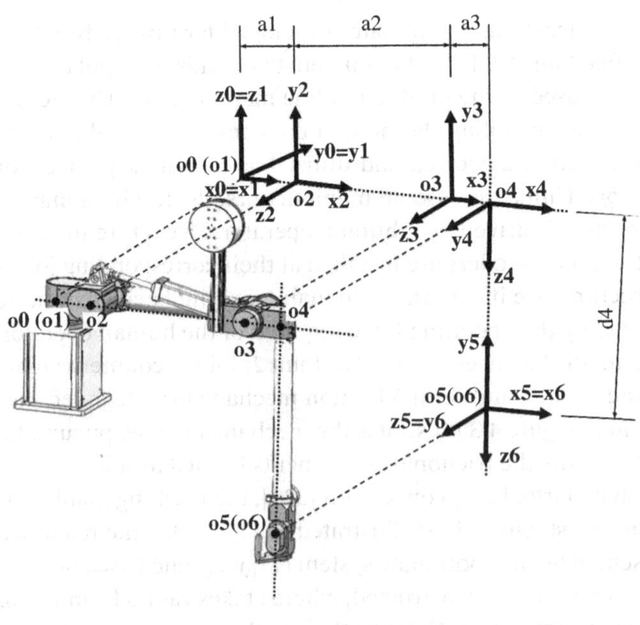

Figure 4.5 Coordinate systems of the teaching manipulator.

Table 4.1 D-H parameters of the 6-DOF teaching manipulator.

Joint$_i$	α_i	a_i	d_i	θ_i
1	$\pi/2$	0.160	0	q_1
2	0	0.790	0	q_2
3	$\pi/2$	0.155	0	q_3
4	$-\pi/2$	0	0.995	q_4
5	$\pi/2$	0	0	q_5
6	0	0	0	q_6

The teaching manipulator's dynamic model is represented using the Lagrangian approach, which is expressed in the equation below:

$$M(q)\ddot{q} + C(q,\dot{q})\dot{q} + G(q) + \tau_{frc}(\dot{q}) = J^T f_{opr}(t) + \tau_b \tag{4.1}$$

where the generalized angular acceleration of the system is represented by $\ddot{q} \in \mathbb{R}^6$. The inertia matrix, denoted by $M(q) \in \mathbb{R}^{6\times6}$, is a symmetric and positive-definite matrix that characterizes the mass distribution and rotational inertia of the manipulator. Gravitational moments acting on the joints are represented by a six-dimensional vector $G(q) = [G_1, G_2, \ldots, G_6]^T$. The matrix $C(q,\dot{q}) \in \mathbb{R}^{6\times6}$ represents the centripetal and Coriolis forces acting on the system. Specifically, the elements of this matrix that depend solely on joint velocities $\dot{q}$ represent the centripetal forces, which arise from the relative motion between the joints. In contrast, the elements that involve both joint positions q and velocities $\dot{q}$ are associated with the Coriolis forces, which occur due to the interactions between the moving links.

In addition, the counterweights' mass and the lengths of the rods connecting them to the joints play a critical role in modifying both the mass distribution and inertia of the manipulator's links, which, in turn, affect the operating forces. The teaching manipulator's Jacobian matrix, with respect to the reference inertial frame, is denoted by $J \in \mathbb{R}^{6\times6}$. The vector $f_{opr}(t) = [F_x, F_y, F_z, n_x, n_y, n_z]$ represents the operating forces and torques applied by the human operator, where F_x, F_y, F_z are the linear force components along the x-, y-, and z-axes, and n_x, n_y, n_z correspond to the torques around each of these axes, respectively. It's important to note that the operating torque is relatively small due to the absence of friction disks in the three joints of the spherical wrist, and because the counterweight A helps maintain the balance of Joint 5. As a result, the focus of the design should be more on the operating force rather than the operating torque. Additionally, the vector $\tau_b \in \mathbb{R}^6$ represents the moment exerted on the manipulator's joints by the balancer, and $\tau_{frc}(\dot{q}) \in \mathbb{R}^6$ corresponds to the friction moment vector, which arises from the interaction of the friction disks with the joints, and depends on the joint velocities $\dot{q}$.

2. Design Variables

The optimization problem takes into account a total of ten design variables, which are specified as follows:

$$x = [m_A, L_A, m_B, L_B, k, H_b, b, T_1, T_2, T_3] \tag{4.2}$$

where m_A and m_B represent the masses of the counterweights that are affixed to Joint 5 and Joint 3, respectively. These counterweights play a significant role in balancing and stabilizing the manipulator. Correspondingly, L_A and L_B represent the lengths of the rods connected to these counterweights. The spring's stiffness coefficient within the balancer is represented by k. The vertical distance, which is measured from Joint 2's rotational axis to the balancer's lower attachment point, is represented by H_b. The force exerted by the pneumatic cylinder, which assists in the balancing process, is represented by b. Additionally, T_i (where $i = 1, 2, 3$) denotes the kinetic friction moments generated by the friction disks at each of the joints. A comprehensive diagram illustrating these variables can be found in Figure 4.6.

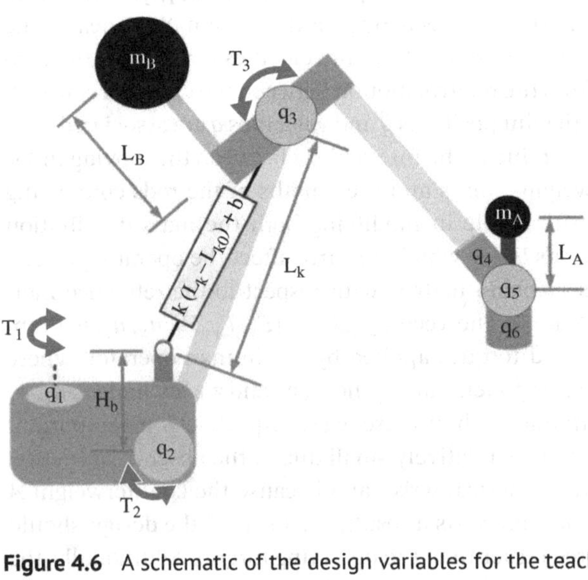

Figure 4.6 A schematic of the design variables for the teaching manipulator.

3. Objective Functions

To ensure that the human operator's perception of motion is not significantly affected during the manipulator's operation, two key objectives are prioritized: first, it is essential to minimize the peak operating force, ensuring that the operator does not encounter excessive resistance; second, the goal is to reduce the variation between the maximum and minimum operating forces, ensuring a smoother and more consistent interaction throughout the operation.

a. Peak Operating Force

In this research, the first goal is to minimize the peak operating force, with the aim of enabling the human operator to control the teaching manipulator smoothly and effortlessly, without facing substantial resistance. The operating force magnitude, which is crucial to this objective, can be calculated using the following equation.

$$f_{mag}(x) = \sqrt{F_x^2 + F_y^2 + F_z^2} \tag{4.3}$$

Therefore, the objective function is expressed in the following equation:

$$f_1(x) = \max_t f_{mag}(x, t) \tag{4.4}$$

b. Difference Between the Maximum and Minimum Operating Forces

Additionally, the second objective of this study focuses on minimizing the variation between the maximum and minimum values of the operating forces. Large fluctuations in these forces can significantly affect the human operator's experience, leading to an inconsistent or uncomfortable perception of motion while controlling the manipulator. This objective is captured by the following function:

$$f_2(x) = \| \max_t f_{mag}(x, t) - \min_t f_{mag}(x, t) \| \tag{4.5}$$

4. Constraints

To guarantee that the teaching manipulator can maintain the angular positions of Joints 1, 2, and 3 at any given point along its trajectory, it is essential for the manipulator to remain stationary at these points. This requirement is achieved by enforcing three specific constraints, as outlined below:

$$\max_t |G_2(q_t, m_A, L_A, m_B, L_B) - \tau_{b,2}(q_t, k, H_b, b)| \leq T_2 \tag{4.6}$$

where $\tau_{b,2}$ refers to the second component of the moment force vector τ_b, which corresponds to the rotational moment applied by the balancer to Joint 2.

$$\max_t |G_3(q_t, m_A, L_A, m_B, L_B)| \leq T_3 \tag{4.7}$$

To maintain the balance of Joint 5, it is necessary to meet the following constraint.

$$\max_{t} |G_5(q_t, m_A, L_A)| \leq \epsilon \tag{4.8}$$

where ϵ is set to 0.1 N · m. For a more in-depth analysis, see the supplementary material in [13].

The design variables, along with their respective ranges, are detailed in Table 4.2, as provided by an experienced engineer. Additionally, Table 4.3 lists

Table 4.2 The initial values and ranges of the design variables.

Variables	Range	Units	Original design
m_A	[0, 20]	kg	1.5
L_A	[0, 0.2]	m	0.18
m_B	[0, 50]	kg	15
L_B	[0, 0.8]	m	0.46
k	[0, 50000]	N/m	3730
H_b	[0.11, 0.50]	m	0.12
b	[450, 4000]	N	2500
T_1	[0, 90]	N·m	75.7
T_2	[0, 90]	N·m	75.7
T_3	[0, 90]	N·m	75.7

Table 4.3 The constant parameters derived from the virtual prototype.

Parameter	Value	Units
m_{L1}	13.0	kg
m_{L2}	11.7	kg
m_{L3}	5.3	kg
m_{L4}	4.5	kg
m_{L5}	2.0	kg
m_{L6}	1.5	kg
ρ_A	1.8	kg/m
ρ_B	3.7	kg/m

other constant parameters, which are derived from the virtual prototype created in SolidWorks. Among these parameters, m_{Li} represents the mass of Link i, while ρ_A and ρ_B correspond to the mass per unit length of the connecting rods for counterweights A and B. These parameters are crucial for accurately modeling the dynamics of the manipulator and its counterweight system.

5. Trajectory Design

To thoroughly assess the performance of a particular design, it is crucial to define a specific trajectory that aligns with the intended application scenario. For this purpose, a trajectory previously employed in [332, 333] is adopted, as it serves as a widely accepted benchmark for evaluating the performance of 6-DOF manipulators. This trajectory is defined in the base coordinate system and mathematically expressed as the following equation:

$$\begin{aligned} X_{ef} &= 1.2 - 0.3 \times \left(1 - \cos\left(\frac{\pi}{10}t\right)\right) \\ Y_{ef} &= 0.8 \sin\left(\frac{\pi}{10}t\right) \\ Z_{ef} &= 0.3 + 0.5 \cos\left(\frac{\pi}{10}t\right) \end{aligned} \tag{4.9}$$

All functions are expressed in meters (denoted as m). The Euler angles for the end-effector are specified as $[0, \pi, \pi]$, which ensures that the end-effector remains vertically oriented always facing downward toward the ground during its entire motion. To analyze the motion, the entire trajectory is divided into $N = 2000$ time segments, each lasting 0.01s. For each segment, it is assumed that the end-effector's movement follows a uniformly accelerated pattern.

With the aid of the Robotics Toolbox in MATLAB [334], a simulation model of the teaching manipulator is developed to determine the operating force at different positions along the trajectory mentioned earlier. Figure 4.7 depicts the teaching manipulator model along with the designed trajectory of its end-effector in the simulation environment.

6. Experimental Study

In order to assess the effectiveness of the proposed PPS-MOEA/D and PPS-M2M approaches, a comparative analysis is conducted with four other CMOEAs, namely MOEA/D-ACDP, MOEA/D-CDP, NSGA-II-CDP, CM2M, and PPS-MOEA/D. The parameters utilized for these algorithms are provided in detail below:

a. The mutation probability P_m is defined as $1/n$, where n represents the dimensionality of the decision vector. In addition, in the context of the polynomial mutation, the distribution index is fixed at a value of 20.

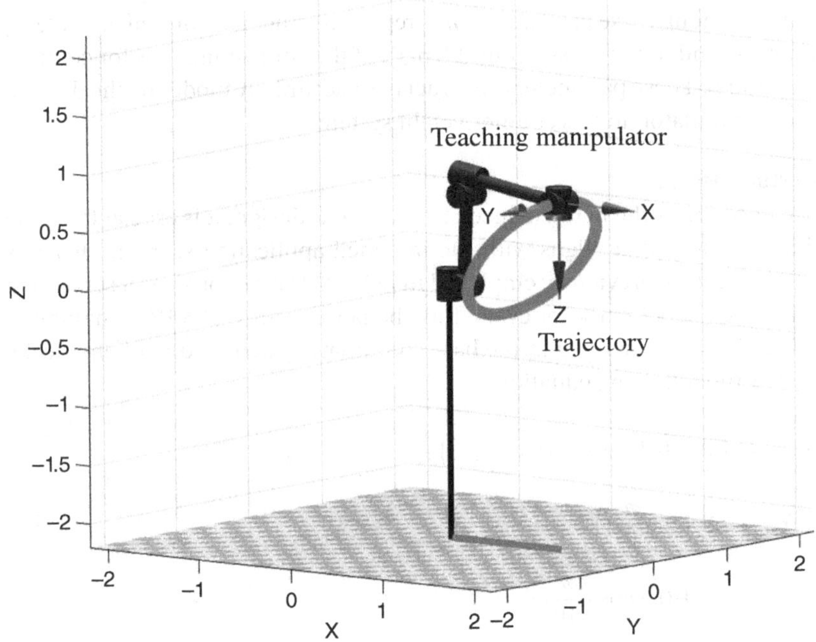

Figure 4.7 The teaching manipulator model and the trajectory of its end effector in MATLAB.

b. The parameter settings for differential evolution are: *CR* is set to 1.0 and *f* is set to 0.5.
c. The population size *N* is set to 200, and the neighborhood size *T* is configured to 20.
d. The halting condition for each algorithm specifies that the process runs independently for 30 iterations and terminates once 400,000 function evaluations have been completed.
e. The probability δ, which determines the likelihood of selecting individuals from within their respective neighborhoods, is set at 0.9.
f. Each child can update a maximum of $n_r = 2$ solutions.
g. For the PPS-MOEA/D algorithm, the basic parameter settings are: $Tc = 1600$, $\alpha = 0.95$, $\tau = 0.05$, $cp = 2$, $l = 20$.
h. For the PPS-M2M algorithm, the basic parameters are identical to those of PPS-MOEA/D. Additionally, the number *K* of subregions is set to 10.
i. For the CM2M approach, the number *K* of subregions is also set to 10.

For each CMOEA, 30 separate runs are performed to address CMOP in designing the teaching manipulator. For each CMOEA, 30 separate executions were performed to address CMOP in designing the teaching manipulator. To further assess the performance of PPS-MOEA/D relative to the other five CMOEAs, a Wilcoxon's rank-sum test was applied at a significance level of 0.05. This statistical test evaluates the differences between the algorithm's results. The outcomes, based on the HV metric, reveal that PPS-MOEA/D significantly surpasses the other algorithms – PPS-M2M, NSGA-II-CDP, MOEA/D-ACDP, MOEA/D-CDP, and CM2M – in effectively addressing the teaching manipulator design optimization problem.

Figure 4.8 displays the distributions of HV values obtained across the 30 independent runs for each of the six algorithms. From the graph, it is evident that PPS-MOEA/D achieves the largest median HV value, along with the lowest interquartile range. This indicates that PPS-MOEA/D outperforms the other five algorithms in terms of stability and effectiveness in solving the teaching manipulator design optimization problem.

The HV results for four different algorithms, PPS-MOEA/D, MOEA/D-CDP, PPS-M2M, and CM2M, are shown in Table 4.4. To evaluate the statistical significance of the observed performance differences, Wilcoxon's rank-sum test was performed at a 0.05 significance level, with comparisons drawn between PPS-MOEA/D and MOEA/D-CDP, as well as between PPS-M2M and CM2M. The results in Table 4.4 clearly indicate that PPS-MOEA/D outperforms MOEA/D-CDP with respect to the HV index. Similarly, PPS-M2M demonstrates a marked improvement over CM2M. A key observation is that both PPS-MOEA/D and

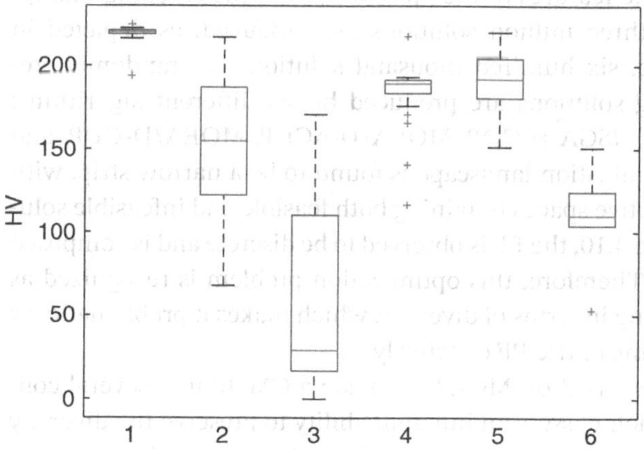

Figure 4.8 Comparison of HV values between PPS-MEEA/D and other five CMOEAs across 30 independent runs for the teaching manipulator design optimization problem. For ease of interpretation, the numbers 1 through 6 correspond to the following algorithms: 1 – PPS-MOEA/D, 2 – PPS-M2M, 3 – NSGA-II-CDP, 4 – MOEA/D-ACDP, 5 – MOEA/D-CDP, and 6 – CM2M.

Table 4.4 Comparison of HV values between the two PPS-based CMOEAs with their counterparts that do not use the PPS framework. The test compares the performance of PPS-MOEA/D with MOEA/D-CDP, as well as PPS-M2M with CM2M.

	PPS-MOEA/D	MOEA/D-CDP		PPS-M2M	CM2M
mean	**217.9**	190.3†	mean	**157.1**	109.7†
std	5.1	18.0	std	36.6	30.0

The symbol † is used to indicate that the performance of the corresponding algorithm is significantly inferior to that of either PPS-MOEA/D or PPS-M2M.

PPS-M2M, which incorporate the PPS framework, consistently deliver superior performance compared to their counterparts, MOEA/D-CDP and CM2M, which do not utilize the PPS approach. This emphasizes the effectiveness and advantage of the PPS framework in enhancing the optimization of the teaching manipulator design, demonstrating its clear superiority in this particular problem.

Figure 4.9 presents the non-dominated solutions derived from each approach during the optimization of the teaching manipulator design, highlighting those that yielded the highest HV results across the 30 separate runs. It is evident that the majority of the solutions produced by PPS-MOEA/D are positioned along the reference PF, demonstrating its ability to find solutions that closely align with the true optimal set. In contrast, the other five CMOEAs tend to produce solutions that are concentrated in local optima.

In order to investigate the features of the optimization of the teaching manipulator design, a total of three million solutions are produced, as depicted in Figure 4.10. Among these, six hundred thousand solutions are randomly created, while the remaining solutions are produced by six different algorithms: PPS-MOEA/D, PPS-M2M, NSGA-II-CDP, MOEA/D-ACDP, MOEA/D-CDP, and CM2M. The resulting optimization landscape is found to be a narrow strip, with certain regions of the objective space containing both feasible and infeasible solutions. As depicted in Figure 4.10, the PF is observed to be discrete and is composed of two distinct segments. Therefore, this optimization problem is recognized as being inherently challenging in terms of diversity, which makes it problematic for CMOEAs to approximate the entire PF effectively.

The CMOEA algorithms based on MOEA/D divide a CMOP into several constrained subproblems, which possess an inherent ability to preserve the diversity of the population. Given that the teaching manipulator design optimization problem is particularly challenging in terms of diversity, MOEA/D-based methods are better suited to address it compared to NSGA-II-based CMOEAs, as evidenced by the results presented in Table 4.5. In the comparison between PPS-MOEA/D and PPS-M2M, a key distinction lies in their underlying frameworks: PPS-MOEA/D

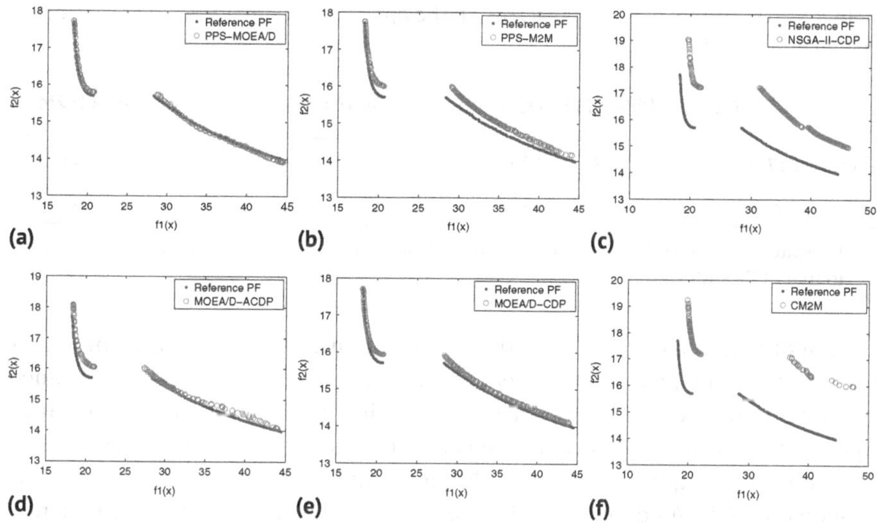

Figure 4.9 The non-dominated solutions with the highest HV values obtained by each approach during the 30 separate runs.

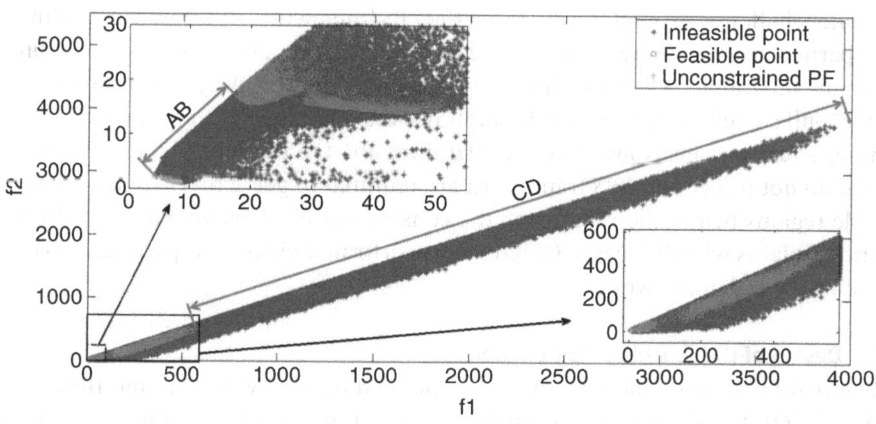

Figure 4.10 The distribution of solutions of the teaching manipulator design optimization problem in the objective space.

employs the MOEA/D framework, while PPS-M2M adopts the NSGA-II framework within each subregion. As shown in the optimization results for the teaching manipulator design, PPS-MOEA/D outperforms PPS-M2M, demonstrating the superior capability of the MOEA/D framework in this context.

Table 4.5 HV results of the six CMOEAs on the optimization problem of the teaching manipulator design.

	PPS-MOEA/D	PPS-M2M	NSGA-II-CDP	MOEA/D-ACDP	MOEA/D-CDP	CM2M
mean	**217.9**	157.2†	61.6†	184.1†	190.3†	109.7†
std	5.1	36.6	58.0	17.7	18.0	30.0

† Indicates that the HV performance of the corresponding algorithm is significantly inferior to that of PPS-MOEA/D.

An additional characteristic of the optimization of the teaching manipulator design is the close proximity between the unconstrained PF and the constrained PF, along with the small ratio of feasible to infeasible solutions, as shown in Figure 4.10. Specifically, the distance between these two fronts is represented by *AB*, while the length of the infeasible region in front of the constrained PF is presented by *CD*. As depicted in Figure 4.10, the distance of *AB* is significantly smaller than *CD*, which means that the PPS-based methods can more easily pull infeasible solutions towards the constrained PF. In contrast, CMOEAs that do not use the PPS framework must traverse a much larger infeasible region to reach the constrained PF. This distinct feature of the optimization problem presents a greater challenge for CMOEAs without the PPS framework. PPS-based algorithms, in particular, follow a two-stage process: the first stage identifies solutions along the unconstrained PF, disregarding constraints, while in the second, known as the "pull search" stage, infeasible solutions situated along the unconstrained PF are gradually pulled towards the constrained PF. On the other hand, CMOEAs that do not utilize the PPS framework are required to get across extensive infeasible regions before they can reach the constrained PF. This fundamental difference explains why PPS-based CMOEAs outperform their counterparts that do not utilize the PPS framework.

7. Design of the Teaching Manipulator

Figure 4.11 presents the reference PF generated by PPS-MOEA/D and the other five CMOEAs, along with the original design proposed by a human expert. It can be observed that the reference PF solutions are superior to the initial design, indicating a substantial improvement in the teaching manipulator's performance when utilizing the proposed PPS-MOEA/D and PPS-M2M algorithms. As the reference PF is composed of two distinct segments, the endpoints of each segment are selected as typical examples, which are depicted in Figure 4.11. This study conducted a comprehensive comparison between the original design by the human expert and four typical designs selected from the reference PF. The detailed results can be found in the original paper.

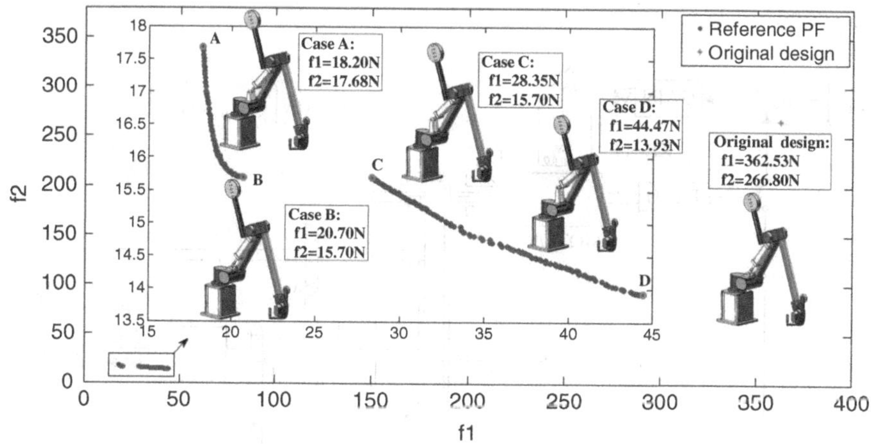

Figure 4.11 The four endpoints of the reference PF are chosen as the typical examples for comparison with the human expert's original design. For each of these selected solutions, CAD models are generated, with distinct appearances that differentiate them from the original design.

4.1.2 Optimal Design of Drive Mechanism for Electric Typewriter

In order to develop an integrated mechatronics design framework, a method combining BG and GP is studied to improve the level of automation design of mechatronics systems to some extent. To enhance the topology search capabilities of GP and offer designers a broader array of potential design options, this section introduces a tailored version of parallel GP, referred to as hierarchical fair competition GP. The practical effectiveness of this approach is demonstrated through a compelling case: the redesign of the drive mechanism for an electric typewriter, as detailed in [37]. By investigating open-topologies within the bond graph framework, multiple feasible design alternatives for the typewriter drive mechanism are produced, effectively showcasing the versatility and robustness of the proposed method in generating diverse and optimized solutions.

1. Problem Formulation

The problem of redesigning the drive mechanism for an electric typewriter was initially introduced in 1972 by C. Denny and W. Oates from IBM, Lexington, KY. As depicted in Figure 4.12, the system under consideration is a closed-loop control system, specifically designed to position a rotational load, which is referred to as inertia and represented by J_L. This load is part of a dynamic system that involves various components working together. The system is made up of several interconnected elements, including an electric voltage source, a motor, and a range of mechanical parts. Given that the system involves multiple domains, namely electrical, mechanical, and control, the use of a bond graph is particularly suitable

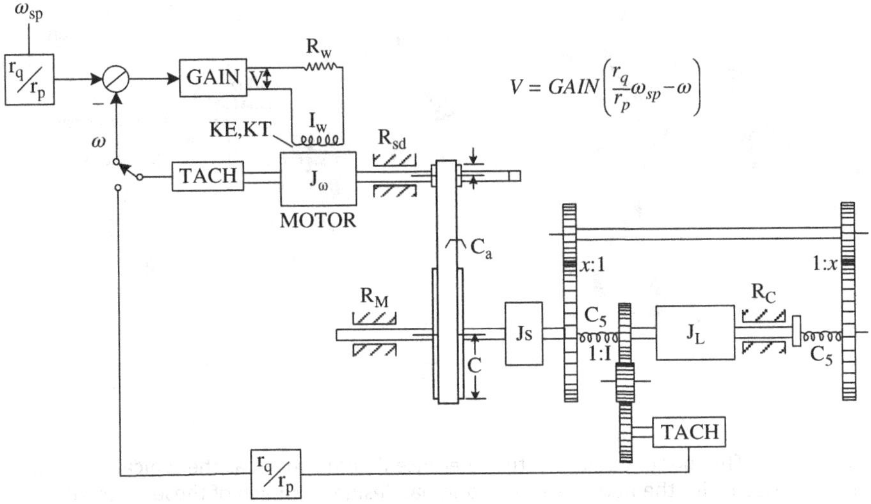

Figure 4.12 Schematic of the typewriter drive system.

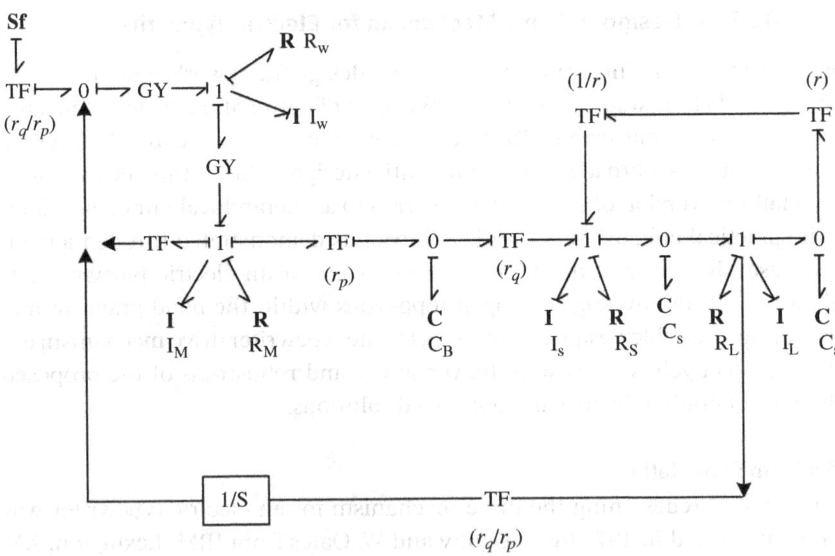

Figure 4.13 BG model.

for modeling the interactions and dynamics within the system, as illustrated in Figure 4.13. A key challenge with the original design is that the load J_L experiences significant vibrations in its position output, as shown in Figure 4.14. The primary goal of the redesign is to minimize the vibrations of the load's position, ensuring they fall within an acceptable range, while also meeting the specified rotational position command conditions. As defined in the design specifications in [37], the system should exhibit a settling time of less than 70 ms when the input voltage is stepped from zero to one. In stark contrast, the original system exhibits a much slower settling time of around 2000 ms, as shown in the time scale of 4000 ms in Figure 4.14. Thus, the challenge lies in significantly improving the system's response time and reducing the oscillations within the load's position.

Through a detailed analysis of the system model, it becomes clear that the key component influencing the design is the subsystem that includes the drive shaft and the load, as illustrated in Figure 4.15. In this configuration, the driving torque, T_d, serves as the input to the subsystem. This torque is produced by the motor and

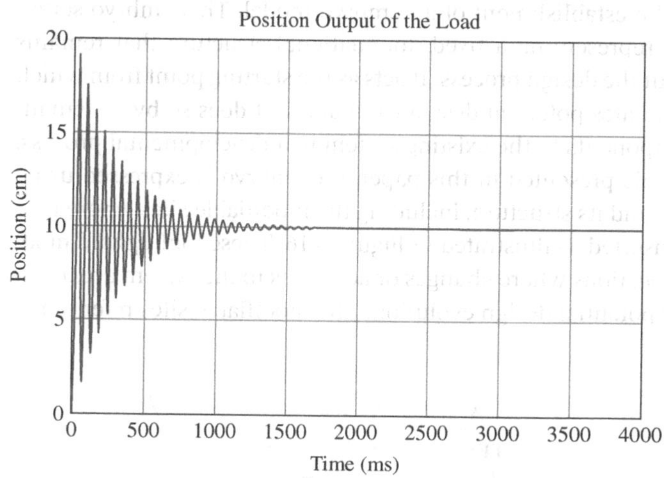

Figure 4.14 Positional vibration of the load.

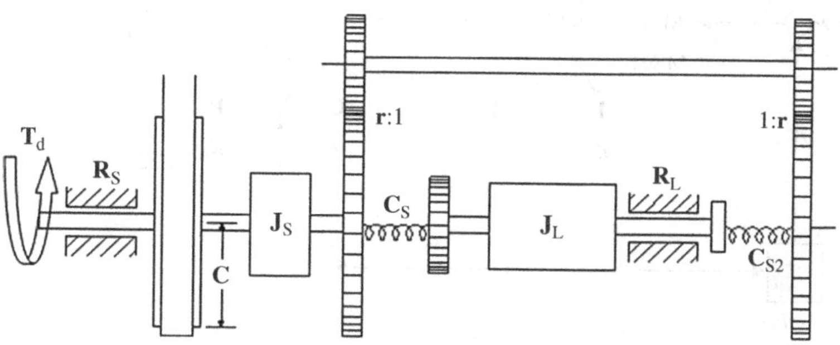

Figure 4.15 The embryo subsystem.

subsequently transferred through a belt coupling mechanism (not shown in the figure).

For the design of this subsystem, the designer should consider the following three aspects: (a) determining the precise locations within the subsystem where new components are to be incorporated, (b) selecting the appropriate types and quantities of components to be added, and deciding on their arrangement, and (c) specifying the parameter values for these newly added components. The methodology presented in this section is capable of autonomously answering all three of these critical questions, provided that the basic framework or "embryo" of the system has already been established.

2. Embryo of Design

In order to explore new design alternatives using the BG/GP methodology, the process begins with the establishment of an embryo model. This embryo serves as the system's core, representing a fixed, foundational structure that remains unchanged throughout the design process. It acts as the starting point from which the GP algorithm generates potential design candidates. It does so by systematically adding new components to the existing system in a developmental process. For the specific example presented in this paper, the embryo is expressed using bond graph language, and its structure, including the modifiable sites where new components can be inserted, is illustrated in Figure 4.16. These highlighted modifiable sites indicate locations where changes or additions to the system are possible, marking areas of potential design evolution. The modifiable sites refer to the

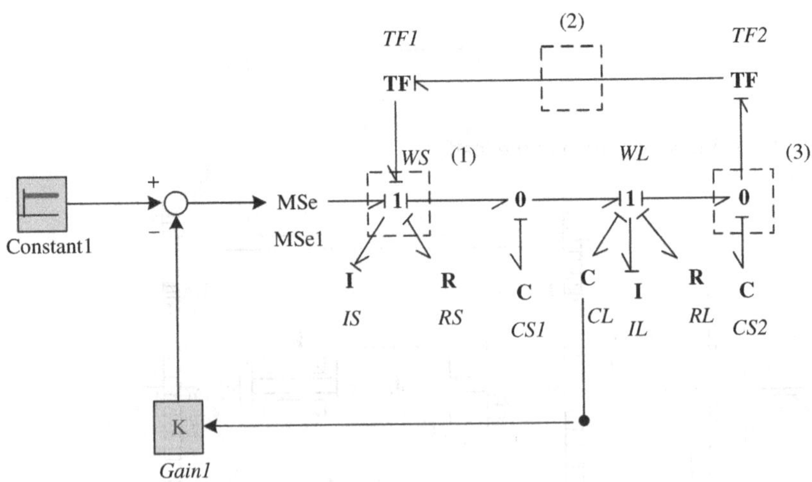

Figure 4.16 The embryo subsystem.

locations in the system where new components can potentially be inserted. Typically, the designer finds it straightforward to identify these sites, as they represent the designated areas for modification. However, it is important to note that while these sites offer the possibility for component insertion, not all of them will necessarily be utilized during the design search process. In the example provided, the designer is not required to predefine the specific locations for adding new components – whether it be at modifiable site (1), site (2), site (3), or any combination of these potential positions. Instead, the algorithm autonomously determines both the optimal locations and configurations for component integration. This eliminates manual input and allows the system to operate independently. Such an approach not only reduces the designer's involvement but also facilitates a more efficient exploration of design alternatives, enabling the algorithm to systematically examine a broader range of possibilities with greater speed and accuracy.

The parameters for the embryo model are:

$$\begin{cases} I_s : 6.7 \times 10^{-6}\ \mathrm{kg} \cdot \mathrm{m}^2 \\ R_s : 0.013 \times 10^{-3}\ \mathrm{N} \cdot \mathrm{m} \cdot \mathrm{sec}/\mathrm{rad} \\ C_{s1} : 0.208\ \mathrm{N} \cdot \mathrm{m/rad} \\ C_{s2} : 0.208\ \mathrm{N} \cdot \mathrm{m/rad} \\ R_L : 0.58 \times 10^{-3}\ \mathrm{N} \cdot \mathrm{m} \cdot \mathrm{sec}/\mathrm{rad} \\ I_L : 84.3 \times 10^{-6}\ \mathrm{kg} \cdot \mathrm{m}^2 \\ C_L : 1.0 \times 10^{6}\ \mathrm{N} \cdot \mathrm{m/rad} \\ \mathrm{TF1} : 0.1,\ \mathrm{TF2} : 10 \end{cases} \tag{4.10}$$

To simplify the analysis without affecting its general applicability, both the gain values K and MSe are assigned a value of one.

3. HFC Model

In this study, a specialized variant of GP (called hierarchical fair competition GP) is utilized to redesign the drive mechanism of an electric typewriter. As illustrated in Figure 4.17, this model organizes multiple subpopulations into a hierarchical structure. Each subpopulation is confined to individuals who fall within a specific fitness range [335]. New individuals are constantly generated in the lower layers of the hierarchy. The HFC model is particularly effective in striking a balance between exploration and exploitation within GP. From our experience, this model not only enhances the diversity of the entire population but also offers the designer

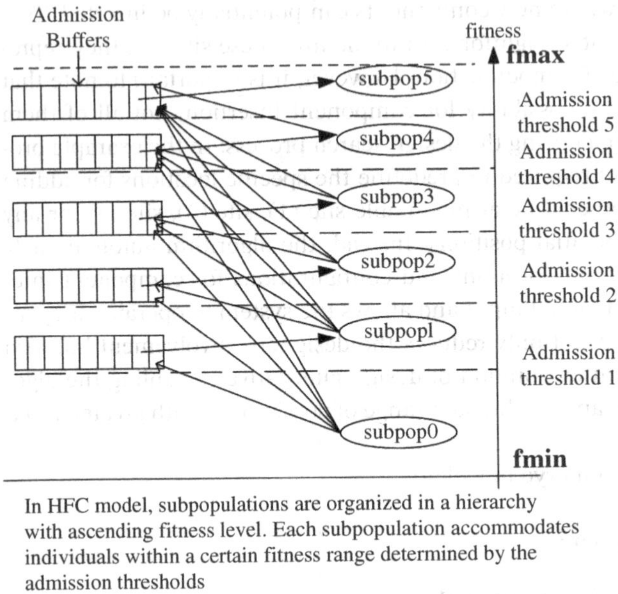

Figure 4.17 The HFC model of GP.

a wide array of competing design alternatives, facilitating better trade-offs during the design process.

4. Definition of Fitness Function

The fitness function for evaluating an individual design is formulated based on the position output response of the load J_L, which is described in Equation (4.11). Over the specified time interval of interest (which in this example spans from 0 to 500 ms), 1000 uniformly distributed points are sampled from the output response. At each of these points, the magnitude of the load's position output is compared to the desired target value (which is set to 1 in this example). The deviation between the actual output and the target is computed, and the squared sum of these differences is calculated, yielding the raw fitness value, denoted as $\text{Fitness}_{\text{raw}}$. Finally, the raw fitness value is normalized using the following formula:

$$\text{Fitness}_{\text{norm}} = 0.5 + \frac{1000}{2000 + \text{Fitness}_{\text{raw}}} \tag{4.11}$$

It can be roughly inferred that a higher fitness value after regularization generally indicates a better design. However, this fitness definition is an approximation for two key reasons: first, it does not directly align with a precise definition of stability time, and second, it overlooks other factors in the system design that extend beyond the output response.

While a more refined fitness function could be defined at a later stage if necessary, the current definition is adequate for this study. It serves to demonstrate the feasibility and validity of the method presented. The design outcomes, as illustrated in Figures 4.18–4.24, show that the performance satisfies the design specifications proposed in this case study.

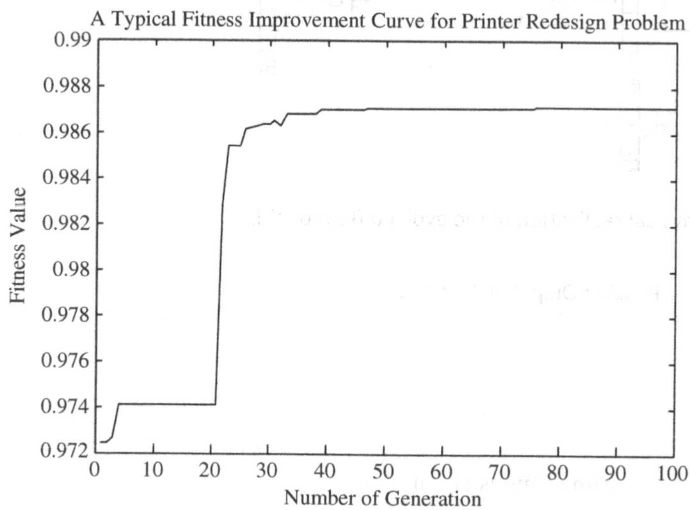

Figure 4.18 Fitness progression for a typical run of the typewriter drive redesign.

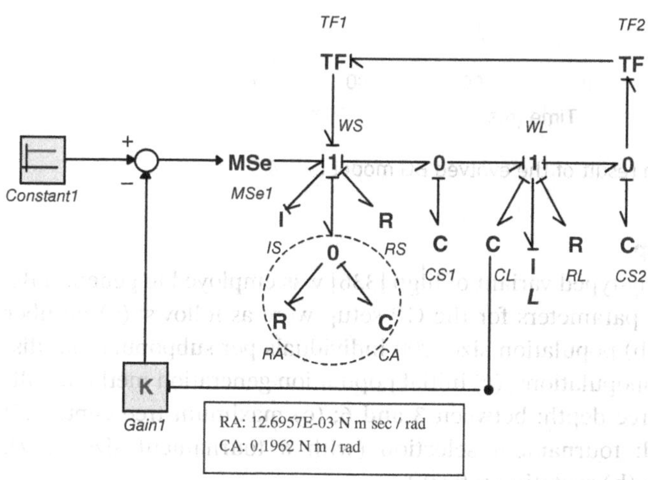

Figure 4.19 The evolved BG model I.

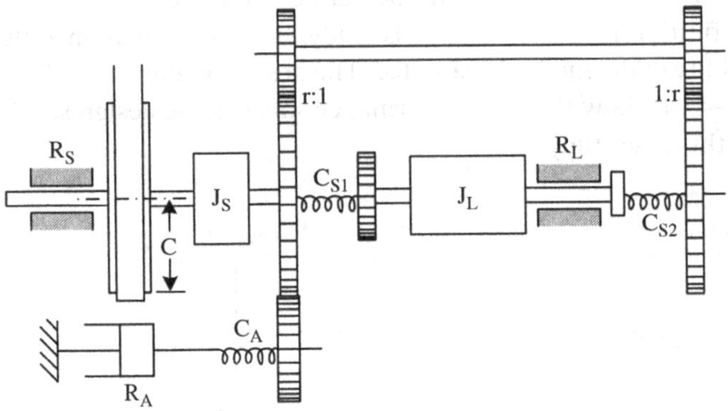

Figure 4.20 The physical realization of the evolved BG model I.

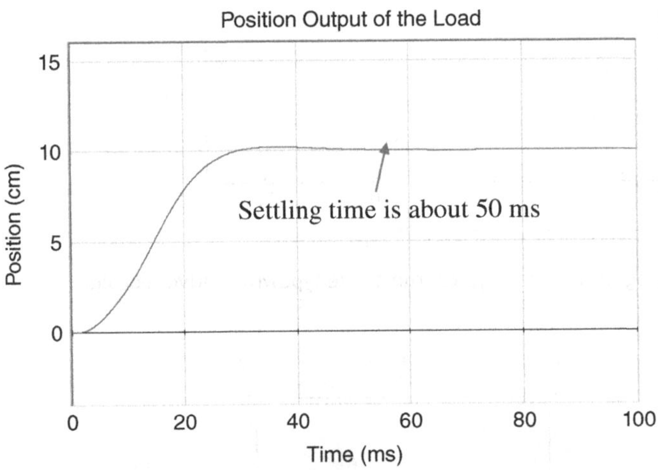

Figure 4.21 Simulation result of the evolved BG model I.

5. Experimental Setup

In this study, a strongly-typed variant of lilgp [336] was employed to generate BG models. The primary parameters for the GP setup were as follows: (a) number of generations: 100; (b) population size: 200 individuals per subpopulation, distributed across 15 subpopulations; (c) initial population generation method: half-and-half; (d) initial tree depth: between 3 and 6; (e) maximum tree depth: 17; (f) selection method: tournament selection (with a tournament size of 7); (g) crossover rate: 0.9; (h) mutation rate: 0.1.

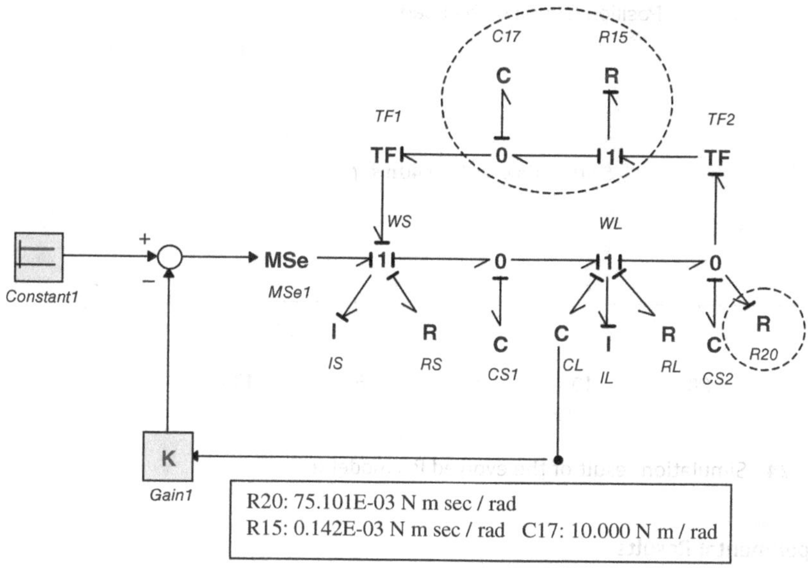

Figure 4.22 The evolved BG model II.

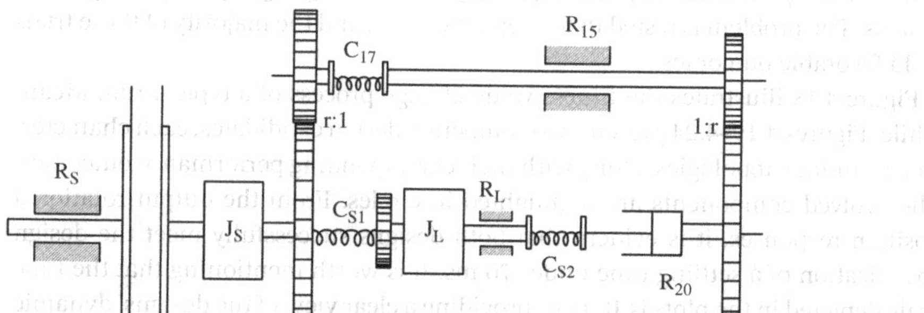

Figure 4.23 The physical realization of the evolved BG model II.

For the implementation, three major code modules were developed. The core algorithm of the hierarchical fair competition GP [37] was derived from a modified version of lilgp. Additionally, a BG class was written in C++, and the fitness evaluation component was implemented in C++ by converting MATLAB code, supplemented by hand-coded functions for seamless integration with other project modules. To validate the dynamic properties of the evolved designs, the commercial software 20-Sim was utilized for simulation and analysis.

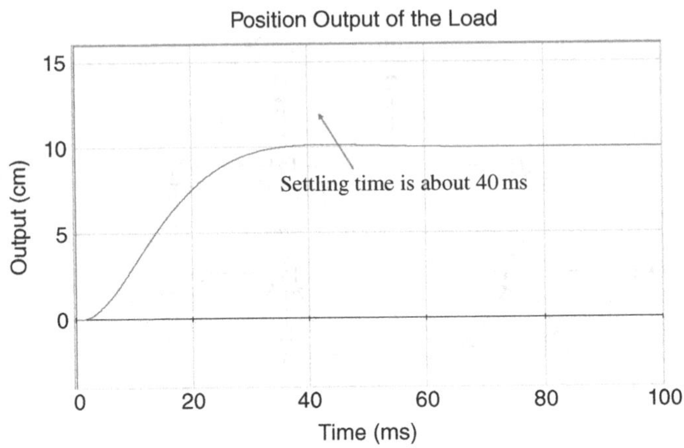

Figure 4.24 Simulation result of the evolved BG model II.

6. Experimental Results

The hierarchical fair competition GP algorithm successfully produces satisfactory results within a time range of 5 to 15 minutes on a Pentium IV 1 GHz processor, demonstrating its efficiency and capability in searching high-quality design candidates. The problem is tested in ten separate trials, and the majority of these trials yield favorable outcomes.

Figure 4.18 illustrates the fitness value change process of a typical run. Meanwhile, Figures 4.19–4.24 present two competing design candidates, each characterized by unique topologies, along with their corresponding performance outcomes. The evolved components are highlighted in circles. From the output rotational position responses, it is evident that both designs successfully meet the design specification of a settling time under 70 ms. It is worth mentioning that the time scale depicted in the plots is 100 ms, providing a clear view of the designs' dynamic performance.

Figure 4.19 illustrates one of the design solutions, which is developed after only 20 generations of the GP process. In each of the 15 subpopulations, 200 individuals are evaluated, yet the resulting design remains remarkably simple in structure. The key modifications involve the addition of three components – specifically, a 0-junction, a capacitor (C), and a resistor (R) – which are inserted into modifiable site 1 of the embryo model, as illustrated in Figure 4.19. These newly introduced elements are clearly marked with dashed circles in the corresponding bond graph diagrams.

Figure 4.20 presents one potential physical implementation of the bond graph model. In this configuration, a spring and damper have been introduced and integrated into the existing printer subsystem to enhance its dynamic behavior. As illustrated in Figure 4.21, the position response to a step function input stabilizes rapidly, converging within approximately 50 ms, which is deemed an acceptable response time. Furthermore, an alternative design configuration is depicted in Figure 4.22, offering another possible solution.

In the design, three elements are incorporated: a 0-junction with a capacitance (C) and a 1-junction with a resistance (R), which are added to modifiable site (2), while an additional resistor (R) is placed at modifiable site (3). Figure 4.23 illustrates a potential physical representation of this design configuration. The performance outcomes of this model are displayed in Figure 4.24, highlighting its operational characteristics. Additionally, the statistical outcomes from 10 independent runs of the printer drive optimization process are summarized in Table 4.6.

The approach presented in this study demonstrates both efficiency and effectiveness, offering designers a broad range of potential design alternatives. This flexibility empowers designers to explore and evaluate numerous design options, facilitating the creation of diverse and optimized solutions for their specific needs.

Table 4.6 Fitness outcomes for the printer system.

Run No	Distance	Fitness
1	15.08	0.99
2	15.82	0.98
3	15.19	0.99
4	16.7	0.98
5	15.05	0.99
6	14.09	0.99
7	15.12	0.99
8	15.5	0.99
9	15.13	0.99
10	15.88	0.98
Best	14.09	0.99
Worst	16.72	0.98
Average	15.36	0.99
Standard Deviation	0.69	0.0007

The ability to evolve and evaluate a wide array of design alternatives enhances the decision-making process and supports more informed design choices.

4.2 Controller Design Automation

Within the domain of intelligent robot systems, the robot control system plays a crucial and indispensable role, significantly impacting the performance, stability, and operational capacity of the entire robotic setup. Its primary function is to manage and modify the robot's actions, ensuring seamless and precise task execution to meet predefined goals.

The optimization of controller design is of utmost importance. It boosts robots' efficiency and adaptability and strengthens their resilience and decision-making in complex environments. Consequently, the design optimization of controllers represents a fundamental and central task in the advancement of intelligent robot systems. In subsequent sections, two case studies will be presented to illustrate the efficacy and practicality of the MODENA framework in enhancing control system design. The initial case centers on the optimization of an FSA controller for a hybrid mechatronic system, whereas the second case delves into the optimization of a vehicle suspension system.

Even though these two cases do not directly pertain to the optimization of robot controller systems, they are both rooted in the core algorithms of the MODENA framework. These case studies yield valuable theoretical perspectives on the application of sophisticated optimization methods in control systems and can be extrapolated to the design optimization of intelligent robot control systems. The outcomes of these cases not only showcase the practical viability of the MODENA framework across diverse engineering scenarios but also lay the groundwork for future research endeavors focused on optimizing complex control systems for robotic applications.

4.2.1 Discrete Controller Design for Hybrid Mechatronic Systems

As modern mechatronic systems experience rapid development, the integration of mechanical, electronic, and information processing technologies has become increasingly pronounced. This trend has led to a significant increase in the complexity of system design. Hybrid mechatronic systems are characterized by the coexistence of continuous-time dynamics and discrete event-driven behaviors. This hybrid nature presents substantial challenges for controller design. Traditional design techniques, which depend on accurate mathematical models and manual analysis, are not only time-consuming and labor-intensive but also ill-equipped to handle the dynamic changes and high complexity inherent in these systems.

In recent years, evolutionary computation has emerged as a potent optimization technique, opening up new avenues for the automated design of hybrid systems. By integrating evolutionary algorithms, such as genetic algorithms, into controller design, it is possible to search the entire design space of discrete controllers, thus enabling the automated optimization of control strategies. Nevertheless, most current research efforts are concentrated on time-driven systems, and there is a dearth of studies on the design of discrete controllers for hybrid systems. Additionally, the representation of discrete controllers has a profound impact on both design efficiency and implementation performance. Selecting an appropriate representation that strikes a balance between interpretability and computational efficiency remains an important area that requires further exploration.

FSA has shown great potential as a representation for discrete controllers due to its straightforward structure and its ease of extension into statecharts, which are widely used in industrial settings. Meanwhile, HBG serves as an effective modeling tool capable of capturing both discrete and continuous dynamic behaviors in hybrid mechatronic systems. Consequently, integrating FSA with HBG and utilizing evolutionary computation for controller optimization could offer a viable solution to the control problems in hybrid systems.

Against this backdrop, this subsection presents a case study of a two-tank system to explore the evolutionary design approach for discrete controllers in hybrid systems based on FSA [139]. The proposed method is also compared with alternative control techniques, such as the lookahead controller, to assess their respective advantages and applicability across different scenarios.

1. Case Study of Two-tank System

Multi-tank systems are often cited in research literature because of their distinctive research significance and extensive applications in nonlinear multivariable feedback control and fault diagnosis [337, 338]. Researchers are drawn to these systems due to their mechanical simplicity and the ease with which their physical behavior can be understood, which helps in gaining in-depth knowledge of system dynamics. Moreover, multi-tank systems have an appropriate degree of control complexity, presenting a meaningful challenge for research and acting as an excellent experimental setup for validating complex control algorithms and diagnostic approaches.

In this subsection, a typical two-tank system is chosen as a case study to verify the proposed controller design approach. The design of this system not only accurately emulates the essential features of multi-tank problems but also offers a straightforward yet representative environment for evaluating the performance of controllers in nonlinear, multivariable control scenarios. In particular, in the two-tank system, water is continuously pumped into the left tank at a steady flow rate, and a series of controllable valves enable flexible fluid management. These valves can independently regulate the outflow of each tank and also allow for two-way fluid transfer between the two tanks, thereby creating a testing ground that can mimic a variety of dynamic behaviors.

The detailed configuration of the system is depicted in Figure 4.25. A pump maintains a constant flow rate to continuously supply water to the left tank, and the open-closed states of the valves govern the tank drainage processes and the fluid transfer between the tanks. Through precise control of the valve states, different control strategies can be evaluated based on aspects such as system stability, response time, and control accuracy.

The HBG of the two-tank system is shown in Figure 4.26, where the various components of the system are represented by different modeling elements. The flow restrictions of the pipes and valves are modeled as resistive components, *R*, while the tanks are represented as capacitive components, *C*, which reflect their capacity to store fluid. The input from the pump is modeled as a flow source, S_f, providing a continuous inflow of liquid. The function of the valves is represented by switch components, S_w, which can operate as either a 0-flow source or a 0-effort source, depending on their state. When a switch is in the 0-flow source state, it enforces

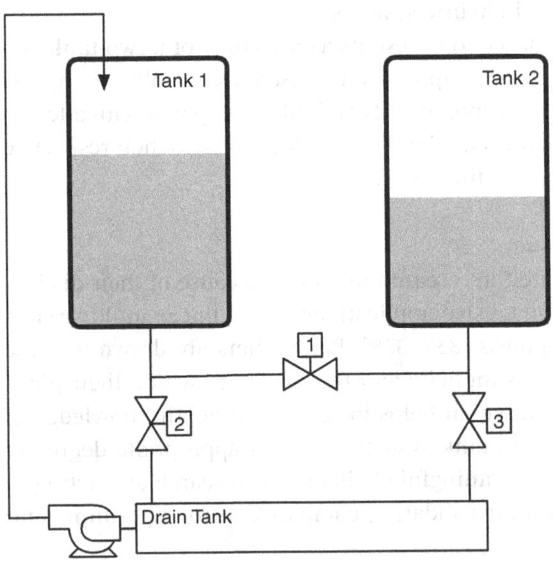

Figure 4.25 A typical two-tank system.

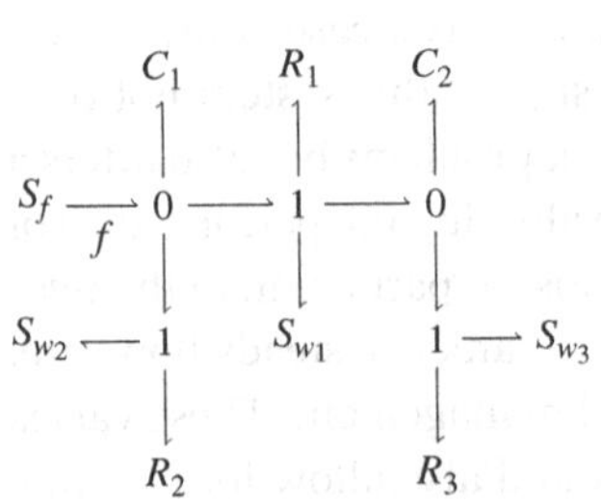

Figure 4.26 The hybrid bond graph of the two-tank system.

a flow of zero at the connected junction. This indicates that the valve is closed, preventing any liquid from passing through and effectively interrupting the flow path. Conversely, when the switch is in the 0-effort source state, it imposes no restriction on the flow. In this state, the valve is considered fully open, allowing liquid to flow freely. Through this modeling approach, the hybrid bond graph provides a clear representation of the interaction between continuous dynamic behaviors and discrete states in the system.

The dynamic characteristics of the two-tank system are obtained based on its HBG model. Accordingly, the system can be formulated as a series of differential equations that elucidate the connections among the state variables, inputs, and outputs. The vector state equation that characterizes the two-tank system is presented as follows:

$$\begin{aligned}
\dot{x}_1(t) &= f - \frac{u_1(t)}{R_1}\left(\frac{x_1(t)}{C_1} - \frac{x_2(t)}{C_2}\right) - \frac{u_2(t)}{R_2}\frac{x_1(t)}{C_1} \\
&= g_1(\mathbf{x}(t), \mathbf{u}(t)) \\
\dot{x}_2(t) &= \frac{u_1(t)}{R_1}\left(\frac{x_1(t)}{C_1} - \frac{x_2(t)}{C_2}\right) - \frac{u_3(t)x_2(t)}{R_3C_2} \\
&= g_2(\mathbf{x}(t), \mathbf{u}(t))
\end{aligned} \tag{4.12}$$

where $x(t)$ denotes the state vector, $u(t)$ represents the input vector, and f is the external inflow. The functions g_1 and g_2 represent the nonlinear dynamics of the two tanks, determined by the parameters R_1, R_2, R_3 (resistances) and C_1, C_2 (capacitances).

The levels $y_i(t)$ of each tank, corresponding to the heights of the liquid, can be directly obtained from the state variables. The relationship is given by:

$$y_i = \frac{x_i(t)}{C_i \rho g} \tag{4.13}$$

where ρ is the fluid density and g is the gravitational acceleration. This equation relates the state variables $x_i(t)$ to the physical quantities representing the tank levels.

In order to make the system representation more concise, the previously mentioned equations can be rewritten in a matrix-based format. Consequently, the state equation of the two-tank system can be presented as:

$$\begin{aligned}
\dot{\mathbf{x}}(t) &= \begin{bmatrix} -\frac{u_1(t)}{R_1C_1} - \frac{u_2(t)}{R_2C_1} & \frac{u_1(t)}{R_1C_2} \\ \frac{u_1(t)}{R_1C_1} & -\frac{u_1(t)}{R_1C_2} - \frac{u_3(t)}{R_3C_2} \end{bmatrix} \mathbf{x}(t) + \begin{bmatrix} f \\ 0 \end{bmatrix} \\
\mathbf{y}(t) &= \mathrm{diag}\left(\frac{1}{C_1\rho g}, \frac{1}{C_2\rho g}\right)\mathbf{x}(t)
\end{aligned} \tag{4.14}$$

where the first equation describes how the state vector $\mathbf{x}(t) = [x_1(t), x_2(t)]^T$ evolves under the influence of the control inputs $\mathbf{u}(t) = [u_1(t), u_2(t), u_3(t)]^T$, as well as the external inflow f. The system matrix $A(\mathbf{u}(t))$ contains the coefficients determined by the resistances and capacitances, while B represents the input term from the inflow.

Finally, the entire system can be compactly written in state-space form as:

$$\begin{aligned}\dot{\mathbf{x}}(t) &= A(\mathbf{u}(t))\mathbf{x}(t) + B \\ \mathbf{y}(t) &= C\mathbf{x}(t)\end{aligned} \tag{4.15}$$

where $A(\mathbf{u}(t))$ is the state matrix, B is the input vector, and C is the output matrix. The matrix C is a diagonal matrix derived from the physical properties of the system, specifically the capacitances, fluid density, and gravitational constant.

2. Evolved FSA Controller

a. Encoding Scheme

Evolutionary computation has found extensive applications across diverse areas of controller design, vividly illustrating its remarkable ability to tackle intricate optimization challenges [339–341]. In the present research, a genetic algorithm is harnessed to evolve the controller, adopting an encoding approach inspired by the methodology outlined in [342]. The controller devised in this study is an FSA characterized by a predefined number of states, denoted as N. Each state within this FSA corresponds to a distinct switch configuration within the hybrid bond graph. Intriguingly, every bit in the binary representation of an FSA state's numerical value has a direct and one-to-one correspondence with a specific switch.

In this study, the state transition rules of the FSA are extremely simple, lacking any additional action logic. They merely determine the next state given the received input symbol. The current state of the FSA directly dictates the "ON" or "OFF" status of each switch, thereby controlling the behavior of the actuators connected to these switches.

To formally characterize the controller's behavior, the FSA can be represented as a state transition matrix, where row indices correspond to current states and column indices denote received input symbols. Each matrix element specifies the next state the FSA should transition to for a given input. Moreover, an initial state must be separately defined outside the matrix to serve as the controller's starting point.

For ease of encoding and optimization by the genetic algorithm, the state transition matrix is encoded as a bit string. In this bit string, each matrix element is represented using ceil($\log_2 N$) bits. Additionally, an extra set of bits is appended to the end of the bit string to define the encoding of the initial state. If the controller has N states and M input symbols, the total length of the evolved bit string can be expressed as:

$$\text{ceil}(\log_2 N)(NM + 1). \tag{4.16}$$

Figure 4.27 offers a clear and accessible depiction of the proposed approach when applied to a more straightforward situation. Here, the system comprises merely four states and two input symbols, enabling each state to be encoded with a mere two bits. This simplified instance serves as an excellent aid for grasping the encoding procedure of the proposed method.

In detail, all entries of the state transition matrix are inserted into the corresponding genotype in a row-major fashion. That is, beginning with the first row, the elements of each row are added to the genotype sequentially until the entire mapping between states and input symbols is completely represented. Moreover, to fully characterize the system's operation, the final two bits of the genotype are designated to define the initial state. This specification is crucial as it sets the starting point for the controller's operation, providing the essential basis for its initial functioning.

The controller for this example is engineered to oversee a system featuring two switches. By examining the state transition diagram depicted in Figure 4.27c, it becomes evident that in state 1 (01), the first switch is in the "OFF" state, whereas the second switch is "ON." When the system occupies this state, regardless of the input symbol received, it will shift to state 3 (11). In state 3, both switches are activated, placing the system in a fully operational mode. A more in-depth examination reveals that when the system is in state 3 (11) and encounters input symbol 1,

	0	1
0	2	1
1	3	3
2	0	1
3	3	0
Initial = 1		

(a) Transition table.

10 01 11 11 00 01 11 00 01

(b) Genotype

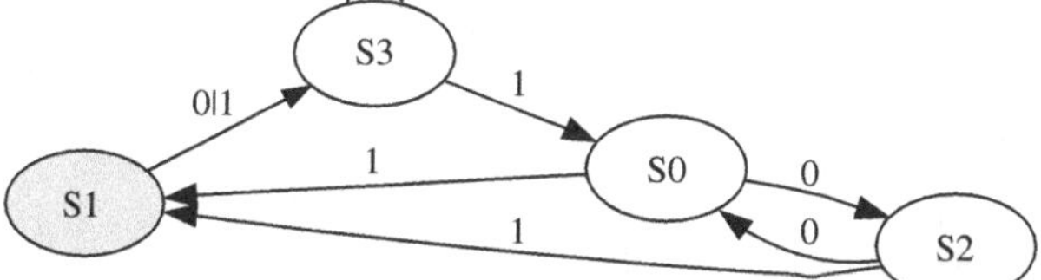

(c) Phenotype

Figure 4.27 Encoding example: (a) transition table, (b) genotype, and (c) phenotype.

it transitions to state 0 (00). In state 0, with both switches deactivated, the system enters a completely idle state.

To develop the controller through an evolutionary approach, this research implemented a basic genetic algorithm, as shown in Figure 4.28. The algorithm incorporates conventional methods including one-point crossover, bit-flipping mutation, and tournament selection. These techniques strike a balance, enabling the algorithm to efficiently search the solution space while keeping computational complexity in check. Notably, elitism was deliberately excluded from this study. Experimental findings indicated that the inclusion of elitism often resulted in premature convergence. This phenomenon significantly diminished the diversity within the population and hampered the algorithm's capacity to thoroughly explore the entire solution space.

Table 4.7 presents a comprehensive summary of the evolutionary algorithm's specific parameters. The core goal of this algorithm is to evolve a finite state controller that effectively reduces the errors in tank levels. Fitness assessment is conducted by calculating the average of the squared errors of the most deviant output from the target value, across six distinct training scenarios.

For the selection process, tournament selection with a size of 2 was utilized. This approach promotes competition among individuals in the population while simultaneously preventing overfitting. The algorithm's termination condition was

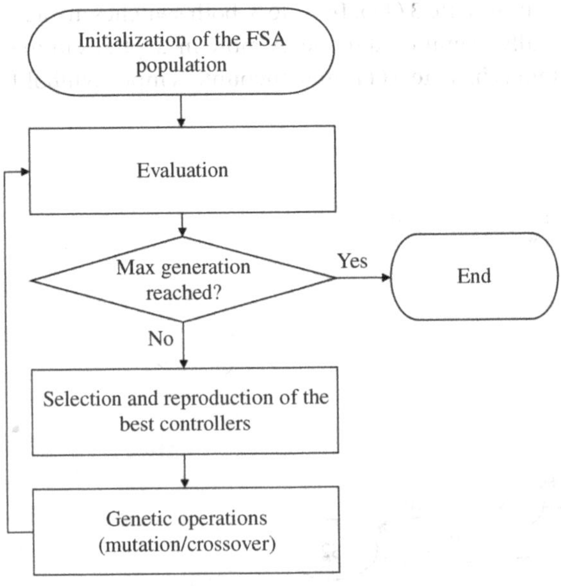

Figure 4.28 Genetic algorithm flow chart.

Table 4.7 Parameters used to evolve the FSA controller.

Objective:	Develop a finite state controller to minimize tank level errors.
Fitness:	For six training cases, the average of the sum of squared errors from the target exhibited by the worst output.
Selection:	Tournament consisting of 2 participants.
Termination:	Maximum number of generations attained.
Parameters:	Population size of 300, with a 0.3 one-point crossover rate and 0.01 bit-flip mutation rate.

defined as reaching the maximum number of generations, a strategy that optimally balances computational efficiency and the comprehensive exploration of the solution space.

Among the crucial parameters are a population size of 300, a one-point crossover rate of 0.3, and a bit-flipping mutation rate of 0.01. The genetic algorithm was implemented using the Open BEAGLE C++ framework [343], a robust platform for evolutionary computation. To boost computational performance, fitness evaluations were parallelized and distributed across a computer cluster using MPI technology, enabling concurrent execution of these evaluations.

b. FSA for Two-tank System

The FSA controller of the two-tank system operates with eight distinct states, and each state is associated with three switches. Input symbols are produced through an interface mechanism. This mechanism triggers the generation of symbols whenever the state variables intersect pre-determined surfaces within the state space. These surfaces are divided into two distinct sets, with each set corresponding to one of the two tanks. Details are as follows:

$$\begin{aligned} h_{i1} &= y_i - T_1 + \beta \\ h_{i2} &= y_i - T_1 - \beta \end{aligned} \tag{4.17}$$

In this context, T_1 denotes the intended target level for tank i, and β represents the acceptable tolerance margin around this target. For each tank, the state space is partitioned into three distinct regions: above the target level, below the target level, or precisely at the target level. These regions are demarcated by the relationship between the actual tank level and the target value, with the tolerance β establishing the boundaries of each region. Subsequently, a total of 9 distinct symbols are generated. These symbols are derived from the logical combination, or conjunction, of the two sets of regions (one set for each tank). Given that there are 9 input symbols and 8 possible states, with each state represented by 3 bits, the total length of the bit string to be evolved is computed as follows: $(8 \times 9 + 1) \times 3$, resulting in a bit string length of 219 bits. It is crucial to note that this bit string must also include a specification for the initial state.

- Fitness assessment

Within the framework of the two-tank system, the FSA controller was developed to govern the fluid levels in each tank, ensuring they consistently track predefined target profiles. For evaluating the fitness of the evolved controllers, every simulation is executed over a 15-second interval. At the end of this 15-second duration, an objective function $f(i)$ is computed for each tank to gauge its performance. This function is derived from the square of the error between the actual fluid level and the target level, thus quantifying how much each tank's fluid level deviates from the desired profile. The function is shown as follows:

$$f(i) = \int_0^{15} (y_i - T_i)^2 \, dt, \ i \in \{1, 2\} \tag{4.18}$$

The overall fitness for the simulation case is determined by taking the maximum value of the two objective functions, which corresponds to the larger error between the two tanks. In essence, the system's fitness is dictated by the tank that exhibits the greater deviation from its target level.

$$F = \frac{1}{\max\{f(i) \mid i \in \{1, 2\}\}} \tag{4.19}$$

Through observations and practical insights, it has been found that defining the fitness function using a maximization or minimization strategy (depending on the objective) yields superior results compared to using the average error of the two tanks as the fitness criterion. When the average error is employed, a notable issue arises: in most scenarios, the system tends to prioritize the performance of one tank at the expense of the other. This occurs because the evolved controller often focuses on optimizing one tank's performance while neglecting the other, leading to no improvement in the second tank's performance. The underlying reason is that the tank with compromised performance can "compensate" via the other tank's better performance, resulting in a deceptively high average fitness that does not reflect genuine overall system improvement.

In contrast, defining fitness based on the poorer-performing tank mitigates this undesirable behavior. Specifically, by selecting the tank with the worst performance as the evaluation criterion, this approach ensures that the evolution process optimizes both tanks' performance simultaneously. This strategy prevents the trade-off between tanks and drives holistic improvement in system performance.

- Training and testing

In the process of evolving controller design, it is essential to partition the experimental procedure into a training phase and a testing phase. The training phase predominantly relies on specific training objectives to direct the controller's evolution,

Table 4.8 Training targets employed in evolving the FSA.

Case	$t = 0$		$t = 5$		$t = 10$	
	Tank 1	Tank 2	Tank 1	Tank 2	Tank 1	Tank 2
a	0.2	0.4	0.25	0.35	0.35	0.2
b	0.48	0.3	0.4	0.2	0.32	0.25
c	0.35	0.2	0.4	0.45	0.2	0.4
d	0.2	0.4	0.2	0.2	0.4	0.2
e	0.4	0.4	0.2	0.05	0.3	0.3
f	0.4	0.2	0.1	0.2	0.4	0.2

whereas the testing phase evaluates the evolved controller's generalization capability when faced with additional objectives not addressed during training. This separation ensures the controller performs effectively and adapts to diverse scenarios. To enable the evolved FSA controller to adapt proficiently to a broad range of goals, multi-objective training is commonly utilized. This approach helps the controller maintain robust performance across a wider spectrum of objectives. In the experiment, six randomly generated objectives listed in Table 4.8 were selected to validate the evolution process, with their average performance serving as the training basis. The diversity of these objectives ensures the evolved controller exhibits robustness and adaptability across various contexts. All experiments were conducted on an HP ProLiant SL165z G7 server equipped with an AMD Opteron 6168 processor (12 cores, 1.9 GHz, 12 MB L3 cache) and 64 GB of memory, ensuring experimental efficiency and stability. It is important to note that six objectives may not be the optimal number for training; the quantity and type of objectives may require further adjustment and optimization depending on the specific application. However, as the primary goal of this work is to demonstrate the feasibility of the evolutionary approach in controller design, further optimization of the controller is reserved for future research.

During the training phase, the progress of the evolutionary process can be visually tracked via the fitness curve in Figure 4.29. By comparing the performance of the best controllers obtained at intermediate and final evolutionary stages, key characteristics of the process are revealed. The controller's behavior at generation 150 is shown in Figure 4.30, while its behavior at the evolution's conclusion is depicted in Figure 4.31. The comparison clearly shows that ripples observed in mid-evolution stages are nearly eliminated in later stages. This indicates that the proposed evolutionary design method effectively guides FSA controller evolution, progressively enhancing performance and ensuring the final controller fully meets

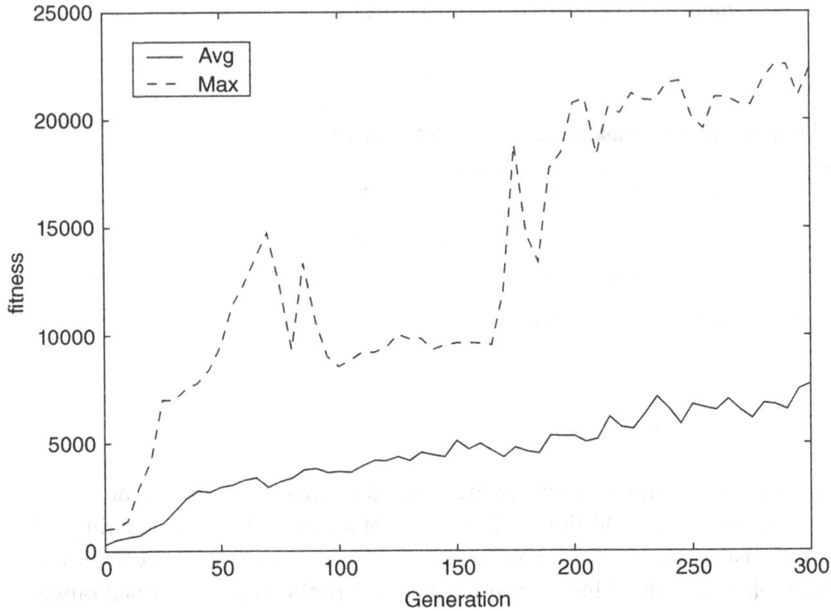

Figure 4.29 Progression of the population fitness during the evolution of the FSA controller for the two-tank system.

predefined design goals. Figure 4.32 presents the best FSA controller obtained at evolution's end, and Table 4.9 lists its fitness values for each training case, further validating its superior performance.

During the testing phase, to verify the evolved FSA controller's generalization ability, it was applied to a set of 20 randomly generated targets not used in training, with results reflecting its performance on unseen objectives. The controller's best evolved phenotype is shown in Figure 4.32. Figure 4.33 illustrates the controller's behavior on the 20 validation targets, demonstrating excellent generalization: the controller effectively handled most randomly generated targets, failing only on the 13th case out of 20.

3. Lookahead Controller

To further assess the control performance, a one-step predictive controller was employed to manage both the two-tank system and the DC-DC converter. This controller was chosen for its simplicity and manageable computational load, making it suitable for real-time implementation. At each time step, the controller operates by simulating the system to predict future state variable values $x_{S_{w_i}}$ for all possible switch configurations $SW = \{S_{w_i} | i = 1, \ldots, n\}$. Using these simulation

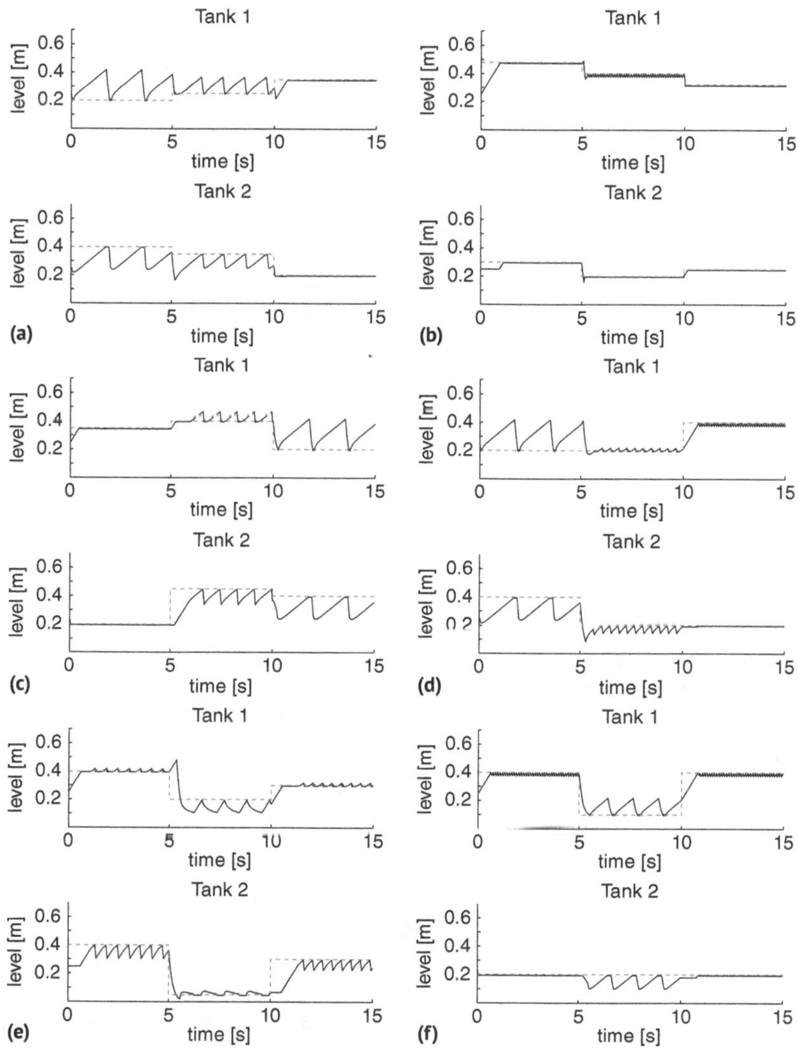

Figure 4.30 FSA controller behavior on the training set for the two-tank system of the best individual at the 150th generation.

results, it calculates the future states for each configuration and selects the optimal switch configuration for the next step based on the predictions. Specifically, the selected configuration is the one whose resulting trajectory aligns most closely with the target point x_d, ensuring the system progresses steadily toward the desired direction. The control objective is to minimize the angle between two vectors: one

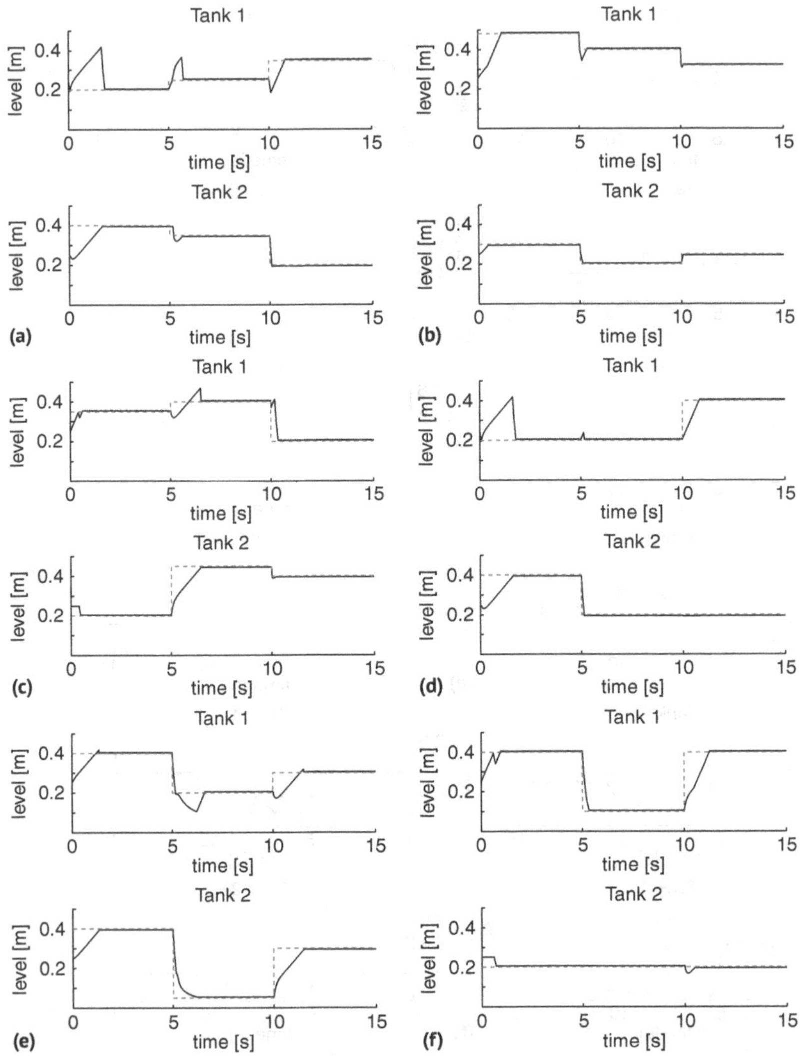

Figure 4.31 FSA controller behavior on the training set of the two-tank system. The tanks' levels and desired targets are represented by solid and dashed lines, respectively.

formed by the current state x and the target x_d, and the other connecting the current state to the predicted system trajectory $x_{s_{w_i}}$. This is formalized as:

$$\underset{s_{w_i} \in SW}{\arg\min} \left(\arccos \frac{(x - x_d)(x - x_{s_{w_i}})}{|x - x_d||x - x_{s_{w_i}}|} \right) \tag{4.20}$$

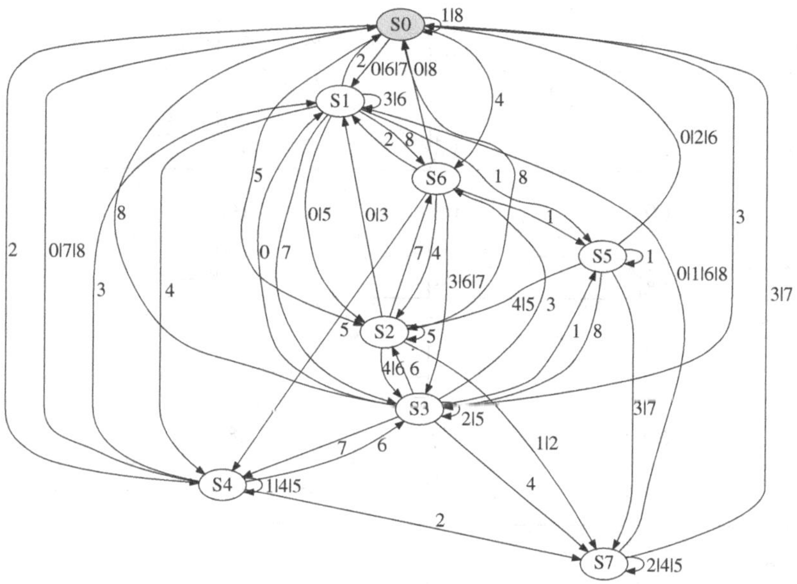

Figure 4.32 FSA controller for the two-tank system obtained at the end of the evolution.

Table 4.9 Fitness obtained on the training cases.

Case	FSA	Lookahead
a	23656	6314
b	40867	55102
c	48888	32965
d	24291	8887
e	24352	22476
f	22376	20479

An important characteristic of this type of controller is its lack of requirement for training or redesign for a specific system, which enhances its flexibility and adaptability. Unlike traditional control methods, the sole prerequisite for using this controller is an accurate system model to predict future states during operation. Free from complex training processes or application-specific adjustments, it can rapidly adapt to diverse control tasks. Additionally, the controller has only one adjustable parameter: the lookahead horizon, which determines the "depth"

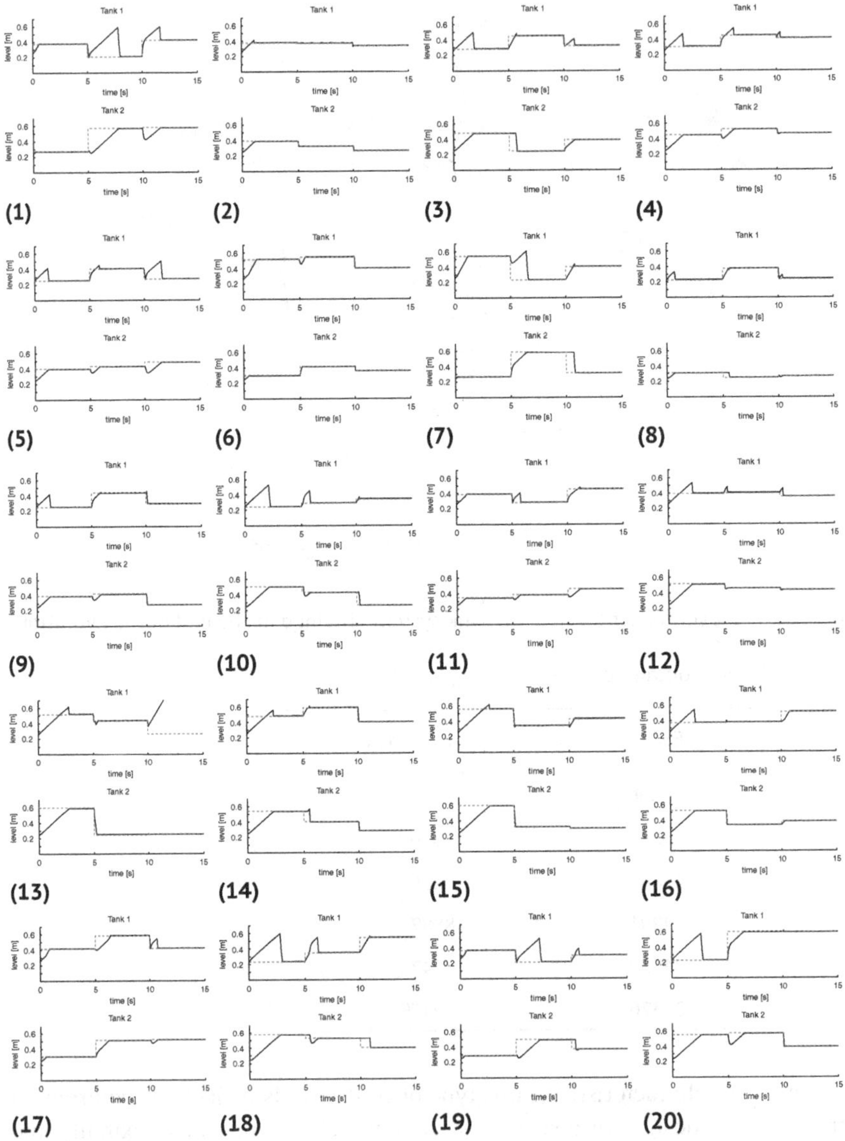

Figure 4.33 The behavior of the evolved FSA controller on the testing set.

of future state predictions at each time step. To achieve efficient control, this horizon should be adjusted according to available computational resources and sampling time. In this study, the lookahead horizon was set to one step to minimize computation time.

Table 4.9 presents the fitness values of the predictive controller across training cases. As evident in Figure 4.34, the lookahead controller struggles to meet the target for the second tank in the first three cases (illustrated in Figure 4.34c, d, f).

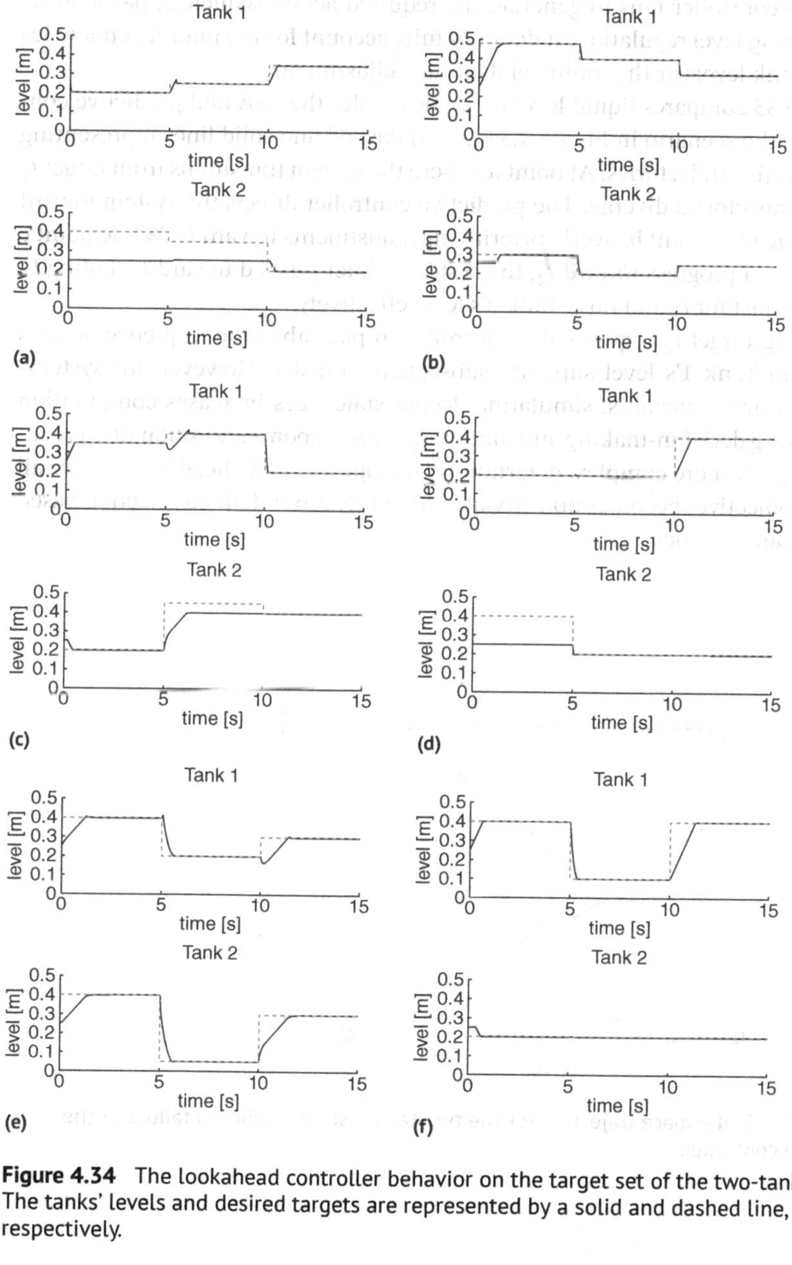

Figure 4.34 The lookahead controller behavior on the target set of the two-tank system. The tanks' levels and desired targets are represented by a solid and dashed line, respectively.

These cases require Tank 2's liquid level to exceed Tank 1's, posing a challenge for the predictive controller. Analysis of Figure 4.25 reveals that increasing Tank 2's level requires first raising Tank 1's level and then transferring fluid between tanks. This necessitates a controlled decrease in Tank 1's level to facilitate transfer, but the lookahead controller fails to generate the required action sequence, particularly in fine-tuning level regulation. It does not fully account for the interdependencies between tank levels or the timing of dynamic adjustments.

Figure 4.35 compares liquid level dynamics under the FSA and predictive controllers for the scenario in Figure 4.34d, with dashed and solid lines representing their respective trajectories. At point a, where the system transitions from target t_1 to t_2, the trajectories diverge. The predictive controller directs the system toward point c instead of point b, overly prioritizing adjustments toward t_2. While point c appears to aid progress toward t_2, the system is later pushed toward t_3, failing to properly adjust for t_2 and raise Tank 2's level effectively.

Achieving target t_2 requires the controller to plan ahead to sequence actions that ensure Tank 1's level supports subsequent transfer. However, for systems with small time constants, simulating deeper state trees increases computation time, slowing decision-making and hindering rapid response. Additionally, as control tasks grow more complex, determining the optimal lookahead horizon for all potential objectives becomes non-trivial, with no clear pre-defined selection based on task characteristics.

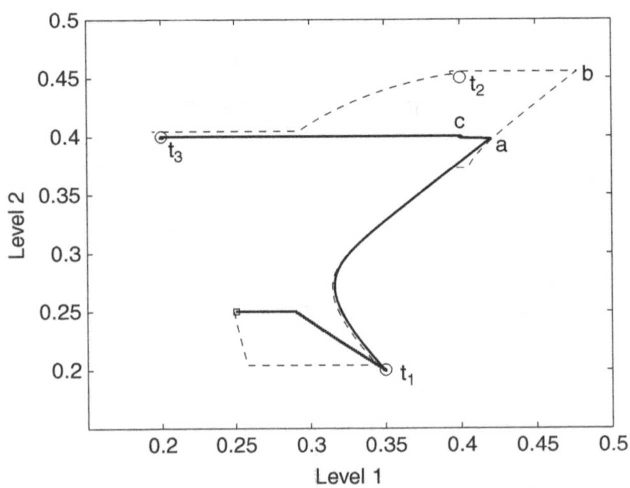

Figure 4.35 State-space trajectory for the two-tank system exhibiting failure of the lookahead controller.

4. Experimental Discussion

a. Capability of Evolutionary Design

Notably, even though the fitness function does not explicitly incorporate a reward mechanism to promote the simultaneous reduction of errors in both tanks, the evolutionary algorithm successfully evolves an efficient FSA controller. This controller generates precise control sequences that effectively coordinate the complex dynamic interactions between the two tanks, enabling the system to achieve the targets for both tanks concurrently.

More significantly, the experiments show that the control sequence does not rigidly adhere to a fixed path to reach the targets. Instead, it permits the system to temporarily deviate from the target state. This strategic flexibility allows the system to make necessary trade-offs to optimize global performance, ultimately achieving an optimal solution.

This result not only confirms the practicality of the proposed method in solving complex control tasks but also underscores its adaptability to the dynamic characteristics of various systems. A core advantage of this approach lies in its capability to design intelligent controllers that can adapt to complex environments through an automated evolutionary process, eliminating the need for human intervention or extensive domain-specific knowledge.

b. The Issue of Optimal Design for FSA Controllers

While the FSA controllers designed in this case study achieved notable success, it is essential to recognize that they do not represent globally optimal solutions. This limitation stems from the fundamental design of FSA controllers, which inherently involve two critical components: symbol generation and actuator control, as illustrated in Figure 4.36. In essence, FSA controllers can only respond when specific predefined hyperplanes in the state space are crossed, prompting the execution of corresponding control strategies.

In our research, to streamline the problem, we chose several very simple hyperplanes. The definition of these hyperplanes depends on the targets' positions within the hyperspace and their corresponding tolerance ranges, with the aim of guiding the system's trajectory toward the targets. However, such predefined hyperplanes do not always constitute the optimal solution. To further elaborate on this point, consider a simple double integrator system as an example. The double integrator system is a classic dynamic system, and its state-space representation is as follows:

$$\dot{\mathbf{x}}(t) = \begin{bmatrix} 0 & 1 \\ 0 & 0 \end{bmatrix} \mathbf{x}(t) + \begin{bmatrix} 0 \\ 1 \end{bmatrix} \mathbf{u}(t) \tag{4.21}$$

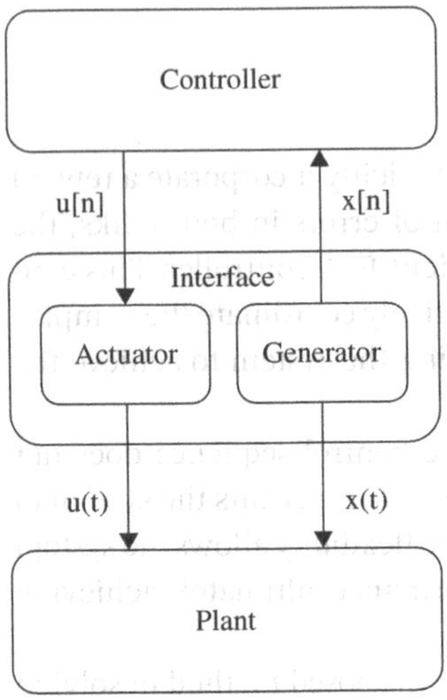

Figure 4.36 Representation of a general hybrid control system.

In this system, the control strategy $\mathbf{u}(t)$ is restricted to $u(t) \in \{-1, 0, 1\}$. Due to the system's dynamic nature, the controller must adjust the input $\mathbf{u}(t)$ precisely according to the current state to steer the system toward the target. However, if the controller can only make decisions based on simple predefined hyperplanes, it may cause the system to deviate from the globally optimal trajectory or even fail to reach the target in some scenarios.

Figure 4.37 demonstrates two different state-space partitioning methods. The first method, derived via the invariant analysis approach proposed in [344], is depicted by solid lines in the figure. This partitioning effectively captures the system's dynamic characteristics and is regarded as the theoretical optimal solution. In contrast, the partitioning method used in this study's experiments is shown by dashed lines in Figure 4.37a. A direct comparison highlights substantial differences between the experimental partitioning and the optimal one. This approach fails to adequately reflect the system's key dynamic features, as input symbols may be generated even when minimal changes occur in the vector field. This leads to wasted symbol-generation potential and reduces the controller's overall efficiency.

In Figure 4.37b, an adjusted partitioning approach is presented, where the partitioning hyperplanes are rotated at an angle to optimize the method. This modification aligns the partitioning more closely with the theoretically optimal configuration, enabling the generated input symbols to more accurately capture relevant changes in the vector field. This improvement highlights that optimizing the symbol generation process can significantly enhance the performance of the FSA controller, as more precise symbol generation provides more reliable input information for the control sequence.

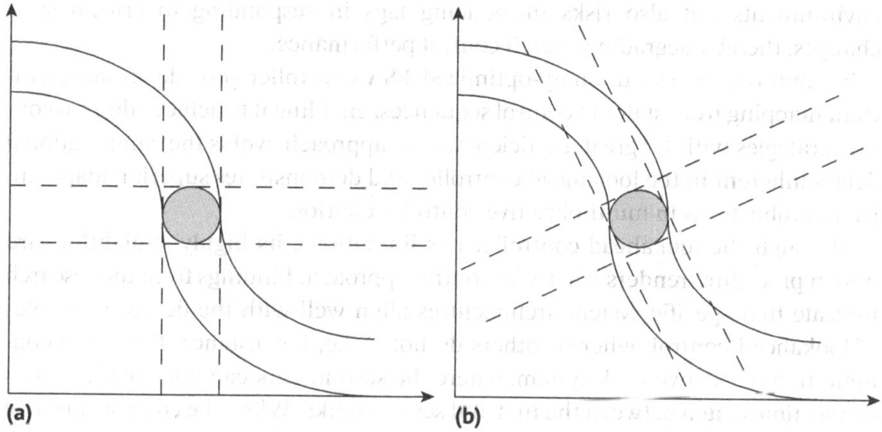

Figure 4.37 Double integrator partitioning.

Figure 4.38 Lookahead control with limited horizon.

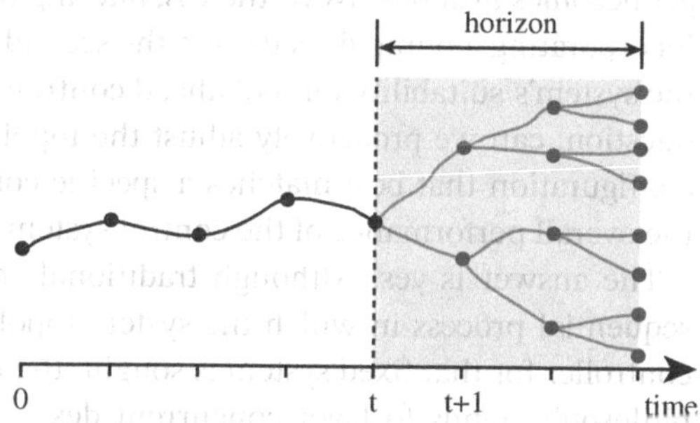

c. Potential of Lookahead Controller

In this research, a lookahead controller was deployed to enable a performance contrast with the evolved FSA controller (illustrated in Figure 4.38). The outcomes of the experiments demonstrate that the evolved FSA controller proficiently attains all six control objectives established during the training process. In contrast, the lookahead controller falls short in fulfilling three of these objectives. This divergence is particularly evident in situations where the water level of the second tank must surpass that of the first tank. These distinctions underscore the evolved FSA controller's enhanced aptitude for managing intricate control situations.

Notably, the lookahead controller exhibited significant limitations during real-time execution, particularly regarding its practical applicability. The need to simulate and compute a lookahead tree of possible states often leads to excessively long execution times, making it difficult to meet real-time control requirements even in fast-response systems. This delay not only limits its usability in dynamic environments but also risks introducing lags in responding to critical state changes, thereby degrading overall control performance.

In contrast, the evolutionary-optimized FSA controller provides a more efficient mapping from states to control sequences, enabling it to achieve desired control strategies with far greater efficiency. This approach avoids the computational delays inherent in the lookahead controller and demonstrates superior adaptability and robustness in multi-objective control scenarios.

Although the lookahead controller has limitations, its highly straightforward design procedure renders it a useful control approach. Findings from the research indicate that specific system architectures align well with the design principles of lookahead control, whereas others do not. Take, for instance, the initial configuration of the two-tank system, where the second tank can only be filled via a connection switch between the first and second tanks. When the control objective demands that the water level in the second tank exceeds that of the first, achieving this goal for the second tank without negatively impacting the first tank's target becomes arduous. Nevertheless, altering the system architecture – such as by incorporating a pump directly into the second tank – could substantially enhance the system's suitability for lookahead controller design. This raises an interesting question: can we proactively adjust the topology of the control system to find a configuration that best matches a specific controller design, thereby optimizing the overall performance of the control system.

The answer is yes. Although traditional controller design typically follows a sequential process in which the system topology is first fixed and then the best controller for that fixed system is sought, the modern mechatronic system design philosophy tends to favor concurrent design. This concurrent design approach advocates treating the system topology as a variable parameter that can be optimized alongside the controller design, allowing the system and controller to achieve the best possible match.

While it is theoretically possible to optimize FSA controllers and hybrid bond graph structures simultaneously, the lookahead controller is more likely to be a practical choice for representing the controller component in real-world scenarios. The reason is that each system topology generates a large number of potential FSA controller candidates, all of which need to be explored via evolutionary computation to determine the most suitable controller for a specific system. Meanwhile, the design space of system topologies must also be explored, leading to a search space that becomes the Cartesian product of the controller space and the system space. Such a vast search space is computationally intractable and may even render the problem unsolvable. Moreover, evolutionary algorithms may waste significant computational resources optimizing controllers for poorly designed system topologies, resulting in suboptimal controllers and further increasing problem complexity.

In contrast, by selecting a lookahead controller for the controller design process, the problem of navigating the Cartesian product of spaces (as in the previous approach) can be avoided. This is because the design process for a lookahead controller remains consistent across all plant candidates. Once the lookahead horizon is predefined, each plant candidate is uniquely associated with a single lookahead controller. Consequently, the exploration is limited to the plant design space alone, eliminating the need to explore the controller design space simultaneously and greatly simplifying the overall process.

4.2.2 Evolution Design for Vehicle Suspension System

Vehicle suspension systems are critical subsystems in wheeled vehicles. From the perspective of system design, thc two main types of disturbances that affect vehicles are road and load disturbances. The characteristics of road disturbances are high intensity at low frequencies (e.g., hills) and low intensity at high frequencies (e.g., uneven road surfaces). Load disturbances arise from changes in loads due to acceleration, braking, and cornering. An effective suspension design seeks to mitigate both types of disturbances, ensuring minimal impact on the vehicle's performance, such as its vertical position. Typically, a suspension system nccds to be "soft" enough to absorb road irregularities for comfort and to damp high-frequency disturbances, while also being "stiff" enough to minimize the effects of load disturbances [345].

Suspension systems are widely used in vehicles to minimize body vibrations caused by road and load disturbances. These systems can either be based on passive physical designs or incorporate active control mechanisms. Commonly, suspension systems are classified as active, passive, or semi-active, depending on the amount of external power required for their operation.

Figure 4.39 shows a schematic diagram of a quarter-car model, which describes the dynamic behavior of a suspension system. The sprung mass m_s (kg) represents the vehicle's main body supported by the suspension, while the unsprung mass m_u (kg) includes the tire, hub, and wheel. The tire is modeled as a spring with

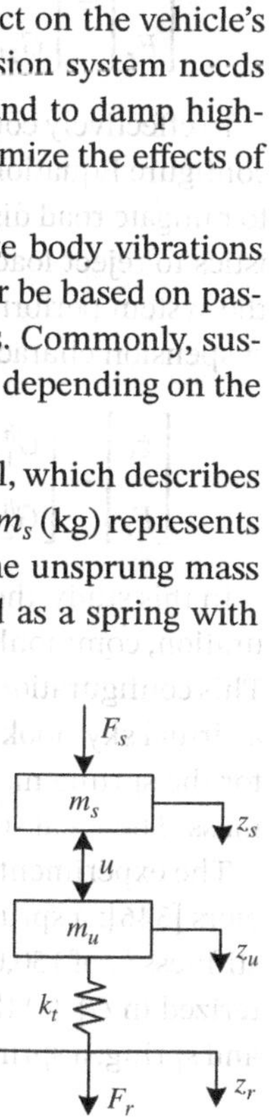

Figure 4.39 Quarter-car schematic model.

stiffness k_t (N/m). Vertical positions z_r, z_u, and z_s denote the positions of the road disturbance input, unsprung mass, and sprung mass, respectively. The interaction force between the road and the tire is denoted by F_r; u stands for any applied suspension force; the load disturbance is represented by force F_S. This model is based on the study by Smith [346].

The system's motion can be described by the following equations:

$$m_s\ddot{z}_s = -u + F_s \tag{4.22}$$

$$m_u\ddot{z}_u = u + F_r$$

$$F_r = k_t(z_r - z_u) \tag{4.23}$$

Viewed as part of a multiport mechatronics network, the quarter-car suspension system can be represented as a two-port network. The associated mixed immittance matrix, denoted as G, is characterized by the following definition:

$$\begin{bmatrix} \dot{z}_s \\ F_r \end{bmatrix} = \begin{bmatrix} G_{11} & G_{12} \\ G_{21} & G_{22} \end{bmatrix} \begin{bmatrix} F_s \\ \dot{z}_r \end{bmatrix} \tag{4.24}$$

To effectively counteract disturbances from both road and load, it is necessary to configure Equation (4.24) such that $G_{12}(s)$ and $G_{22}(s)$ exhibit "soft" characteristics to mitigate road disturbances, whereas $G_{11}(s)$ and $G_{21}(s)$ display "hard" characteristics to reject load disturbances. This desired behavior is achieved by specifying the system performance in the immittance matrix, combining both soft and hard suspension characteristics as follows:

$$\begin{bmatrix} \dot{z}_s \\ F_r \end{bmatrix} = \begin{bmatrix} G^h_{11} & G^s_{12} \\ G^h_{21} & G^s_{22} \end{bmatrix} \begin{bmatrix} F_s \\ \dot{z}_r \end{bmatrix} \tag{4.25}$$

In this study, the target system is designed as an ideal "double skyhook" configuration, commonly used for target suspension forces, as illustrated in Figure 4.40. This configuration is represented by the additional dashed system, which includes a virtual sky-hook damper c_u for the unsprung mass, a virtual sky-hook damper c_s for the sprung mass, and a spring k_s between the sprung mass and the unsprung mass. The ideal suspension force is given by $u = k_s(z_s - z_u) + c_s\dot{z}_s - c_u\dot{z}_u$.

The experimental setup for the quarter-car model utilizes the following parameters [346]: a sprung mass m_s of 250 kg, an unsprung mass m_u of 35 kg, and a tire stiffness k_t of 150,000 N/m. The frequency response to road disturbances is characterized in $G^s_{12}(s)$ through a double skyhook configuration featuring a soft damper and spring: a spring constant k^s_s of 10,000 N/m, a sprung mass damping coefficient

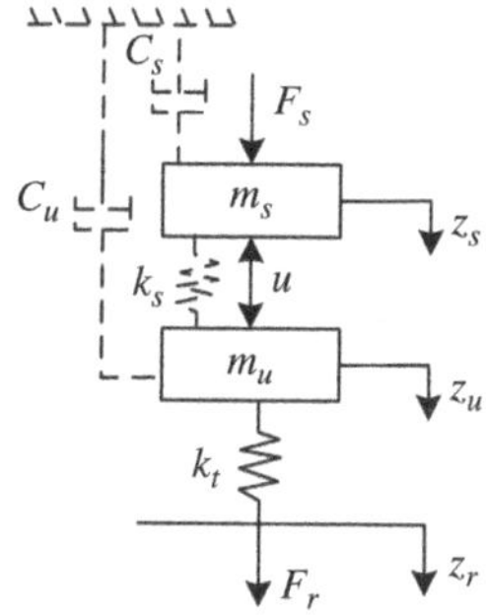

Figure 4.40 Quarter-car with double skyhook suspension configuration.

c_s^s of 4000 Ns/m, and an unsprung mass damping coefficient c_u^s of 2000 Ns/m. The desired frequency response to load disturbances is outlined in $G_{11}^h(s)$ via a distinct double skyhook configuration that incorporates a stiff damper and spring: a spring constant k_s^h of 150,000 N/m, a sprung mass damping coefficient c_s^h of 12,000 Ns/m, and an unsprung mass damping coefficient c_u^h of 6000 Ns/m.

The desired $G_{11}^h(s)$ and $G_{12}^s(s)$ can be computed using the following equations:

$$
\begin{aligned}
G_{11}^h(s) = \frac{\dot{z}_s}{F_s} &= \frac{\left(m_u s^2 + C_u^h s + k_t + k_s^h\right) s}{\begin{array}{l} m_s m_u s^4 + \left(c_s^h m_u + C_u^h m_s\right) + \left(k_s^h m_u + k_s^h m_s + k_t + k_s\right) s^2 \\ + c_s^h k_t s + k_s^h k_t \end{array}} \\
G_{12}^s(s) = \frac{\dot{z}_s}{\dot{z}_s} &= \frac{c_u k_t s + k_s^s + k_t}{\begin{array}{l} m_s m_u s^4 + \left(c_s^h m_u + C_u^h m_s\right) s^3 + \left(k_s^h m_u + k_s^h m_s + k_t m_s\right) s^2 \\ + c_s^h k_t s + k_s^h k_t \end{array}}
\end{aligned} \tag{4.26}
$$

Figure 4.41 displays their Bode plots.

The system possesses a single degree of freedom that enables it to respond to both road and load disturbances. By utilizing two appropriate feedback measurements, such as suspension deflection and sprung mass velocity, this degree of freedom can be independently controlled [346]. The suspension design, based on these feedback signals, is visually represented in the bond graph shown in Figure 4.42.

The control law is represented by the following equation:

$$
u = \begin{bmatrix} u_1 & u_2 \end{bmatrix} = \begin{bmatrix} k_1(s) & k_2(s) \end{bmatrix} \begin{bmatrix} s z_s - s z_u \\ s z_s \end{bmatrix} \tag{4.27}
$$

Here, $k_1(s)$ represents a collocated controller that utilizes relative velocity feedback, while $k_2(s)$ denotes a noncollocated controller that employs absolute

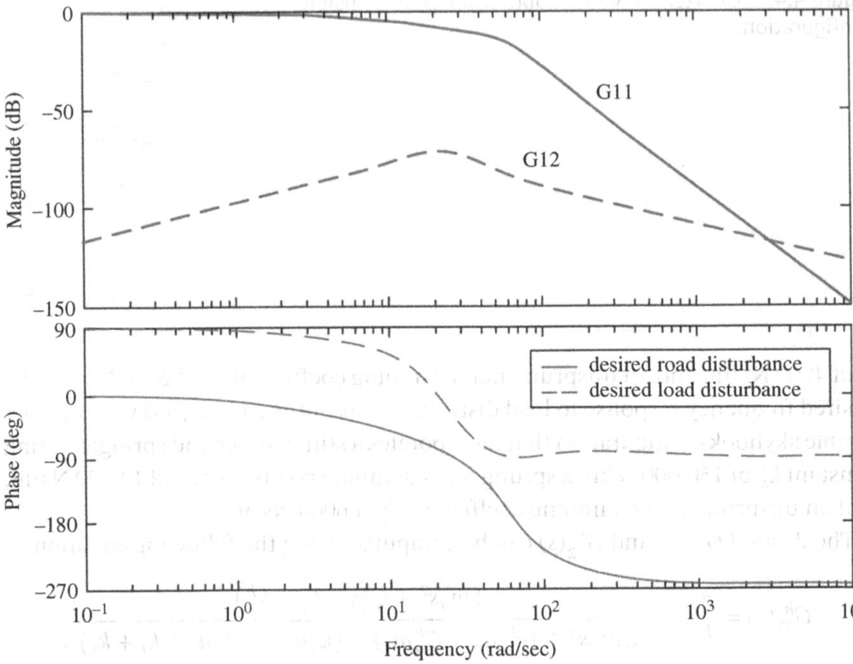

Figure 4.41 Desired frequency response of road and load disturbances.

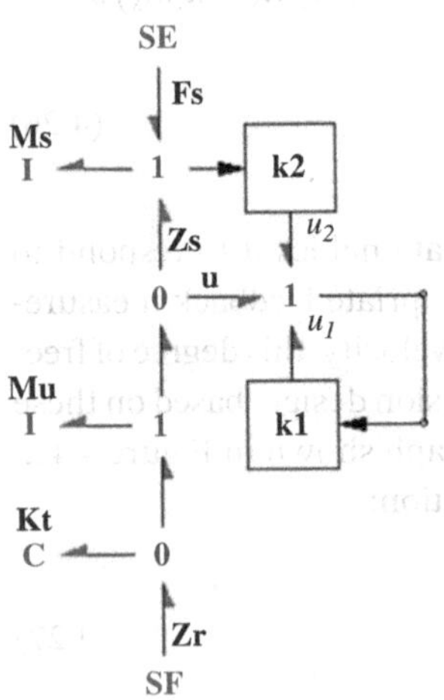

Figure 4.42 Quarter-car suspension control with both road and load disturbances.

velocity feedback. Utilizing both $k_1(s)$ and $k_2(s)$, the actual responses to road and load disturbances are given by:

$$
\begin{aligned}
G_{11} &= \frac{\dot{z}_s}{F_s} \\
&= \frac{(m_u s^2 + k_1 s + k_t)}{m_s m_u s^3 + (k_1 m_s + k_1 m_u + k_2 m_u) s^2 + k_t m_s s + k_1 k_t + k_2 k_t} \\
G_{12}^h(s) &= \frac{\dot{z}_s}{\dot{z}_r} \\
&= \frac{k_1 k_t}{m_s m_u s^3 + (k_1 m_s + k_1 m_u + k_2 m_u) s^2 + k_t m_s s + k_1 k_t + k_2 k_t}
\end{aligned}
\tag{4.28}
$$

Due to the conflicting performance requirements for road and load disturbances, a passive suspension system alone cannot achieve the desired results. An active suspension [347] is necessary to introduce additional energy. From a system design perspective, it is advantageous to concurrently investigate passive and active control strategies to enhance overall performance and energy efficiency.

1. Coevolution of Controllers

The controllers $k_1(s)$ and $k_2(s)$ are modeled using BG and encoded with GP. They belong to two coevolving GP species that work together to provide a complete solution for the quarter-car suspension design. Detailed descriptions of the encoding and decoding programs for BG/GP can be found in Ref. [5].

A superior solution discovered through coevolutionary GP yielded the outcomes $k_1(s)$ and $k_2(s)$ depicted in Figure 4.43 and 4.44, respectively. These BG models are manually simplified according to their corresponding genotypes.

As illustrated in Figure 4.42, $k_1(s)$ is used to measure the velocity difference between the sprung and unsprung masses, generating u_1, which is a portion of the force u acting between them. It should be emphasized that components R8, C12, R13, C9, R10, and C11 possess negative values, requiring active implementation.

The genotype of controller $k_1(s)$ is:

$$
\begin{aligned}
&f1(f1(f1(f0(R(-2536.01), C(-2.91685e-05)), C(7.29305e-06)),\\
&f1(f0(R(-28144.44), C(-1.23919e-05)), f0(R(-465.871),\\
&C(-9.565442e-05)))), R(2104.298))
\end{aligned}
\tag{4.29}
$$

A one-port BG configuration can be expressed in impedance form, which can then be converted into a transfer function.

$$
k_1(s) = \frac{2104s^4 + 93380s^3 + 2238000s^2 + 30470000s + 11930000}{s^4 + 38.83s^3 + 406.5s^2 + 869.8s} \tag{4.30}
$$

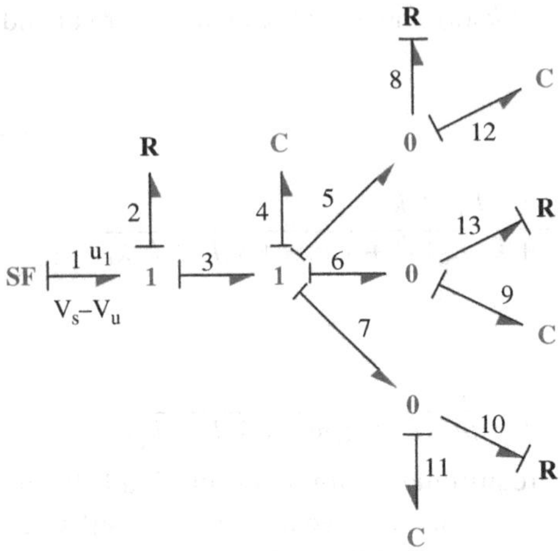

Figure 4.43 BG model of the coevolved controller $k_1(s)$.

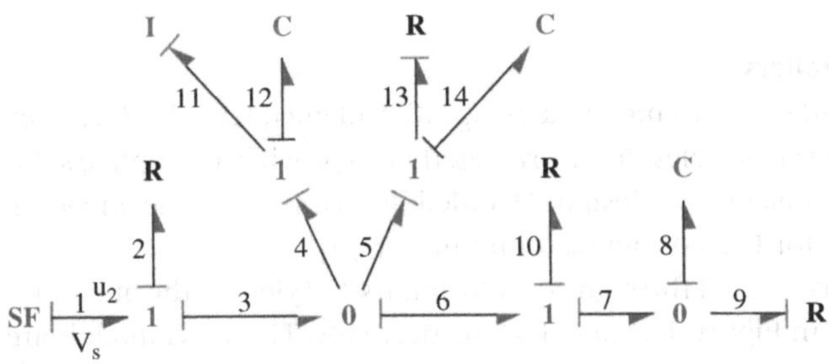

Figure 4.44 BG model of the coevolved controller $k_2(s)$.

The input for controller $k_2(s)$ is the velocity of the sprung mass, while the output u_2 represents another portion of the force u.

The genotype of controller $k_2(s)$ is:

$$\begin{aligned}&f1(f0(f1(f0(R(317.927), C(5.33948e-07)), R(28639.9)),\\&f1(f0(C(8.3877e-06), R(10490.4)), f1(I(48.0437),\\&C(5.9036e-06)))), R(1890.49))\end{aligned} \tag{4.31}$$

$$k_2(s) = \frac{9569s^4 + 5.735e007s^3 + 2.39e009s^2 + 2.216e001s + 1.952e012}{s^4 + 6102s^3 + 9.752e005s^2 + 3.177e007s + 6.328e007} \tag{4.32}$$

In this case, $k_1(s)$ and $k_2(s)$ can also be computed algebraically by employing traditional control techniques to match the desired responses for load and road disturbances, resulting in the following outcomes [346]:

$$k_1(s) = \frac{2000\left(s^5 + 224.4s^4 + 10270s^3 + 251600s^2 + 3600000s + 12860000\right)}{s^5 + 187.4s^4 + 8611.4s^3 + 96000s^2 + 171428.6s} \tag{4.33}$$

$$k_2(s) = \frac{10000\left(s^4 + 57.886s^3 + 4577.14s^2 + 101142.9s + 514285.7\right)}{s^4 + 187.4s^3 + 8611.4s^2 + 96000s + 171428.6} \tag{4.34}$$

The comparison of the two outcomes reveals that the controller $k_1(s)$ has a lower order and is simpler than those achieved through algebraic methods. This illustrates that using GP to coevolve controller structures represented by BG can lead to the discovery of control strategies that are equal to or better than those derived using traditional approaches. The Bode plots for the coevolved $k_1(s)$ and $k_2(s)$, shown in Figure 4.45, are nearly identical to those of the algebraically calculated controllers. However, controllers expressed solely as transfer functions lack physical insight, especially when it comes to determining if specific components of the controller can be realized through passive means.

In this method, the controllers are evolved within the physical domain using BG representations. The resulting BG structures provide valuable insights for designers, allowing them to choose among various physical implementations involving either active or passive subsystems. Based on the analysis of the parallel controller $k_1(s)$, it is found that R2 and C4, which are connected by a 1-junction, have positive real values and can therefore be passively realized as a parallel spring-damper pair. In contrast, components R8, C12, R13, C9, R10, and C11 possess negative

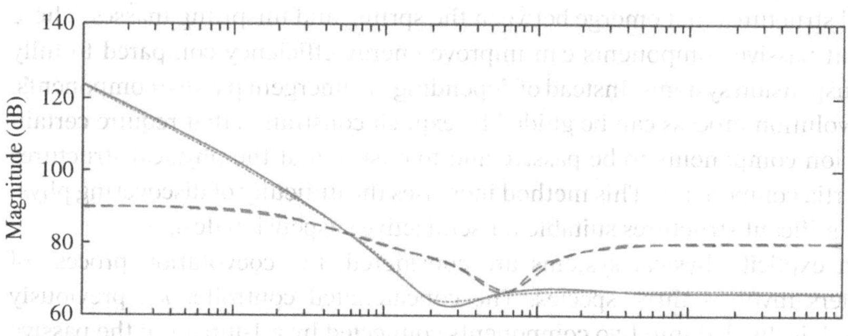

Figure 4.45 Comparison between evolved k_1 and k_2 and calculated ones.

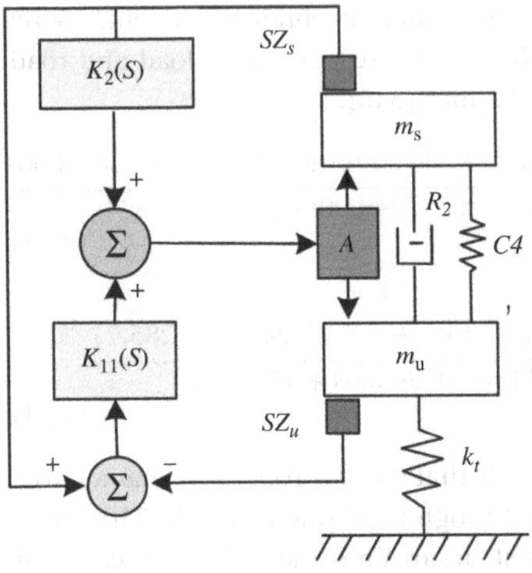

Figure 4.46 Physical realization of suspension control with road and load disturbances.

real values, indicating a necessity for active implementation. This is illustrated in Figure 4.46, where *R*2 is 2104.298 Ns/m and C4 is 137116.9 N/m.

$$k_{11}(s) = \frac{125400s^2 + 3941000s + 27090000}{s^3 + 38.83s^2 + 406.5s + 869.8} \tag{4.35}$$

2. Incorporating Physical System Consideration

BGs offer a unique advantage in mechatronic system design by allowing the simultaneous exploration of passive and active configurations, facilitating coevolutionary synthesis. In prior experiments, the coevolved controllers were not restricted to either passive or active implementations. Coevolutionary computation was utilized to design beneficial controller structures, which may include passive physical structures that emerge between the sprung and unsprung masses. These emergent passive components can improve energy efficiency compared to fully active suspension systems. Instead of depending on emergent passive components, the coevolution process can be guided by explicit constraints that require certain suspension components to be passive and to ensure that the physical structures lack inertia components. This method increases the difficulty of discovering physically significant structures suitable for semiactive suspension design.

When explicit physical systems are considered, the coevolution process of controllers involves three species. The concatenated controller k_1, previously discussed, is divided into two components connected by a 1-junction: the passive component k_{1p}, which corresponds to the physically realizable part of the suspension, and the active component k_{1a}, which represents the active controller k_{11}. These components are all modeled using bond graphs.

In the case of the same parameter settings as before, three simplified optimal structures are obtained through coevolution calculation, all of which have the same active control configuration, as depicted in Figure 4.46. In addition, the BG models and corresponding mechanical implementations of these three design alternatives are shown in Figures 4.47–4.49.

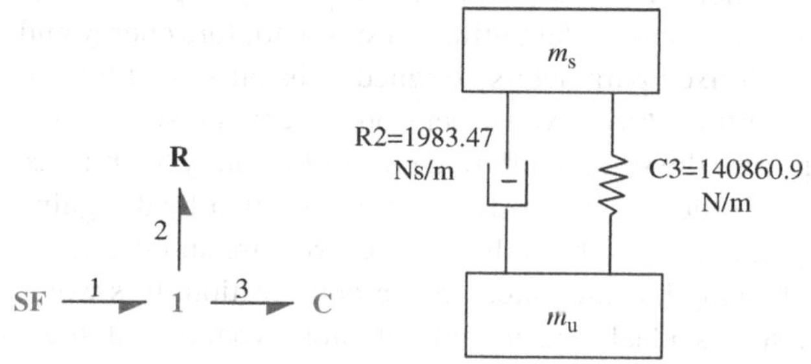

Figure 4.47 Passive physical structure design alternative 1 for suspension controller.

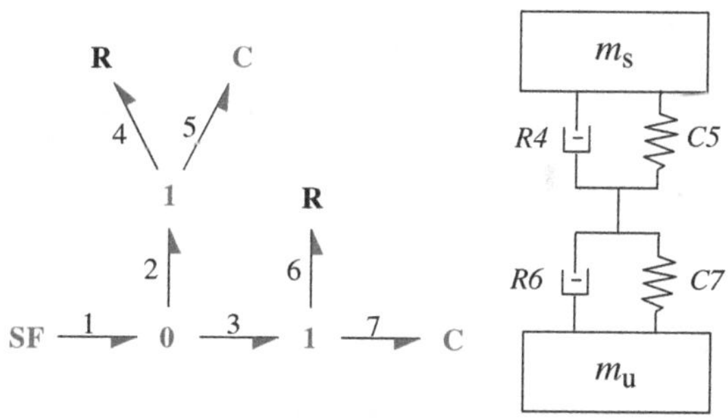

R4	2181.92 Ns/m
C5	119192.6 N/m
R6	4687.44 Ns/m
C7	47258085.9 N/m

Figure 4.48 Passive physical structure design alternative 2 for suspension controller.

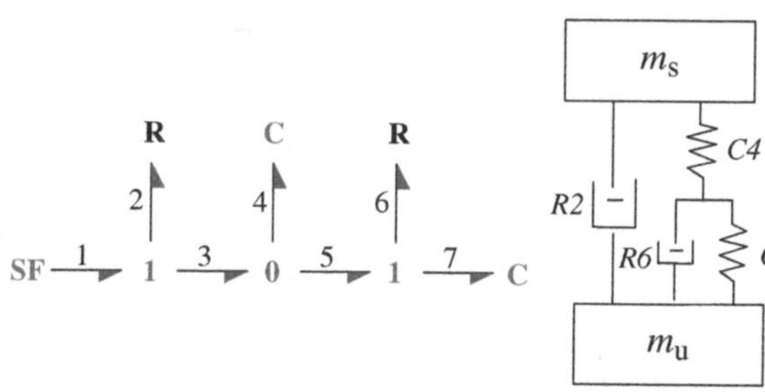

R2	2005.16 Ns/m
C4	192040.7 N/m
R6	1207.97 Ns/m
C7	378034.7 N/m

Figure 4.49 Passive physical structure design alternative 3 for suspension controller.

The suspension controllers in these three coevolving design alternatives contain different passive physical implementations. Comparing the three design alternatives, it can be seen that design alternative 1 shown in Figure 4.47 has the simplest physical structure. By dividing the controller k_1 into two parts, k_{1p} and k_{1a}, Figure 4.50 compares the Bode plots of the coevolutionary controller k_1 and the algebraic computation controller k_1. Both show similar frequency responses.

To summarize, a passive suspension system utilizes a spring to store energy and a damper to dissipate it, with fixed parameters designed to balance road following and load carrying. In contrast, an active suspension system can store, dissipate, and introduce energy into the system, offering greater flexibility to enhance the performance of design solutions. This study demonstrates that by designing controllers within the physical domain, both the passive structure and the active controller can coevolve, allowing for integrated system optimization. It is worth noting that this study assumes ideal sensor and actuator dynamics; if these assumptions do not hold, the suspension design would need significant adjustments.

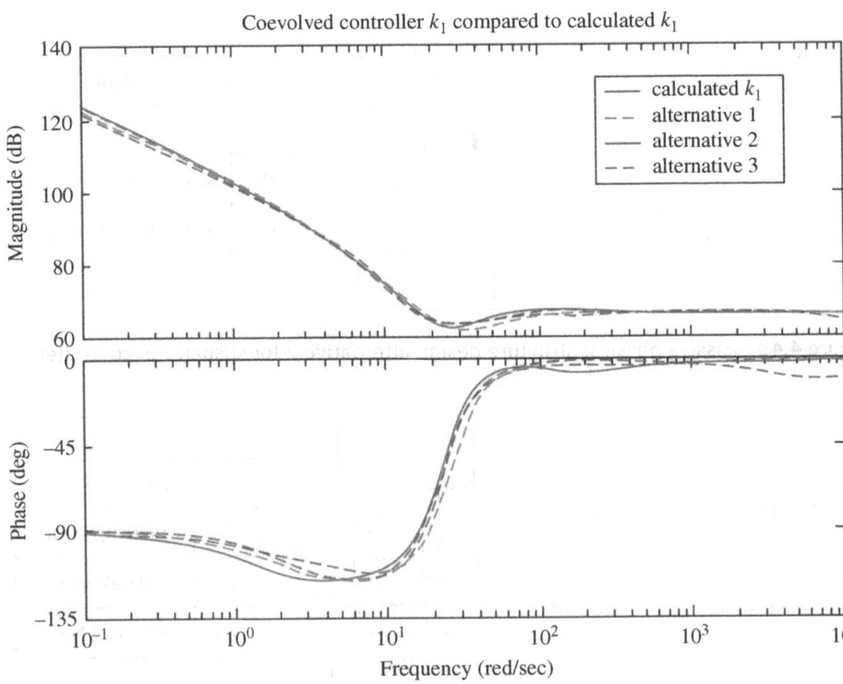

Figure 4.50 Coevolved controller k_1 compared to calculated k_1.

3. Experimental Analysis

In the experimental setup, two species, controllers k_1 and k_2, are involved. For each species, two individuals are selected for fitness evaluation, consisting of the best-performing individual and one randomly chosen from the last generation. The coevolutionary process terminates when either species reaches a fitness value of 0.99. The strong interdependence between the two species leads to significant, dynamic fluctuations in their fitness improvements, in contrast to the more gradual changes typically seen in single-species evolution. The mean and maximum fitness improvement curves for a typical coevolutionary operation are displayed in Figures 4.51 and 4.52.

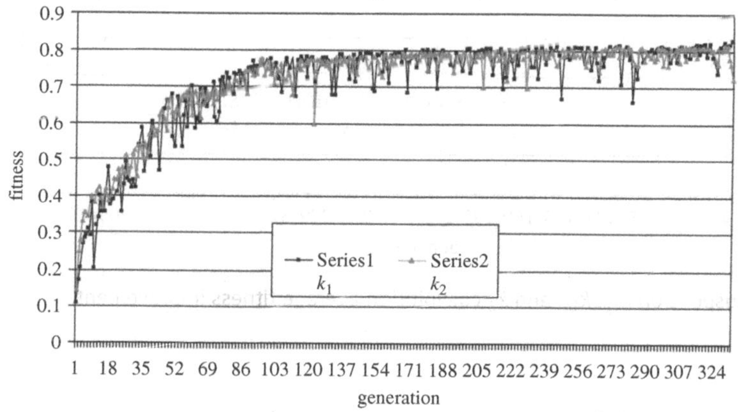

Figure 4.51 Suspension controller k_1 and k_2 coevolution average fitness improvement.

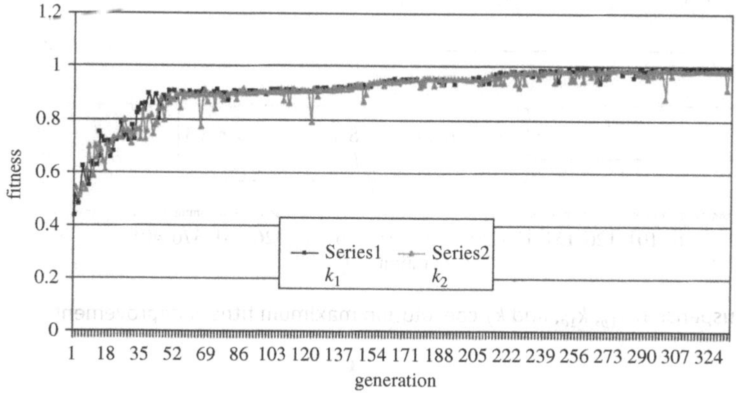

Figure 4.52 Suspension controller k_1 and k_2 coevolution maximum fitness improvement.

In this case, three distinct species are considered: the passive physical system k_{1p}, the collocated active controller k_{1a}, and the noncollocated controller k_2. The experimental setup is similar to that of the two-species coevolution. Figures 4.53 and 4.54 display the mean and maximum fitness improvement curves for a typical coevolutionary operation involving these three species.

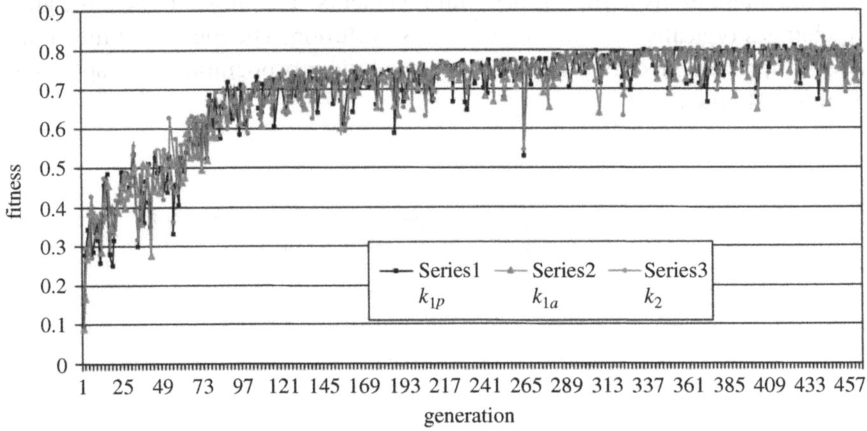

Figure 4.53 Suspension k_{1p}, k_{1a}, and k_2 coevolution average fitness improvement.

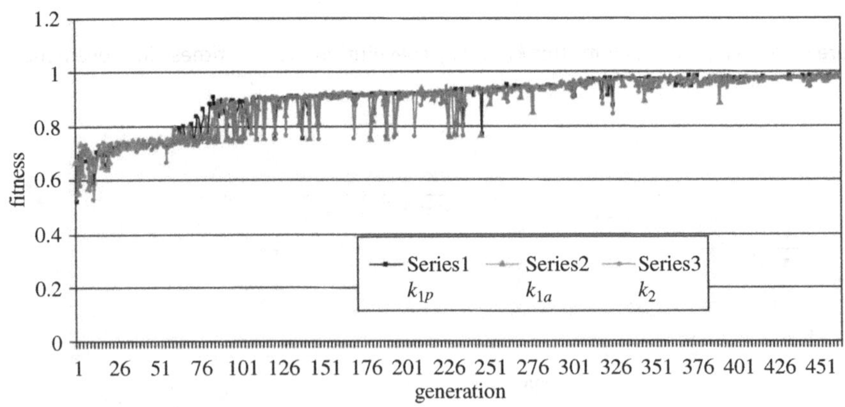

Figure 4.54 Suspension k_{1p}, k_{1a}, and k_2 coevolution maximum fitness improvement.

4.3 Vision System Design Automation

In the following sections, two case studies are presented: (i) automated design of a retinal vessel segmentation network for the retinal vascular system, and (ii) construction of a strabismus-disease detection network. For retinal vessel segmentation, an improved GA is used to optimize a U-Net shaped architecture. For strabismus detection, causal feature learning is introduced to build a high-precision graph neural network. Although not focused directly on robotic vision systems per se, these cases demonstrate techniques that can be applied to improve the design and performance of robotic vision systems.

4.3.1 Optimization Design for the Retinal Vascular System

The complex morphology of the retinal vasculature poses substantial challenges for manual segmentation, which often demands considerable time and clinical expertise, and may still be influenced by individual variability and subjectivity [315, 317]. As a result, developing robust and automated retinal vessel segmentation algorithms has long been recognized as a critical research direction in medical image analysis. Such algorithms have the potential to enhance diagnostic consistency and efficiency, providing ophthalmologists with reliable computational support in clinical decision-making.

1. Method and Experiments

To tackle the challenges outlined previously, we introduce Genetic U-Net, a novel automated architecture design framework tailored for retinal vessel segmentation. This method integrates evolutionary strategies with the classic U-shaped encoder-decoder structure to enhance segmentation accuracy and adaptability. The design emphasizes structural optimization to address the fine-grained complexity of retinal vasculature. A comprehensive description of the Genetic U-Net framework can be found in Section 3.5 of Chapter 3. The experimental protocols, datasets, and settings used in this study remain aligned with those detailed in that section. The subsequent part of this chapter presents a thorough examination of the experimental outcomes, along with corresponding analysis and discussions surrounding the performance and implications of the proposed method.

2. Experimental Results

a. Comparison With the Baseline Models

Given that our proposed method is grounded in a U-shaped encoder-decoder framework, we conducted an extensive performance comparison against two well-established reference models: the original U-Net [348] and FC-DenseNet

Table 4.10 Comparison with U-Net and FC-Densent on four datasets.

Datasets	Models	ACC	SE	SP	F1	AUROC
DRIVE	U-Net	0.96	0.81	0.97	0.82	0.98
	FC-Densent	0.96	0.82	0.97	0.82	0.98
	Genetic U-Net	**0.96**	**0.83**	**0.98**	**0.83**	**0.98**
STARE	U-Net	0.97	0.82	0.98	0.83	0.99
	FC-Densent	0.97	0.83	0.98	0.84	0.99
	Genetic U-Net	**0.97**	**0.87**	**0.99**	**0.86**	**0.99**
CHASE_DB1	U-Net	0.97	0.83	0.98	0.81	0.99
	FC-Densent	0.97	0.83	0.98	0.81	0.99
	Genetic U-Net	**0.97**	**0.85**	**0.98**	**0.82**	**0.99**
HRF	U-Net	0.96	0.82	0.98	0.81	0.99
	FC-Densent	0.97	0.81	0.98	0.81	0.99
	Genetic U-Net	**0.97**	**0.82**	**0.98**	**0.82**	**0.99**

[280], the latter of which also adopts a U-shaped design. To ensure a fair evaluation, all models were trained using identical hyperparameter settings and datasets. As reported in Table 4.10, the architecture identified through our automated approach consistently outperforms both baselines across four benchmark datasets, highlighting its improved capability in retinal vessel segmentation tasks.

Qualitative results further support these findings, as shown in Figure 4.55, 4.56, and 4.58. In these visualizations, the overlaid highlighted regions indicate false negatives—vessel structures that were missed. Both globally and in the magnified subregions, it is evident that the proposed architecture yields fewer undetected vessel pixels compared to U-Net and FC-DenseNet. This suggests a superior ability to capture and represent complex vessel morphologies, which is critical for accurate segmentation in clinical settings.

In addition to accuracy, the computational profile of the proposed model was rigorously evaluated, as detailed in Table 4.11. The discovered architecture comprises approximately 0.27 million parameters – an impressive reduction of over 100-fold and 30-fold relative to U-Net (31.03M) and FC-DenseNet (9.32M), respectively. The corresponding model size is only 1.2 MB, significantly smaller than the 120 MB and 38 MB of the respective reference models.

Efficiency in terms of multiply–accumulate operations (MACs) and inference time was also assessed using an input size of $3 \times 565 \times 584$ on the DRIVE dataset.

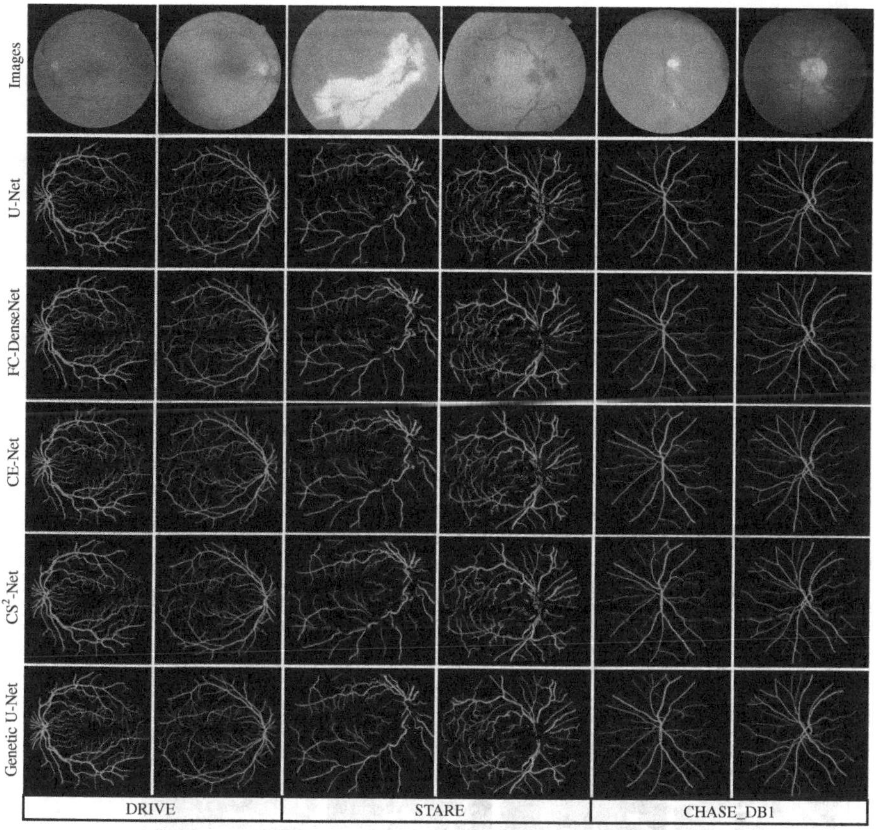

Figure 4.55 Comprehensive view of segmentation outputs across DRIVE, STARE, and CHASE_DB1. The visualizations show each model's predicted vessel maps and allow comparison of correctly segmented vessels, spurious vessel responses (false positives), and missed vessel branches (false negatives)."

The proposed model completes a forward pass with roughly 41 billion MACs, representing a 3.2 × and 2.0 × reduction compared to U-Net and FC-DenseNet. Furthermore, runtime measurements indicate a speedup of approximately 1.3 × over U-Net and 2.9 × over FC-DenseNet, reinforcing the model's suitability for real-time deployment and resource-constrained clinical environments.

b. Systematic Experimental Anatomy of Deep Neural Network Architecture Search

The outcomes of the evolutionary algorithm typically hold certain valuable insights that can assist us in further enhancing our work. In order to accurately

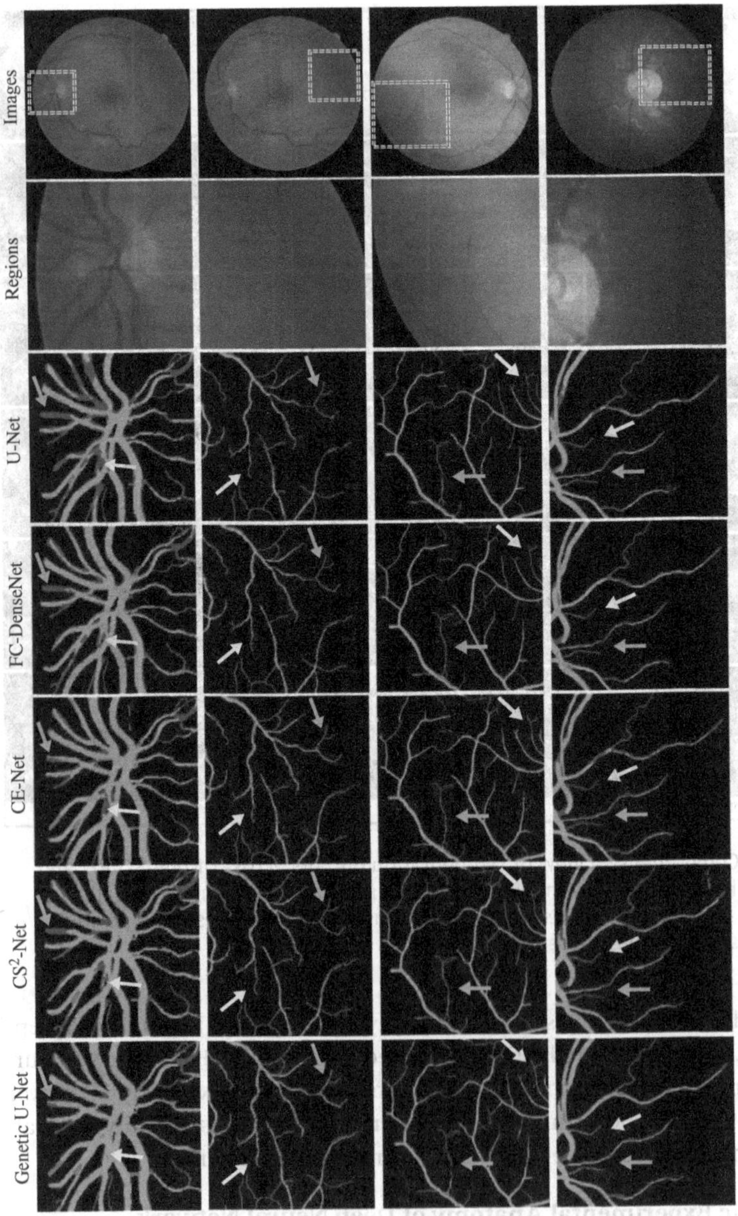

Figure 4.56 Detailed local visualization of the segmentation output.

Table 4.11 The detailed model size, numerous parameters, compute metrics MAC, and execution time are all comprehensively compared to the classic U-Net architecture. The MAC and execution time are precisely calculated based on specific 3 × 565 × 584 input size data to enable in-depth exploration of the differences and characteristics between the two under the same conditions.

Methods	Model size(MB)	Params(M)	MACs(B)	Processing duration(ms)
U-Net	120	31.03	142	35.4
FC-DenseNet	38	9.32	83	80.7
Genetic U-Net	**1.2**	**0.27**	**41**	**27.5**

and comprehensively identify trends in a series of more innovative and efficient architecture designs, we scrutinized and deeply dissected the five main architectures of the last generation that were most representative and influential in the field with a high degree of rigor and nuance.

When analyzing the topology of these architectures in depth, we focused on the structural characteristics of each component (see the supplementary materials and documentation for details). It was found that almost without exception these modules used the maximum number of nodes allowed, showing a highly complex internal structure. The connections between nodes are rich and deep, showing diverse and complex layout patterns. This level of complexity requires a lot of effort to understand the design logic and operation mechanism behind their construction. In addition, multiple paths are prevalent in these modules for flow and fusion. It is worth noting that all modules generally contain two or three parallel branch structures, in a form highly similar to the module design in InceptionNet [283].

To gain a more precise understanding of how specific operation sequences influence architectural performance, we implemented a structured and rigorous data collection protocol. By statistically analyzing the frequency of all operation sequences listed in Table 4.12, we aimed to uncover hidden patterns and correlations that govern the behavior and effectiveness of these sequences within the evolved architectures. As illustrated in Figure 4.57, one particular operation sequence – identified as Sequence 11 – demonstrated a strikingly high frequency of occurrence. It formed a dominant peak in the operation frequency spectrum, suggesting that it may hold a strategic advantage or play a critical role in the overall processing flow of the architecture.

To evaluate its practical contribution, we substituted the standard convolutional unit in U-Net (3×3 convolution followed by ReLU) with Sequence 11. This modification had a negligible impact on model parameters and multiply–accumulate

Table 4.12 Available operation sequences assigned to node IDs.

ID	Sequence	ID	Sequence
0	Conv (3 × 3) / ReLU	8	ReLU / Conv (3 × 3)
1	Conv (3 × 3) / Mish	9	Mish / Conv (3 × 3)
2	Conv (3 × 3) / IN / ReLU	10	IN / ReLU / Conv (3 × 3)
3	Conv (3 × 3) / IN / Mish	11	IN / Mish / Conv (3 × 3)
4	Conv (5 × 5) / ReLU	12	ReLU / Conv (5 × 5)
5	Conv (5 × 5) / Mish	13	Mish / Conv (5 × 5)
6	Conv (5 × 5) / IN / ReLU	14	IN / ReLU / Conv (5 × 5)
7	Conv (5 × 5) / IN / Mish	15	IN / Mish / Conv (5 × 5)

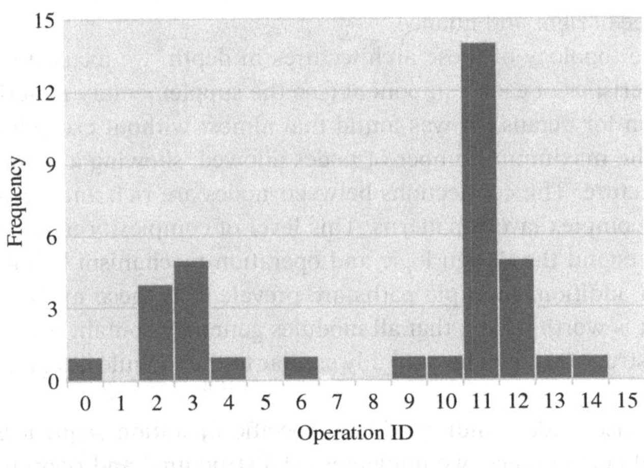

Figure 4.57 Frequency of the operation sequences.

operations (MAC), but, as shown in the eighth row of Table 4.13 (based on experiments on the DRIVE dataset), led to a substantial improvement in segmentation accuracy. The modified U-Net significantly outperformed most baseline methods, ranking just below the best-performing architecture derived from our proposed evolutionary approach.

What is particularly noteworthy is that this improvement arose from localized alterations rather than a fundamental overhaul of the network structure. Sequence 11 introduces three main differences: it incorporates the Mish activation function [288], utilizes a pre-activation layout, and applies instance normalization. To isolate the contribution of each of these components, six additional experiments were

Table 4.13 Empirical evaluation of the designed operations and corresponding sequences.

No.	Setting	Params(M)	F1-score	AUROC
1	U-Net/ReLU	31.0	0.82	0.99
2	U-Net/Mish	31.0	0.79	0.98
3	U-Net/IN/ReLU	31.0	0.83	0.99
4	U-Net/ReLU(P)	31.0	0.83	0.99
5	U-Net/Mish(P)	31.0	0.79	0.98
6	U-Net/IN/Mish	31.0	0.83	0.99
7	U-Net/IN/ReLU(P)	31.0	0.83	0.99
8	U-Net/IN/Mish(P)	31.0	0.83	0.99
9	Genetic U-Net	**0.3**	**0.83**	**0.99**

1 In the U-Net block, "ReLU" and "Mish" denote the activation functions, while "P" stands for pre-activation. The term "IN" refers to the inclusion of instance normalization.

2 The boldface in the "Params(M)" denotes that the metric is computed excluding the FOV regions.

conducted, as summarized in the second to seventh rows of Table 4.13. The results indicate that both instance normalization and pre-activation contribute positively to performance, with instance normalization having a more pronounced effect. Conversely, merely replacing the activation function with Mish, without altering other aspects, did not yield notable improvements.

Further comparisons between rows five through eight confirm that combining pre-activation with instance normalization results in the most consistent gains. Among these, the configuration using Mish in the pre-activation setting slightly outperforms its ReLU counterpart. These findings underscore the potential of fine-grained operation-level modifications in enhancing network performance and provide practical insights into architectural refinement strategies.

The evolutionary process employed by Genetic U-Net has yielded a diverse set of high-performing architectures, each demonstrating strong potential in retinal vessel segmentation tasks. By systematically analyzing these final-generation networks, we are able to distill valuable architectural patterns and recurring design strategies. These insights contribute to the formulation of foundational design principles that can inform future model development. Rather than focusing solely on performance metrics, our statistical and structural evaluations highlight transferable features – such as branching patterns, normalization strategies, and activation configurations – that are closely tied to improved segmentation outcomes. As such, the architectures identified in this study serve not only as high-performing models but also as a rich source of empirical knowledge, offering practical guidance for subsequent advances in neural architecture design within the medical imaging domain.

c. Module-wise Evaluation

Difference-guided Crossover: To investigate the contribution of the disparity-guided crossover approach, a comparative experiment was conducted between two search strategies – one incorporating the crossover and one without it – under identical experimental conditions. For each generation, we recorded the fitness value of the highest-performing individual within the population.

Environmental Selection: To balance search effectiveness with population diversity and reduce the risk of premature convergence, we implemented an environmental selection strategy that integrates both elitism (with a pool size of 5) and diversity preservation. To evaluate the efficacy of this approach, we compared it against two established methods: top-N selection [349] and a strict elitist selection with a single elite [350]. All algorithms were tested under identical settings, including the same initial population and crossover operations.

3. Discussion

a. Evaluation in Complex Cases

As illustrated in columns three and four of Figure 4.55 and columns two of Figure 4.58, several examples contain pathological lesions and, in particular, the image in the third column of Figure 4.55 exhibits pronounced non-uniform illumination. In these challenging scenarios, the segmentation output generated by the discovered architecture exhibits noticeably fewer false positives (represented by red pixels) compared to those produced by the baseline models, suggesting a greater robustness to pathological disruptions and illumination variance. Furthermore, in the detailed visualization of Figure 4.56, the number of false negatives (blue pixels), especially in regions with cross-linked or low-contrast microvessels, is significantly reduced in the results of the discovered architecture. This reflects its enhanced capacity to detect fine vascular structures and complex branching patterns. Taken together, the vascular segmentation maps derived from the proposed model show the highest fidelity to the ground truth across Figures 4.55, 4.56, and 4.58, highlighting its effectiveness in dealing with structurally complex and diagnostically critical regions.

b. Effectiveness in High-resolution Fundus Image Analysis

In contrast to DRIVE and CHASE_DB1, the HRF dataset contains fundus images with significantly higher resolution. Representative experimental outcomes are presented in Figure 4.58, while the quantitative comparisons are listed in Table 4.10. On the HRF dataset, both qualitative and quantitative evaluations

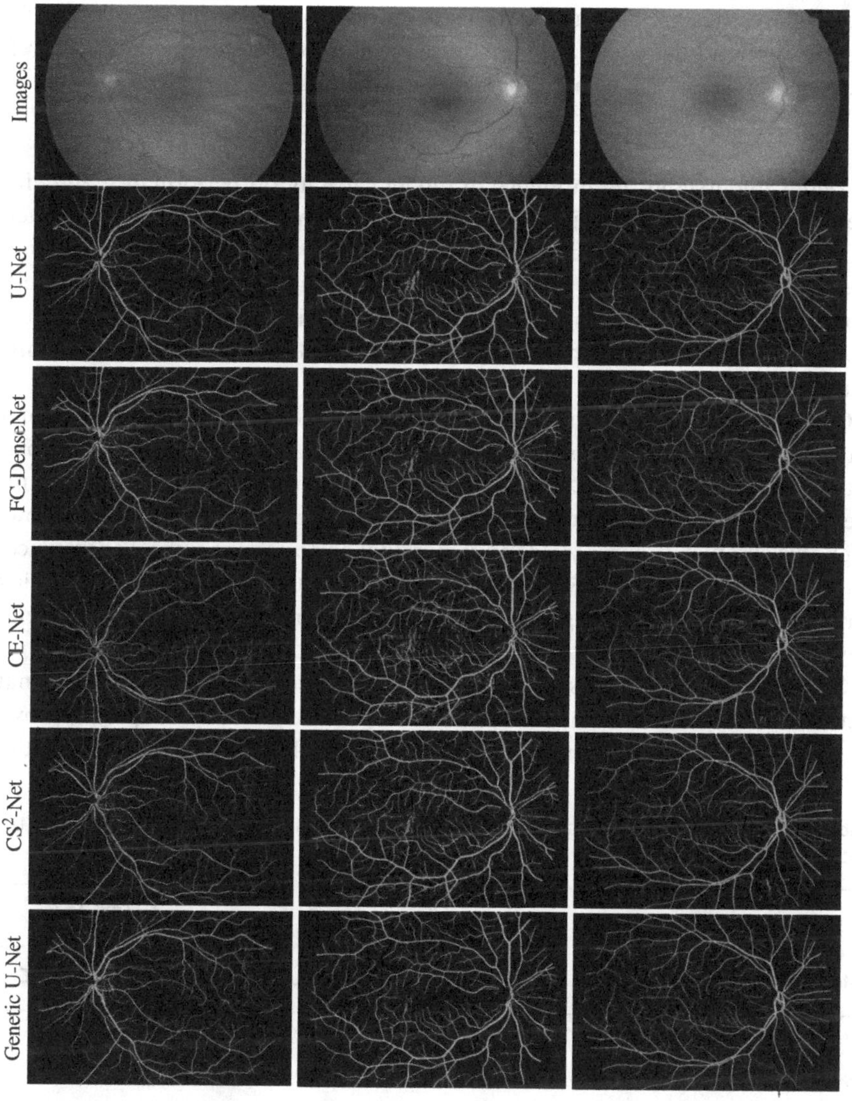

Figure 4.58 Visualization of segmentation performance on the HRF dataset. The predicted vessel maps illustrate correctly detected vessel pixels (true positives), spurious vessel responses (false positives), and missed vessel structures (false negatives), enabling qualitative comparison across models."

indicate that the proposed architecture surpasses top-performing approaches as well as baseline networks (original U-Net and FC-DenseNet) in segmentation performance. These findings confirm the strong adaptability and reliability of the proposed design, demonstrating its capability to deliver superior results for high-resolution fundus imagery and offering valuable insights and technical references for related image analysis and processing tasks.

c. Modifying Quantity of Architecture-derived Parameters

To better understand the influence of model capacity on performance and efficiency, we systematically adjusted the number of convolutional channels (denoted as parameter c) within the nodes of the discovered architecture to a range of values: 2, 3, 4, 5, 10, and 30. These variations were designed to control the model's parameter count and computational load. The modified models were retrained from scratch and evaluated on the DRIVE and CHASE_DB1 datasets, with results summarized in Table 4.14.

Our analysis centers on the role of c in shaping both model expressiveness and efficiency. When c increases from 20 to 30, there is a substantial rise – approximately 125% – in both MACs (multiply–accumulate operations) and total parameters, indicating a corresponding increase in representational capacity. However, this complexity surge yields only marginal performance gains. For instance, on the DRIVE dataset, the F1 score improves by just 0.04%, and AUROC remains almost unchanged. This suggests diminishing returns from increasing capacity beyond a certain threshold.

Conversely, reducing c significantly lowers model size and computational demands. When c is set to 5 or 10, MACs and parameter counts drop by approximately 94% and 75%, respectively, relative to the baseline model. Despite this, the

Table 4.14 Evaluation results after tuning the quantity of parameters determined by the architecture.

			DRIVE		CHASE_DB1	
Model settings	**Params(M)**	**MACs(B)**	**F1-score**	**AUROC**	**F1-score**	**AUROC**
channel = 20	0.27	41.35	0.83	0.98	0.82	0.99
channel = 30	0.61	92.77	0.83	0.98	0.82	0.99
channel = 10	0.07	10.42	0.83	0.98	0.82	0.99
channel = 5	0.02	2.65	0.83	0.98	0.82	0.99
channel = 4	0.01	1.71	0.83	0.98	0.81	0.99
channel = 3	0.01	0.97	0.82	0.98	0.80	0.98
channel = 2	0.01	0.44	0.81	0.97	0.79	0.98

The calculation of MACs is performed for an input shape of $3 \times 565 \times 584$.

performance degradation remains minimal. On DRIVE, F1 scores only decrease by 0.20% to 0.91%, and AUROC by just 0.13% to 0.15%, indicating a high degree of resilience to parameter reduction.

Notably, when c equals 5, the optimized architecture retains strong segmentation performance with only a fraction of the parameters – approximately 0.06% and 0.19% of U-Net and FC-DenseNet, respectively. Even with c reduced to 2, resulting in an ultra-lightweight model with just 2.8K parameters, the network maintains competitive performance when compared to established baselines.

These findings collectively demonstrate that the proposed architecture possesses exceptional parameter efficiency and exhibits strong robustness under aggressive model compression. This makes it highly suitable for deployment in resource-limited environments or real-time clinical applications, where computational and memory constraints are critical.

d. Model Performance When Amount of Training Data Is Gradually Reduced

To evaluate the impact of reduced training data on model performance and architectural robustness, experiments were conducted using the DRIVE and CHASE_DB1 datasets. Originally, each dataset comprised 20 training images, but these were systematically reduced to 2, 5, and 10 images to assess the identified architecture alongside baseline models, namely U-Net and FC-DenseNet. The variation in F1 scores with respect to the number of training images is presented in Figure 4.59 and Table 4.15.

When evaluated on the DRIVE and CHASE_DB1 datasets, both the identified architecture and the baseline models exhibited performance declines as training data decreased. However, a detailed analysis reveals that the identified architecture demonstrates notably greater resilience compared to the baseline models. Specifically, in a demanding scenario with only five training images per dataset, the identified architecture's F1 score on DRIVE/CHASE_DB1 decreased marginally from 0.8314/0.8223 to 0.8270/0.8063, reflecting modest reductions of 0.52%/1.94%. In contrast, the baseline U-Net experienced a more pronounced decline, with F1 scores dropping from 0.8191/0.8092 to 0.8043/0.7881, corresponding to reductions of 1.80%/2.61%. Similarly, FC-DenseNet's F1 scores fell from 0.8235/0.8144 to 0.8073/0.7827, marking declines of 1.97%/3.89%.

These results, derived from rigorous quantitative analysis, underscore the identified architecture's superior stability and adaptability under conditions of limited training data. This robustness is particularly valuable in clinical settings, where annotated medical imaging data, such as those used for retinal vascular segmentation, are often scarce. The architecture's ability to maintain high performance despite constrained datasets highlights its potential to provide critical technical support for advancing precision medicine applications.

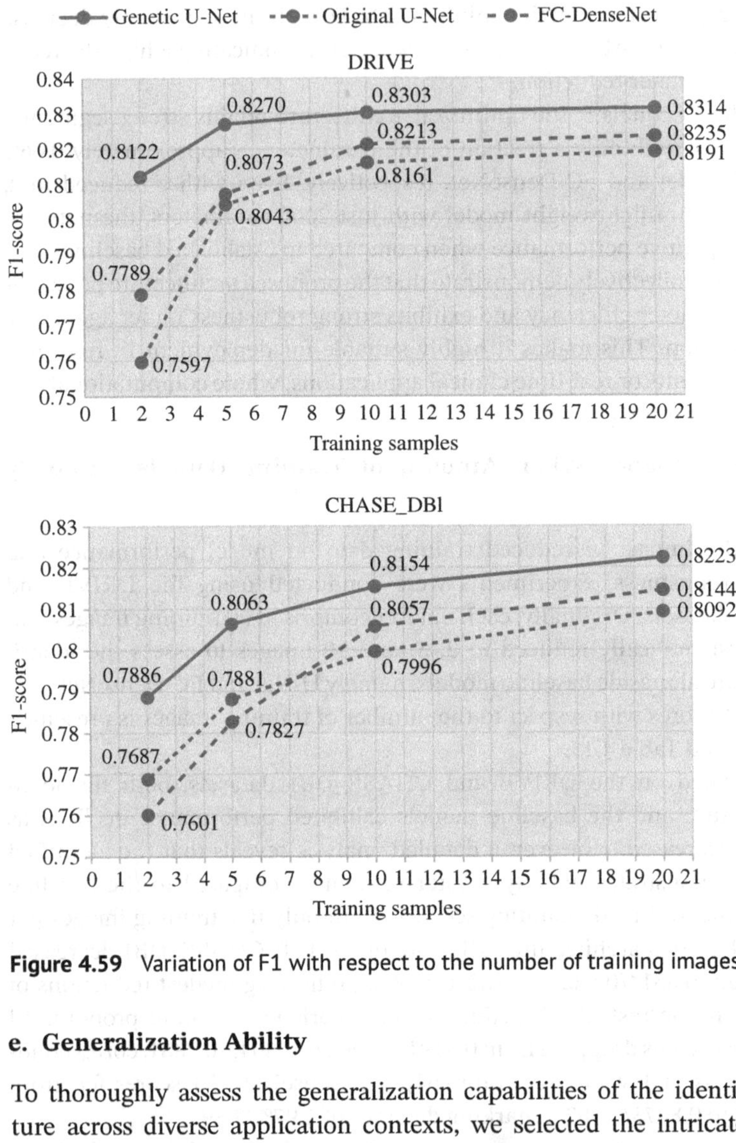

Figure 4.59 Variation of F1 with respect to the number of training images.

e. Generalization Ability

To thoroughly assess the generalization capabilities of the identified architecture across diverse application contexts, we selected the intricate task of cell membrane (boundary) segmentation for an in-depth evaluation. This experiment leverages the DRIVE dataset, noting that cell membrane segmentation shares substantial similarities with vascular segmentation in terms of features and processing methodologies. In this carefully designed experimental framework, the cell membrane's boundary region is designated as the target for segmentation (the positive region). The integrated EM dataset [351], comprising microscopy and electron microscopy data, includes 30 images of 512×512 pixels for training and

Table 4.15 Variation of F1 with respect to the number of training images.

Models	Training instances	DRIVE F1-score	DRIVE AUROC	CHASE_DB1 F1-score	CHASE_DB1 AUROC
Genetic U-Net	20	0.83	0.98	0.82	0.99
	10	0.83	0.98	0.82	0.99
	5	0.83	0.98	0.81	0.98
	2	0.81	0.97	0.79	0.98
FC-DenseNet	20	0.82	0.98	0.81	0.99
	10	0.82	0.98	0.81	0.98
	5	0.81	0.97	0.78	0.98
	2	0.78	0.96	0.76	0.96
U-Net	20	0.82	0.98	0.81	0.99
	10	0.82	0.98	0.80	0.98
	5	0.80	0.97	0.79	0.98
	2	0.76	0.95	0.77	0.97

testing. As the ground truth for test images is unavailable, the 30 training images were meticulously divided into 20 for model training and 10 for performance validation through random sampling.

Without altering the original image dimensions, the experiments utilized the full-scale EM dataset in a computational environment that mirrors real-world conditions. The identified architecture was compared against established models, including the baseline U-Net, the sophisticated FC-DenseNet, the novel CE-Net, and the uniquely designed CS2-Net. Their performance was rigorously evaluated on the 10-image test subset using widely accepted metrics: accuracy (ACC) for overall correctness, sensitivity (SE) for detection sensitivity, specificity (SP) for precise negative identification, F1 score for balancing precision and recall, and the area under the receiver operating characteristic curve (AUROC) for assessing discriminative power. These metrics enabled a comprehensive and objective comparison of model performance.

The results, detailed in Table 4.16, demonstrate that the identified architecture consistently surpasses its counterparts across all five key metrics. Visual evidence in Figure 4.60 further underscores its superior performance, particularly in handling complex scenarios with irregular shapes or subtle boundaries. This highlights the architecture's exceptional adaptability and robustness, affirming its outstanding generalization potential for cell membrane segmentation.

Table 4.16 Comparison of methods for cell membrane boundary segmentation.

Methods	Accuracy	Sensitivity	Specificity	F1-score	AUROC
U-Net	0.91	0.82	0.94	0.81	0.96
FC-DenseNet	0.91	0.83	0.94	0.82	0.96
CE-Net	0.91	0.83	0.92	0.82	0.97
CS^2-Net	0.91	0.83	0.93	0.82	0.96
Genetic U-Net	0.92	0.84	0.94	0.82	0.97

The architecture's stable and effective performance across varied data distributions and task demands positions it as a reliable and advanced solution for related research and practical applications.

4.3.2 Causal Feature Selection for Strabismus Diagnosis Using GNN

At present, causal inference methodologies stand out as the only approach capable of discerning genuine causal relationships among variables within the vast datasets of intricate systems, without the need for extensive expert knowledge. They are adept at identifying causal effects between variables and determining the causal direction. Beyond this, causal inference methodologies surpass statistical methods by accounting for confounding variables, thus delivering a more precise analysis. A key advantage is that causal inference relies primarily on empirical data, which minimizes the likelihood of errors stemming from subjective expertise. Furthermore, it offers a suite of quantitative techniques to measure the intensity or magnitude of causal impacts between variables. Extensive research has demonstrated that causal inference methodologies consistently achieve higher accuracy in root-cause analysis compared to traditional statistics. Consequently, employing causal inference methodologies enables a more precise mapping of causal connections among observed variables within complex systems, facilitating the identification and elucidation of the underlying true causes. This offers a more precise and potent methodology for addressing root cause analysis challenges within complex systems.

Strabismus is a prevalent eye condition in which the eyes do not properly align with each other while focusing on an object. This misalignment is particularly common among children and, if untreated, can lead to serious binocular vision problems. Beyond its effects on vision, strabismus can also have significant psychological effects on those affected. Fortunately, strabismus can be effectively treated

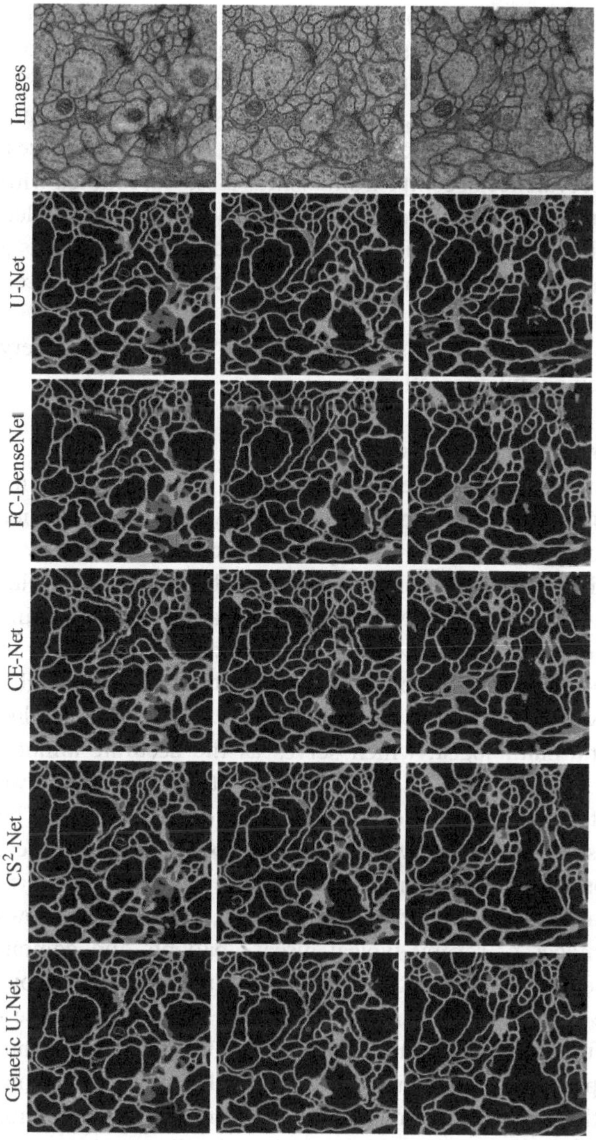

Figure 4.60 Variation of F1 with respect to the number of training images.

with surgery, but the specific surgical approach depends on an accurate diagnosis. The diagnosis of strabismus typically relies on specialized eye examinations such as the Cover test and the Hirschberg test, which require expertise and specific equipment, often found only in larger hospitals or specialized clinics.

With the progress in digital image technology, a growing number of researchers are leveraging images of strabismus patients' eye positions for diagnosing the condition. In recent times, various classification algorithms, including AHT, CCLRR, HOCORC, and U-Net for iris segmentation, have been proposed and have demonstrated high accuracy in distinguishing between normal and strabismus cases. However, due to the low prevalence of complex strabismus diseases, coupled with ethical medical regulations and the high costs associated with obtaining labeled data, collecting data for these complex conditions is extremely challenging. As a result, none of the aforementioned studies have managed to achieve more nuanced classifications of strabismus diseases.

This case is a refined strabismus diagnosis method based on causal discovery using GNNs, and the specific implementation steps are:

1. **Constructing a Strabismus Disease Dataset**
The current strabismus recognition datasets are frequently constrained by limited sample sizes and imprecise annotations. In our study, we aimed to enhance fine-grained strabismus recognition by compiling an extensive dataset comprising 1075 strabismus patients from Xinhua Hospital, which is associated with Shanghai Jiao Tong University School of Medicine. Each patient provided nine high-resolution facial photographs (2592 × 1944 pixels) taken while looking in different directions: upper left, upper, upper right, left, straight ahead, right, lower left, lower, and lower right. Moreover, each patient was given a detailed diagnosis by a senior ophthalmologist, which serves as the accurate annotations for our dataset. For the identification of complex strabismus cases, we stratified the process into two stages based on the pathological features. The initial stage involves distinguishing between esotropia, exotropia, or vertical strabismus, while the subsequent stage identifies whether the pattern is A-pattern, V-pattern, or non-AV pattern. For clarity and illustration, Figure 4.61 shows the nine eye position images of several typical disease patients and demonstrates the distribution of different fine-grained strabismus categories in our dataset. We divided the dataset into training, validation, and testing subsets with a 6:3:1 split ratio. Image preprocessing steps are as follows. In facial photographs, the area occupied by the eyes is relatively small. Applying recognition models directly on the entire facial image may introduce irrelevant details, such as hairstyle, nose, and mouth, which could confuse the diagnostic process and unnecessarily increase computational costs. To address this issue, as shown in Figure 4.62, we perform image preprocessing to extract the main eye area. First, facial landmarks are detected using the dlib library. Based on these landmarks, we calculate the horizontal distance d_h and the vertical distance d_v between the two eyes. Subsequently, a rectangle centered at the midpoint between the eyes is defined as the main eye area, with a length of $r_h \times r_v$ and a width of

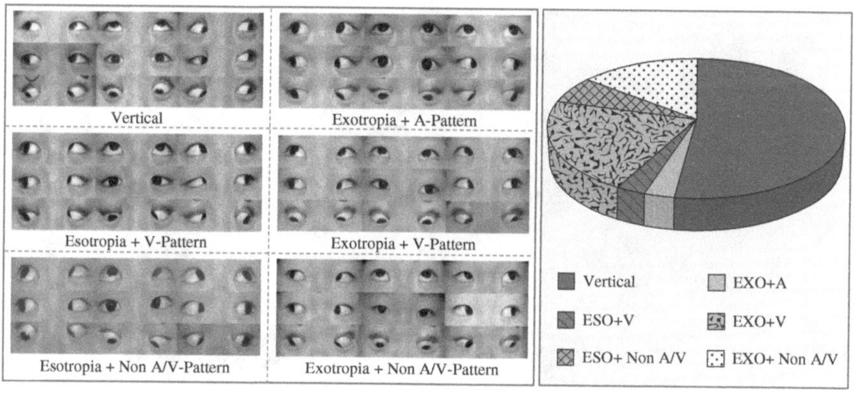

Figure 4.61 Nine gaze position images and the category distribution of the dataset.

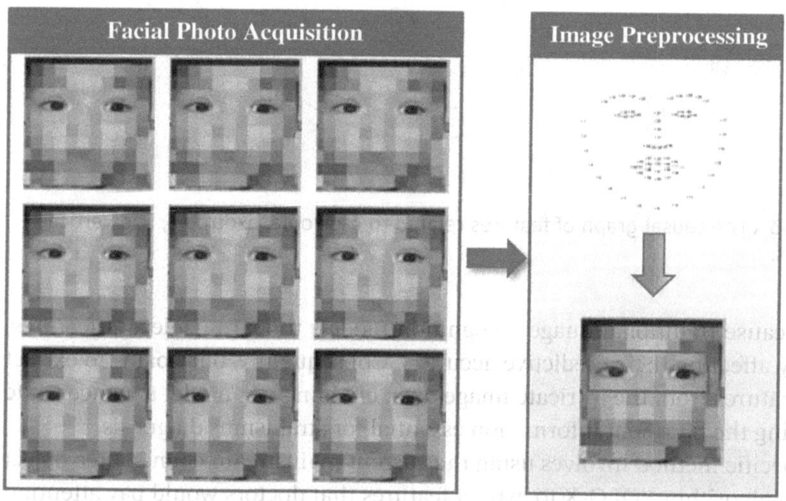

Figure 4.62 The dlib library detects facial landmarks and extracts the eye area.

$r_v \times d_v$, Empirically, we set the relaxation coefficients $r_h = 1.2$ and $r_v = 1.4$ to adjust the size accordingly.

2. Causal Feature Selection Algorithm Filters Key Causal Variables

To accurately identify complex strabismus patterns, our approach involves a two-stage classification based on pathological features. The initial stage identifies whether the condition is esotropia, exotropia, or vertical strabismus, while the second stage determines if it follows an A-pattern, V-pattern, or is non-AV pattern. Given that the images contain substantial interfering or confounding information,

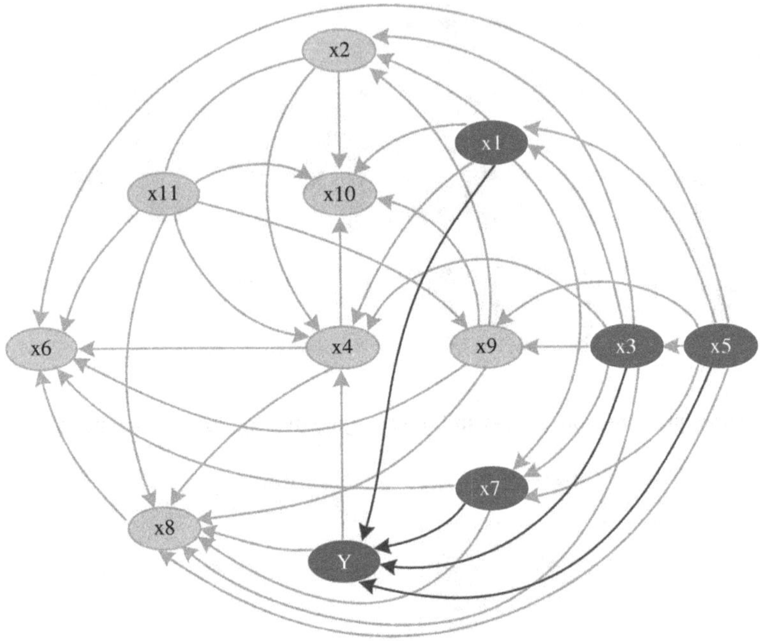

Figure 4.63 The causal graph of features related to esotropia, exotropia, and vertical strabismus.

this can cause traditional image recognition models to learn irrelevant features, adversely affecting their predictive accuracy. Consequently, our goal is to extract salient features from the intricate image data, enabling our model to concentrate on learning the pertinent information essential for strabismus diagnosis.

The specific method involves using the current mainstream open-source object detection algorithm YOLO-X to extract features that doctors would pay attention to, such as the iris and orbital cavity of children, which may be related to the identification of complex disease signs.

We utilize the HCM causal discovery algorithm to sift through the extracted features associated with strabismus. The HCM algorithm, a well-established and open-source method for causal feature selection, can be broken down into three main components:

a. Causal skeleton learning: Combine the PC-Stable algorithm with the mixed-type randomized conditional independence test to determine an undirected causal skeleton graph with edges.
b. Based on the causal skeleton graph obtained in step 1), integrate a greedy search strategy with the cross-validation based mixed information criterion

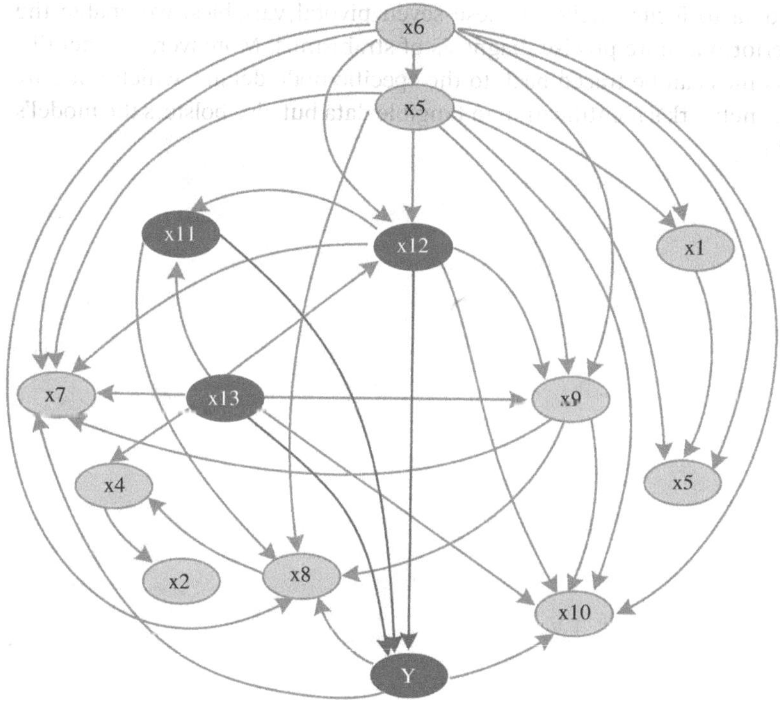

Figure 4.64 Causal graph of features related to AV sign strabismus.

to determine the direction of the edges on the skeleton graph, resulting in a causal graph.

c. On the basis of the causal graph obtained in step 2), perform the mixed-type randomized conditional independence test to remove some causal edges with low credibility, ultimately resulting in a more refined causal graph.

After applying the HCM algorithm for feature selection on strabismus-related variables, we obtained two causal graphs, in which the direct parent nodes of the target variable Y (the highlighted nodes in Figures 4.63 and 4.64) are the causal features for strabismus diagnosis. That is, we believe that only these seven variables are needed to complete the refined diagnosis of strabismus. After applying the HCM algorithm, the causal graph for esotropia, exotropia and vertical strabismus is shown in Figure 4.63, while the causal graph for AV signs related features is shown in Figure 4.64.

Using the key variables identified through causal feature screening for strabismus disease signs, we integrate them as nodes within a GNN. GNNs are adept at capitalizing on the inherent properties of graph structures. By structuring the input data to focus solely on these seven pivotal variables, we enable the GNN to perform a more precise diagnosis of strabismus. Moreover, the identification outcomes can be traced back to the specific node details, which not only grounds the network's identification in tangible data but also bolsters the model's interpretability.

5

MODENA in Swarm Robotics

5.1 Introduction

As tasks and environments become increasingly complex, single-robot systems often struggle to meet application demands. In contrast, swarm robots, as emerging intelligent system, are gradually demonstrating their significant potential for handling complex tasks. Compared to single-robot systems, swarm robots generate complex collective behavior through local interactions among simple individuals [352, 353]. This bottom-up emergence of behavior gives swarm robots unique advantages in adapting to complex environments. However, traditional swarm robot control strategies are often designed using a trial-and-error approach, which requires designers to continuously adjust control parameters and conduct numerous experiments to find a suitable strategy for a specific task. This process not only heavily relies on the designer's expertise and experience but also has significant limitations: it is time-consuming, inefficient, and often lacks adaptability in complex and dynamic environments. Furthermore, due to the limitations of human cognitive abilities, the trial-and-error method generally struggles to find a globally optimal solution when faced with large-scale tasks or highly complex environmental conditions.

Modular design automation (MODENA) has emerged as an effective framework for swarm robot design, addressing key challenges in the field [18]. This approach treats basic control strategies as modular units, which are automatically combined and refined using artificial intelligence (AI) technologies such as genetic programming (GP), evolutionary computation, deep learning, reinforcement learning, and causal reasoning [8, 150]. By automating this process, MODENA reduces reliance on manual parameter tuning while significantly improving design efficiency, quality, adaptability, and robustness in swarm robotic systems [354]. A key strength of MODENA is its ability to generate new modular units during the design

Modular Design Automation of Intelligent Robot Systems, First Edition. Zhun Fan et al.

process. These units are stored in a module library and reused in a closed-loop system, enabling systematic and continuous performance improvements. Compared to traditional design methods, MODENA streamlines workflows, eliminates repetitive trial-and-error tasks, and fosters innovative solutions that surpass the limits of human experience and intuition. This efficient and scalable approach supports the development of advanced, reliable, and high-performing swarm robotic systems.

This chapter progresses from basic to advanced concepts, covering theoretical model construction, specific application cases, and experimental validation. Section 5.2 introduces the fundamental concepts of gene regulatory networks (GRNs) and their roles in multicellular collective behavior. It uses specific mathematical models to delve into the internal mechanisms of GRNs and how these mechanisms enable the automated design of swarm control strategies. Building on this, Section 5.3 explores swarm control methods under communication constraints, proposing a distributed control model based on visual information for common communication-limited scenarios in unmanned systems. By integrating deep reinforcement learning (DRL) techniques, Section 5.4 offers a novel solution for the autonomous control of swarm robots in a communication-free environment. Furthermore, Section 5.5 presents a multi-UAV collision avoidance method based on causal representation learning (CRL). By combining CRL with DRL, this approach extracts causal relationships from depth images, enhancing the obstacle avoidance and navigation capabilities of multi-UAV systems in complex three-dimensional spaces.

5.2 Automated Swarm Pattern Generation for Swarm Robots

5.2.1 Background and Problem Formulation

5.2.1.1 Mapping Relationship

In biologically inspired swarm robot systems, they can be categorized into macroscopic and microscopic approaches [355, 356]. Macroscopic methods take inspiration from natural group behaviors. Microscopic methods, though less explored, are based on the collective movements of microscopic organisms like cells, where the GRN plays a key role. The GRN models interactions between genes and their products, describing how gene expression drives biological morphogenesis [357, 358]. Morphogenesis is a self-organizing process where cells autonomously move to target locations, guided by GRN mechanisms and cell-to-cell interactions [17]. In this process, cells release proteins that disperse into their surroundings, creating a concentration field where protein levels decrease with distance. In Figure 5.1a, motion control relies on the gradient of this concentration field.

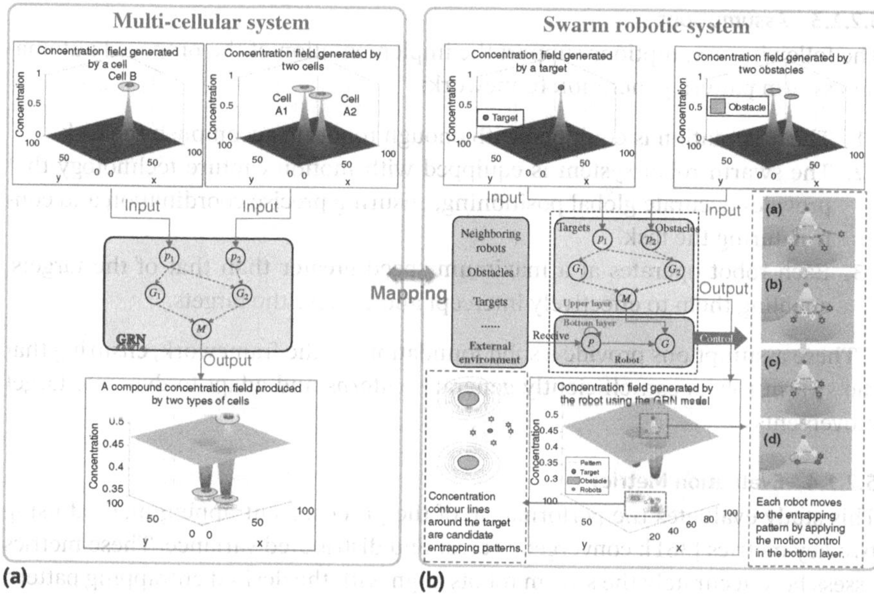

Figure 5.1 Analogy and mapping between swarm robot and multi-cellular.

In morphogenetic swarm robot systems [359], robots mimic the behavior of individual cells. The GRN model maps a robot's position to a corresponding protein concentration within the concentration field, as illustrated in Figure 5.1b. Concentration contours near a target serve as candidate patterns for swarm formation, maintaining a predefined safe distance [360]. By moving along the concentration gradient, robots gradually position themselves into the desired pattern, leading to the emergence of collective encirclement behavior in the swarm system.

5.2.1.2 Swarm Entrapping Task Statement

To entrap dynamic targets using a swarm of robots in environmental constraints, this study proposes a new swarm control method. The method consists of two key stages: swarm entrapment pattern generation and swarm entrapment pattern formation. In the first main stage, swarm entrapping patterns are designed to effectively surround the target while avoiding obstacles in the environment. These patterns provide a clear framework for the robots' movements and positioning. In the swarm pattern formation stage, the robots coordinate and deploy themselves to align with the generated patterns. By moving strategically, they successfully encircle and entrap the target. This two-step method allows swarm robots to efficiently adapt to dynamic targets and complex environments, ensuring obstacle avoidance and effective task execution.

5.2.1.3 Assumptions

The following assumptions support the implementation of the proposed automation swarm pattern generation framework:

1. The base station is equipped with enough robots to accomplish the task.
2. The swarm robot system is equipped with motion-capture technology that provides accurate global positioning, ensuring precise coordination and control during the task.
3. Each robot operates at a maximum speed greater than that of the targets, enabling them to effectively intercept and enclose the targets.

These assumptions provide a solid foundation for the framework, ensuring that the swarm robots can efficiently generate patterns and adapt to dynamic target movements.

5.2.1.4 Evaluation Metrics

This study evaluates the performance of the proposed entrapping method using two key metrics [361]: convergence error and distributed variance. These metrics assess how accurately the swarm robots align with the desired entrapping pattern and how evenly they are distributed within it.

1. Convergence error (C_{e}):

   Convergence error measures the accuracy of the robots' alignment with the target pattern. It is defined as:

$$C_{\mathrm{e}}(\boldsymbol{g}) = \left(\sum_{t=1}^{T}\sum_{i=1}^{n} d_{\min}(f(g_i), i, t)\right)/T \tag{5.1}$$

   where g_i is the position of the i-th robot, and $f(g_i)$ represents the entrapping pattern, such as a unit circle described by $f(g_i) : g_{i,x}^2 + g_{i,y}^2 - 1 = 0$. The term $d_{\min}(f(g_i), i, t)$ calculates the shortest distance between the robot's position and the expected pattern at each time step t. If all robots perfectly align with the entrapping pattern, C_{e} equals zero. Here, n is the number of robots in the swarm robot system and T is the total number of time steps.
2. Distributed variance (D_{v}):

   Distributed variance evaluates how uniformly the robots are distributed across the entrapping pattern. It is given by:

$$D_{\mathrm{v}} = \sum_{t=1}^{T}\left(1 - \sum_{i=1}^{m_t}\frac{N_{i,t}}{m_t} + \sum_{i=1}^{n}\frac{(d_{i,t}^{\min} - \overline{d_t^{\min}})^2}{n}\right)/T \tag{5.2}$$

   where m_t represents the division of the entrapping pattern into m uniform sub-patterns at the t-th time step. And $N_{i,t}$ indicates whether a robot occupies a sub-pattern (1 for occupied, 0 otherwise). $1 - \sum_{i=1}^{m_t}\frac{N_{i,t}}{m_t}$ is zero if all the swarm robots are located on the m sub-patterns in the t-th time step. $d_{i,t}^{\min}$ is the shortest distance between the i-th robot and its neighbors in the t-th time step. $\overline{d_t^{\min}}$ is the mean of all $d_{i,t}^{\min}$ in the t-th time step. n and T denote the number of swarm robots and the total time step, respectively.

5.2.2 Automated GRN Model Structure Design

This subsection presents a constrained multi-objective genetic programming (CMOGP)-based design automation framework for generating entrapping patterns. The framework utilizes CMOGP to automate the design process, creating effective patterns for dynamic target entrapment. The study also introduces ten predefined basic network motifs, which act as structural templates for swarm behavior design. These motifs provide a foundation for generating diverse and adaptable patterns. In addition, some fitness functions are employed to evaluate the performance of the generated patterns.

5.2.2.1 Basic Network Motifs

An essential part of the design automation framework involves defining fundamental network motifs as building blocks, as studies in systems biology have shown that complex networks are typically made up of interconnected modules [362]. For instance, Bowers et al. [363] identified triplet relationships among proteins governed by specific logic rules, a common feature in GRNs of multicellular organisms. Inspired by these findings, ten predefined network motifs (Positive, Negative, AND, NAND, OR, NOR, ANDN, ORN, XOR, and XNOR) are employed as the foundation for constructing GRNs (see Ref. [17]). In these motifs, θ and k act as regulatory parameters and scaling factors for gene expression, respectively. Figure 5.2 illustrates a GRN tree structure example. The terminal set includes

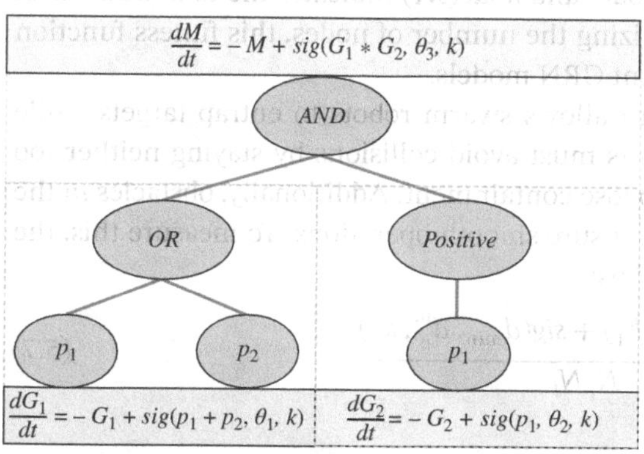

Figure 5.2 A GRN tree structure example.

environmental inputs like p_1 and p_2, which indicate target and obstacle positions, generating corresponding concentration fields. The function set comprises ten predefined motifs. In this example, the GRN model is expressed using the following equations:

$$\frac{dG_1}{dt} = -G_1 + sig(p_1 + p_2, \theta_1, k) \tag{5.3}$$

$$\frac{dG_2}{dt} = -G_2 + sig(p_1, \theta_2, k) \tag{5.4}$$

$$\frac{dG_3}{dt} = -G_3 + sig(G_1 * G_2, \theta_2, k) \tag{5.5}$$

where G_1 combines concentration fields from p_1 (target) and p_2 (obstacle), while G_2 integrates the concentration field from p_1. G_3 further fuses the outputs of G_1 and G_2, generating the final entrapping patterns. The resulting concentration fields produce closed contour lines around the target, which act as candidate entrapping patterns. This process showcases how the GRN model effectively integrates environmental inputs to create adaptive, precise, and dynamic entrapping patterns for swarm robot systems.

5.2.2.2 Fitness Function

A simpler GRN model is favored for its greater clarity and reduced complexity. The simplicity of a GRN is determined by the number of nodes it contains – fewer nodes lead to a less complex and more interpretable model. To measure this, the node count is used as a fitness function, expressed as:

$$f_1 = node(\mathcal{M}) \tag{5.6}$$

where $\mathcal{M}$ represents the GRN model and $node(\mathcal{M})$ indicates the total number of nodes in the network. By minimizing the number of nodes, this fitness function encourages the creation of efficient GRN models.

An effective entrapping pattern allows swarm robots to entrap targets while maintaining safe distances. Robots must avoid collisions by staying neither too close to risk impact nor too far to lose containment. Additionally, obstacles in the environment must be avoided to ensure smooth operations. To measure this, the second fitness function is defined as:

$$f_2 = \Sigma_{i=1}^{N_p} \Sigma_{j=1}^{N_t} \frac{sig(d_{ij}^{pt}, d_{\max}, k_1) + sig(d_{\min}, d_{ij}^{pt}, k_2)}{N_p N_t} \tag{5.7}$$

$$sig(x, \theta, k) = \frac{1}{1 + e^{-k(x-\theta)}} \tag{5.8}$$

where N_p and N_t represent the number of swarm robots and targets, respectively, and d_{ij}^{pt} is the distance between the i-th robot and the j-th target. Parameters d_{min} and d_{max} specify the acceptable minimum and maximum distances, while k_1 and k_2, set to 1 in this study, determine the slope of the sigmoid function. By minimizing f_2, the fitness function ensures that robots maintain an optimal distance, effectively containing the targets without collisions.

When generating an entrapping pattern, it is crucial to account for obstacles and maintain safe distances between neighboring robots to prevent collisions. These considerations are addressed through two key constraints:

1. Obstacle Avoidance Constraint:

$$d_i^{\text{obs}} \geq d_{\text{min}}^{\text{obs}} \tag{5.9}$$

   This constraint ensures that the distance d_i^{obs} between the i-th robot and the nearest obstacle remains greater than the minimum allowable distance $d_{\text{min}}^{\text{obs}}$. Maintaining this buffer helps robots navigate safely without colliding with environmental obstacles.
2. Robot Spacing Constraint:

$$d_{ij}^{\text{robot}} \geq d_{\text{min}}^{\text{robot}} \tag{5.10}$$

   This condition guarantees that the distance d_{ij}^{robot} between any two robots, i and j, is no less than $d_{\text{min}}^{\text{robot}}$, the minimum safe distance. This prevents collisions within the swarm and ensures the stability and integrity of the pattern.

   These constraints are vital for safe and efficient pattern generation. They enable swarm robots to function effectively in environments with obstacles while maintaining proper spacing, ensuring reliable and collision-free operations.

5.2.2.3 Automation Entrapping Pattern Generation Based on CMOGP

The design of a GRN model often relies on a trial-and-error process guided by human experience and intuition. However, this manual approach does not always ensure the model's feasibility or effectiveness for specific applications. To overcome these challenges, the proposed design automation framework provides a systematic and automated process for generating entrapping patterns. As illustrated in Figure 5.3, the framework consists of three main components: (1) hierarchical GRN model construction, (2) automated upper-layer optimization with CMOGP, and (3) application to various scenarios.

1. Hierarchical GRN Model Construction: In Figure 5.3, the new GRN model consists of two distinct layers. The upper layer is entrapping pattern generation layer to focus on designing the optimal trapping patterns based on the dynamic

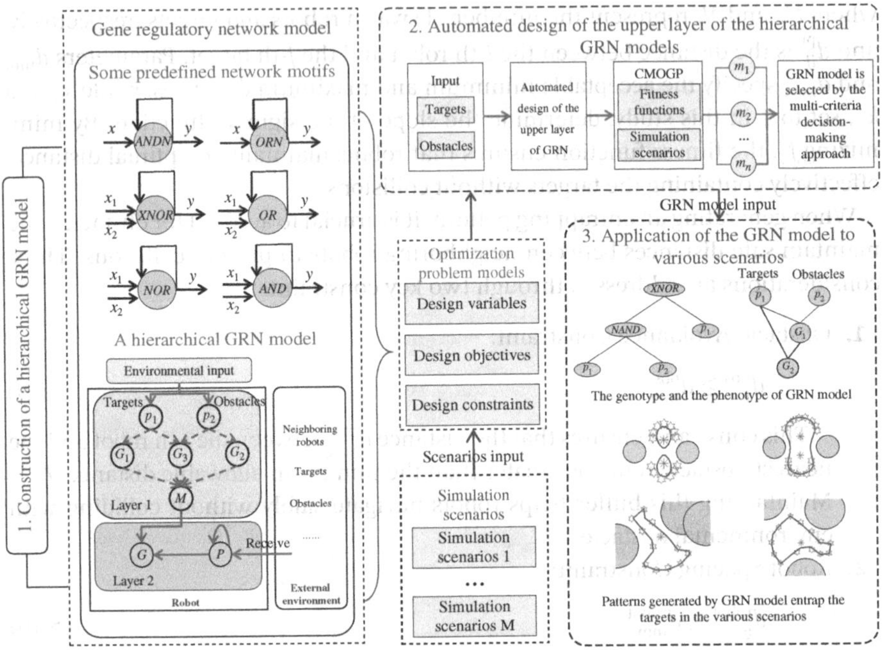

Figure 5.3 The design automation framework.

behavior of the target and environmental conditions. The bottom layer is entrapping pattern formation layer to ensure the coordinated movement and positioning of the swarm robots to execute the generated pattern effectively.

The upper layer focuses on generating adaptable entrapping patterns that respond dynamically to changes in target positions and environmental conditions. Unlike traditional approaches that depend on human intuition and experience, this upper layer leverages CMOGP to automate the design process. The process begins by transforming the target positions (p_1) and obstacle positions (p_2) into a unified morphogen gradient space. Within this space, the GRN model is automatically configured using predefined network motifs as building blocks. These motifs enable the system to construct flexible and efficient patterns tailored to the current environment. From the resulting concentration fields, M is created by combining the individual fields of targets and obstacles. Closed contour lines of concentration values are then extracted from M and transmitted to the lower layer. These contours form the basis for generating precise, collision-free entrapping patterns.

The bottom layer ensures swarm robots align with the entrapping patterns generated by the upper layer. It achieves this by managing robot motion dynamics through two key vectors: ***G*** and [364].

The internal state vector ***P*** processes position data from neighboring robots, targets, and obstacles to prevent collisions. For instance, if a robot gets too close to a target, ***P*** directs it to move away. In contrast, the position vector ***G*** dynamically adjusts based on ***G*** and the generated entrapping pattern, ensuring robots approach their designated positions. The two vectors are governed by the following equations:

$$\frac{d\boldsymbol{G}_i}{dt} = -\boldsymbol{G}_i + \boldsymbol{P}_i + \boldsymbol{C}_i \tag{5.11}$$

$$\frac{d\boldsymbol{P}_i}{dt} = -\boldsymbol{P}_i + \boldsymbol{D}_i + \boldsymbol{T}_i + \boldsymbol{O}_i \tag{5.12}$$

where i denotes the i-th robot and $\boldsymbol{c}_i$ guides it toward the closest position $\boldsymbol{C}_i$ on the entrapping pattern:

$$\boldsymbol{C}_i = -\frac{\boldsymbol{R}_i - \boldsymbol{C}_i}{||\boldsymbol{R}_i - \boldsymbol{C}_i||} \tag{5.13}$$

where $\boldsymbol{R}_i$ is the robot's current position.

To maintain safe operation, the bottom layer incorporates mechanisms to prevent collisions with neighboring robots, targets, and obstacles. The vector $\boldsymbol{D}_i$ ensures robots maintain safe distances from each other:

$$\boldsymbol{D}_i = \sum_{j=1}^{n_i} \boldsymbol{D}_i^j \tag{5.14}$$

$$\boldsymbol{D}_i^j = \frac{\boldsymbol{R}_i - \boldsymbol{R}_j}{||\boldsymbol{R}_i - \boldsymbol{R}_j||} \tag{5.15}$$

where n_i is the number of neighboring robots and $\boldsymbol{D}_i^j$ directs the robot to move away from its neighbors.

The vector $\boldsymbol{T}_i$ prevents robots from colliding with targets while maintaining an appropriate distance:

$$\boldsymbol{T}_i = \sum_{j=1}^{t_i} \boldsymbol{T}_i^j \tag{5.16}$$

$$T_i^j = \begin{cases} -\dfrac{R_i - T_j}{||R_i - T_j||}, & ||R_i - T_j|| > d_{\max} \\ 0, & d_{\min} \leq ||R_i - T_j|| \leq d_{\max} \\ \dfrac{R_i - T_j}{||R_i - T_j||}, & ||R_i - T_j|| < d_{\min} \end{cases} \tag{5.17}$$

where t_i is the number of targets, and $d_{\min}$ and $d_{\max}$ define the allowable distance range. Robots move closer to the targets if they are too far and move away if they get too close.

The vector O_i ensures robots steer clear of obstacles:

$$O_i = \sum_{j=1}^{o_i} O_i^j \tag{5.18}$$

$$O_i^j = \frac{R_i - O_j}{||R_i - O_j||} \tag{5.19}$$

where o_i is the number of obstacles and O_i^j directs the robot to maintain a safe distance from obstacles.

In summary, the hierarchical framework consists of two interconnected layers: the pattern generation layer (upper layer) and the pattern formation layer (bottom layer). The upper layer generates the entrapping patterns that swarm robots must follow to surround and contain dynamic targets. This layer is critical because the effectiveness of the entire framework depends on producing accurate and adaptable patterns. The bottom layer focuses on implementing the generated pattern. It guides the robots to their designated positions using motion dynamics and collision avoidance mechanisms, ensuring precise alignment with the intended pattern. However, the bottom layer's performance relies entirely on the upper layer. If the entrapping pattern produced in the upper layer is ineffective, the robots cannot successfully complete the entrapment task, regardless of their guidance mechanisms. As a result, this study prioritizes improving the pattern generation layer to ensure the creation of robust, flexible, and reliable patterns.

2. Automated Upper Layer Optimization with CMOGP: Designing GRN models manually through trial and error is a time-consuming process that often falls short of delivering optimal solutions for specific tasks. To overcome this limitation, evolutionary algorithms (EAs) are introduced to automate the process, optimizing both the structure and parameters of GRN models. This automation ensures the generation of effective, innovative, and application-specific network designs. The proposed CMOGP method integrates three critical mechanisms: PPS, differential evolution (DE), and multi-criteria decision-making (MCDM). In this framework, GP handles the expression and evolution of GRN structures, while PPS-MOEA/D

identifies the optimal design candidates by balancing multiple objectives. DE optimizes the parameters of these candidates, and MCDM ensures the selection of the most desirable solution based on predefined criteria.

The CMOGP process is outlined in Algorithm 5.1. In line 1, initial population P_0 is generated using the ramped half-and-half method [365], ensuring diversity in the starting solutions. In line 2, DE optimizes the regulatory parameters of each individual in the population. In line 3, fitness functions are evaluated for all individuals to assess their performance. Line 6 is offspring generation. In the line, a new population Q_g is created using crossover and mutation, introducing diversity and exploring new solutions. Lines 7–10 are parameter optimization and fitness evaluation. DE optimizes the parameters for each offspring individual q followed by an evaluation of their fitness based on performance metrics. Lines

---

Algorithm 5.1: Framework of the CMOGP for Entrapping Pattern Generation

Input:
Ten predefined network motifs;
gen_{max}: the maximum generation.
Output: a set of non-dominant and feasible solutions.
1 Initialise: An initial population P_0 of GRN-based models is randomly created using the ramped half-and-half method;
2 Optimal Parameters: The parameters of these models are optimized by DE;
3 Evaluate: The population P_0 is evaluated using fitness functions;
4 Set $gen = 0$;
5 **while** $gen \leq gen_{max}$ **do**
6 The offspring population Q_g is generated by applying genetic operations;
7 **foreach** $q \in Q_g$ **do**
8 The parameters of q are optimised by DE; // q is an individual in Q_g;
9 Each q in Q_g is evaluated using fitness functions;
10 **end**
11 $R_g \leftarrow P_g \cup Q_g$; // P_g is the gth parent population.
12 The new population P_{g+1} from R_g is formed through the PPS mechanism;
13 $gen = gen + 1$;
14 **end**
15 A desired non-dominant and feasible solution is output by applying a MCDM approach.

12–13 are the selection and population update process. Using PPS-MOEA/D, the next-generation P_{g+1} is selected from the combined parent and offspring populations. The generation counter is updated to track progress. In line 15, MCDM is applied to identify the best-performing non-dominated and feasible solutions from the Pareto-optimal front.

3. Application of the GRN Model: To demonstrate the effectiveness of the GRN model designed through the CMOGP framework, the generated entrapping patterns are tested in various challenging environments. These include scenarios with dynamic targets, moving obstacles, and irregular terrain, reflecting real-world complexities. In these experiments, swarm robots are required to form and maintain entrapping patterns that adapt in real time to shifting target positions and environmental changes. The GRN model enables the robots to dynamically adjust their formations while effectively avoiding obstacles and maintaining safe distances between neighboring robots.

5.2.3 Experimental Results

To evaluate the performance and adaptability of the proposed design automation framework, a series of experiments was conducted in different environments. In the first experiment, the framework generated a GRN model (GRN-1) to entrap two moving targets within a predefined environment containing two circular obstacles (OBS-1). This initial scenario tested the GRN model's ability to guide swarm robots in forming accurate entrapping patterns while successfully avoiding collisions with the obstacles. In the second experiment, the same GRN model (GRN-1) was applied to a more complex environment featuring a random distribution of multiple circular obstacles (OBS-3). This scenario evaluated the model's adaptability and robustness when faced with unpredictable changes in both obstacle layout and target movements.

5.2.3.1 Result Analysis in OBS-1

The effectiveness of the proposed design automation framework in the OBS-1 environment is assessed by comparing its performance with two benchmark methods, EH-GRN [364] and TH-GRN [366], using evolutionary process parameters and key fitness function parameters from Ref [17].

The non-dominated solutions generated by the CMOGP are shown in Figure 5.4, where Point A is identified as the knee point, representing the best trade-off between complexity and performance. The corresponding GRN model, GRN-1, is selected for implementation. Figure 5.5 further details this model, with (a) showing the genotype as a GP tree and (b) depicting the phenotype as the GRN structure used in the entrapping pattern generation layer.

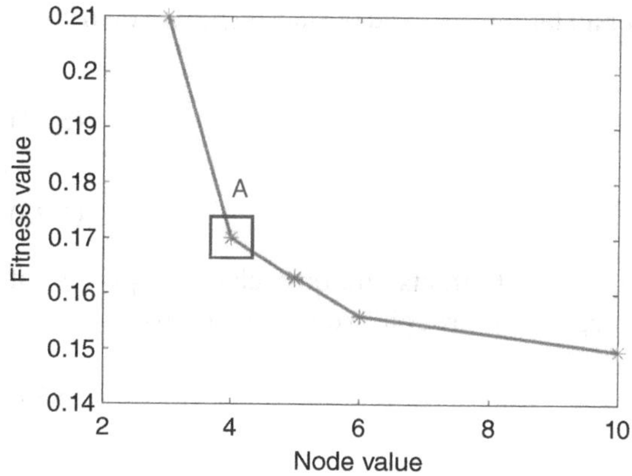

Figure 5.4 Non-dominated solutions generated by CMOGP.

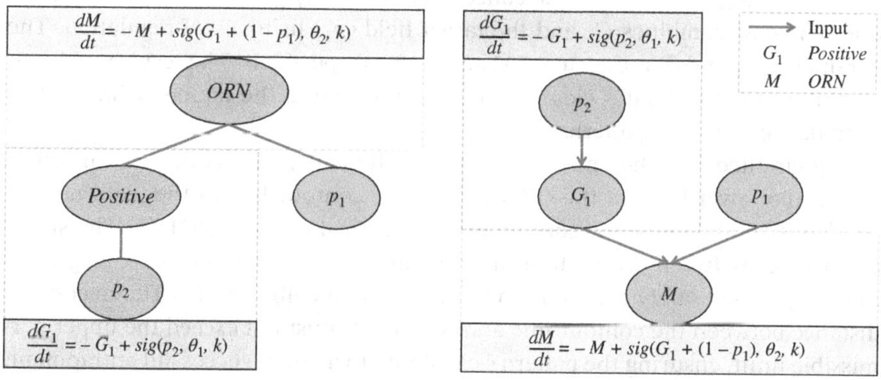

Figure 5.5 Mapping Genotype (GP Tree) to Phenotype (GRN Model) in GRN-1.

GRN-1 can generate entrapment patterns using the following mathematical model. Specifically, it first maps the position of the i-th target (γ_i) and the i-th obstacle (β_i) into concentration fields T_i and O_i, respectively. These fields evolve based on diffusion dynamics defined as:

$$\frac{dT_i}{dt} = \nabla^2 T_i + \gamma_i - T_i \tag{5.20}$$

$$\frac{dO_i}{dt} = \nabla^2 O_i + \beta_i - O_i \tag{5.21}$$

where ∇^2 is the Laplace operator, representing the diffusion effect.

The total concentration fields for all targets (p_1) and obstacles (p_2) are then computed as:

$$p_1 = \Sigma_{i=1}^{N_t} T_i \tag{5.22}$$

$$p_2 = \Sigma_{i=1}^{N_o} O_i \tag{5.23}$$

where N_t and N_o are the total numbers of targets and obstacles. The process of generating entrapping patterns is regulated through two key equations:

$$\frac{dG_1}{dt} = -G_1 + sig(p_2, \theta_1, k) \tag{5.24}$$

$$\frac{dM}{dt} = -M + sig(G_1 + (1 - p_1), \theta_2, k) \tag{5.25}$$

where G_1 integrates the obstacle concentration field (p_2) through *Positive* regulation, while M combines G_1 and the target field (p_1) using *ORN* regulation. The parameters θ_1 and θ_2 are set to 0.702 and 1.189, respectively. The resulting M concentration field generates closed contour lines around the targets, which act as candidate entrapping patterns.

To guarantee that the entrapping pattern efficiently directs the swarm robots within the lower layer of the GRN model, the contour line of the pattern must be identified and chosen according to three critical criteria [367]. (1) The shortest distance between the contour line and any target or obstacle must be greater than a specified safety threshold to avoid potential collisions. (2) The maximum distance between the contour line and the target must not exceed the upper permissible limit, ensuring the pattern's capability to achieve successful entrapment. (3) The contour line should be as centered on the target as possible, ensuring precise and reliable encirclement. Meeting these criteria ensures that the selected contour line serves as a robust entrapping pattern, effectively guiding swarm robots to form safe, accurate, and adaptable formations.

Figure 5.6 presents the EH-GRN model (EH-GRN-1), where all 14 regulatory parameters θ_i ($i = 1, ..., 14$) have been optimized using the CMA-ES. The resulting parameter values are: $\theta_1 = 0.6958$, $\theta_2 = 1$, $\theta_3 = 0.5758$, $\theta_4 = 0.1183$, $\theta_5 = 0.0432$, $\theta_6 = 0$, $\theta_7 = 0.4614$, $\theta_8 = 0$, $\theta_9 = 0.8710$, $\theta_{10} = 0.3903$, $\theta_{11} = 0.5621$, $\theta_{12} = 0.7134$, $\theta_{13} = 0.8934$. Similarly, Figure 5.7 illustrates the TH-GRN model, where three regulatory parameters θ_i ($i = 1, 2, 3$) are also optimized using CMA-ES. The corresponding parameter values are $\theta_1 = 0.5004$, $\theta_2 = 0.0822$, $\theta_3 = 1.9732$.

Figure 5.8 illustrates how GRN-1 dynamically generates adaptive entrapping patterns to effectively entrap two moving targets as they navigate through two circular obstacles. As the targets move closer to the obstacles, GRN-1 initially generates an ellipsoidal pattern, shown in solid line in Figure 5.8a. This pattern

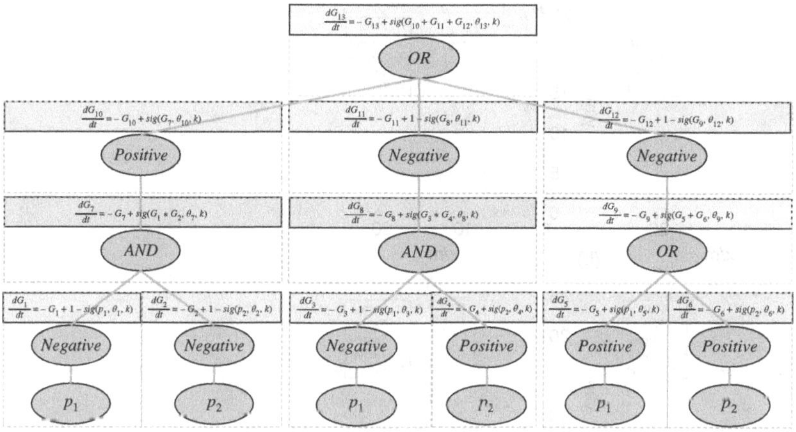

Figure 5.6 The Structure of EH-GRN-1 in OBS-1.

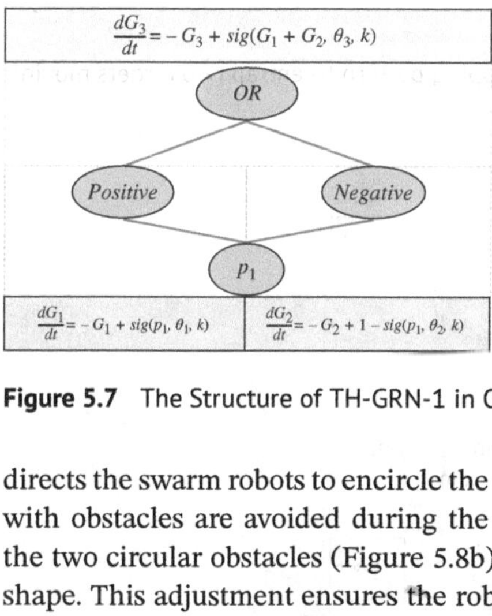

Figure 5.7 The Structure of TH-GRN-1 in OBS-1.

directs the swarm robots to encircle the targets effectively, ensuring that collisions with obstacles are avoided during the process. When the targets pass between the two circular obstacles (Figure 5.8b), the pattern evolves into a dumbbell-like shape. This adjustment ensures the robots maintain entrapment efficiency without colliding with the obstacles. The change occurs because the concentration field created by the obstacles compresses the concentration field around the targets, influencing the pattern's shape. As the targets move away from the obstacles, the entrapping pattern gradually reverts to an ellipsoidal form, as depicted in Figure 5.8c, d.

Figure 5.9 illustrates the entrapping patterns generated by EH-GRN-1 as two targets navigate through two circular obstacles. When the targets approach the obstacles (Figure 5.9a), the generated pattern entraps both the targets and the obstacles. This occurs because the concentration fields of the obstacles merge with those of

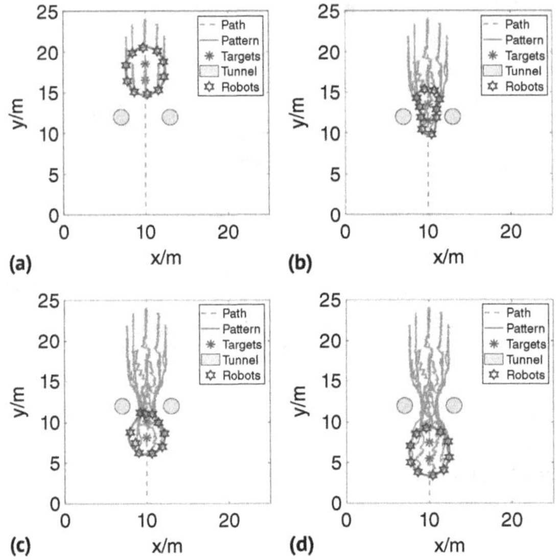

Figure 5.8 GRN-1 generates a dynamic entrapping pattern to entrap two targets moving through a dual-obstacle environment.

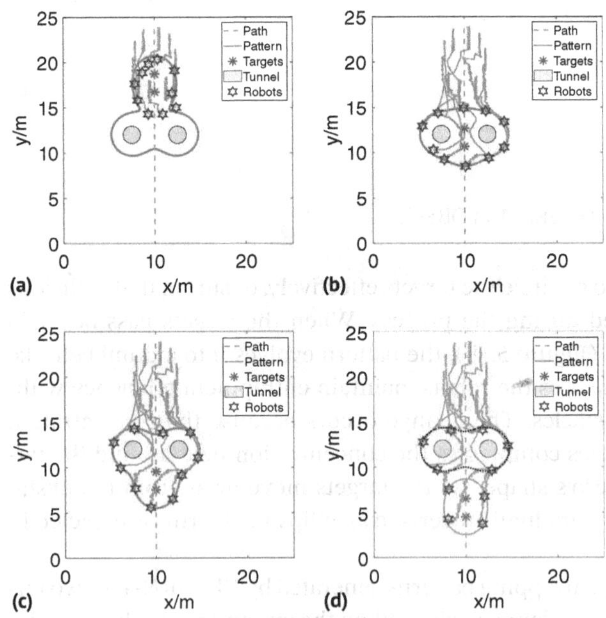

Figure 5.9 The trapping pattern created by EH-GRN-1 for two targets passing through two obstacles.

the targets. While the swarm robots successfully entrap the targets, their distribution is uneven across the pattern. As the targets move between the obstacles (Figure 5.9b), the model produces an ellipsoidal pattern, but the robots cluster on one side, compromising the pattern's efficiency. When the targets move farther away (Figure 5.9c, d), the pattern continues to encompass both the targets and the obstacles. However, some robots remain concentrated near the obstacles instead of realigning to follow the targets, resulting in incomplete entrapment. Although the robots initially achieve partial success, their uneven distribution and misalignment weaken long-term performance. If the targets continue moving under similar conditions, fewer robots will actively participate in the entrapment, increasing the likelihood of task failure. This highlights a limitation in the EH-GRN-1 model, particularly its reduced adaptability and effectiveness in dynamic and obstacle-rich environments.

Figure 5.10 shows the entrapping patterns generated by TH-GRN to guide swarm robots in the OBS-1 environment. As the targets move through the obstacles, the generated pattern remains unchanged and simply follows the targets (Figure 5.10b). This occurs because the upper layer of TH-GRN generates patterns

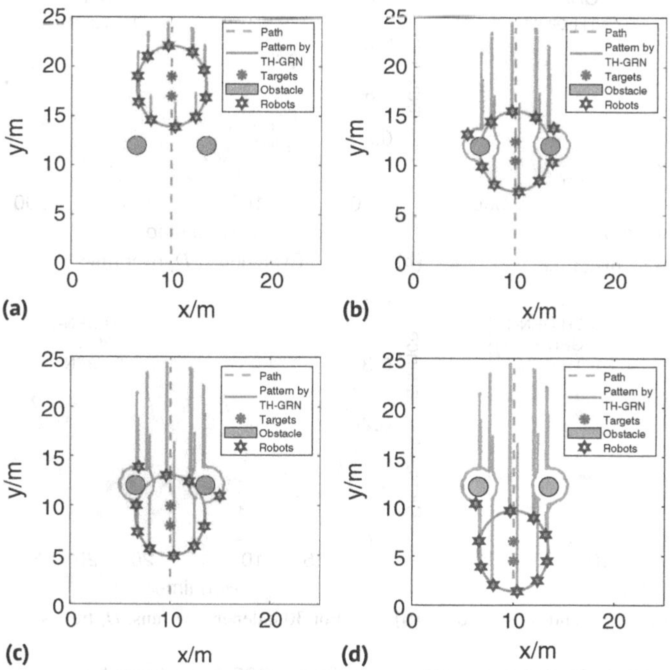

Figure 5.10 While the two targets navigate through the obstacles, the TH-GRN generates a pattern that entraps them.

based solely on target positions, ignoring the influence of obstacles. Meanwhile, the interlayer uses obstacle information to guide the robots, enabling them to navigate efficiently and avoid collisions (Figure 5.10c, d). The resulting pattern consistently maintains a circular-like shape, which does not adapt to changes in the environment. While this unchanged shape temporarily enables entrapment, it proves less effective in complex environments where obstacles obstruct the robots' movements. As demonstrated in Figure 5.15, this rigidity can prevent robots from effectively surrounding the targets, ultimately compromising the entrapment task. This highlights TH-GRN's limitation in handling dynamic and obstacle-constrained scenarios due to its lack of pattern adaptability.

The performance of GRN-1 is compared against EH-GRN-1 and TH-GRN by analyzing the changes in C_e and D_v over time, as shown in Figure 5.11a, b. When no obstacles are present, GRN-1 achieves near-zero values for these metrics, reflecting effective swarm distribution within the entrapping pattern. However, when targets move through obstacles, C_e and D_v fluctuate significantly

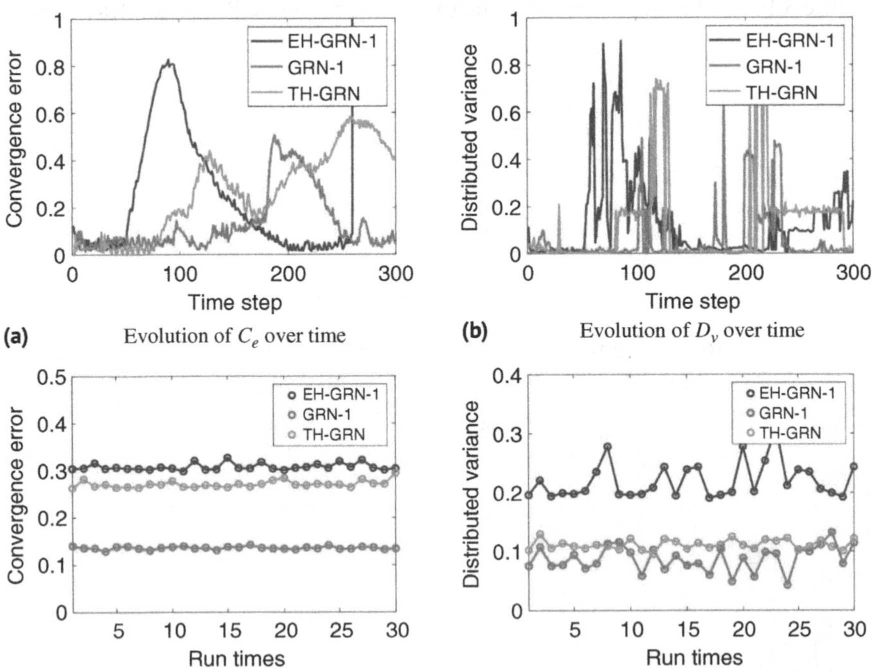

(a) Evolution of C_e over time
(b) Evolution of D_v over time
(c) For 30 independent runs, C_e trends observed
(d) For 30 independent runs, D_v trends

Figure 5.11 Statistical analysis of two evaluation metrics for OBS-1, achieved by EH-GRN-1, TH-GRN, and GRN-1.

during time steps 170–250. This is due to GRN-1's need to adjust its pattern to the obstacles, causing delays before the swarm re-establishes an even distribution (Figure 5.8b, c). In comparison, EH-GRN-1 shows significant fluctuations in C_e and D_v during time steps 70–130 as it adapts its pattern to encircle both targets and obstacles, leading to a larger, more complex formation (Figure 5.9a). Additionally, after time step 265, C_e rises sharply because some robots are blocked by obstacles while the targets move away (Figure 5.9c, d). Statistical evaluations over 30 independent runs, as summarized in Figure 5.11c, d and Table 5.1, reveal that GRN-1 consistently achieves lower values for C_e and D_v than EH-GRN-1 and TH-GRN, demonstrating its superior performance in OBS-1 scenarios.

5.2.3.2 Result Analysis in OBS-3

To assess the transferability of the GRN model developed by the proposed framework, GRN-1 was tested in a new scenario featuring numerous randomly distributed obstacles. For a thorough comparison, EH-GRN-1 and TH-GRN were also evaluated under the same conditions in the OBS-3 environment.

When the GRN model developed by the proposed framework is applied to a new scenario, an intriguing behavior emerges, as illustrated in Figure 5.12. The entrapping pattern dynamically adapts to irregular shapes in response to changes in the environment surrounding the escaping target, enabling the swarm robots to navigate around obstacles and effectively trap the target. This behavior is reminiscent of biological phenomena, such as white blood cells altering their shape to move between crowded cells and engulf bacteria, as shown in Figure 5.13. In contrast, Figure 5.14 reveals that while EH-GRN-1 can merge its entrapping pattern with nearby obstacles to trap the target, this approach often fails when the target escapes. The merged pattern becomes overly large, simultaneously encircling both the target and obstacles, which distracts the swarm and prevents a firm entrapment. Similarly, as shown in Figure 5.15b, c, TH-GRN struggles when the target moves between obstacles, as many robots are blocked and fail to form an effective pattern. This occurs because TH-GRN's entrapping pattern lacks the flexibility to adapt to environmental changes, leaving the swarm unable to respond effectively to obstacles.

The performance of GRN-1, EH-GRN-1, and TH-GRN in the OBS-3 environment is evaluated by analyzing the evolution of C_e and D_v over time, as shown in Figure 5.16a, b. GRN-1, developed through the proposed framework, consistently outperforms EH-GRN-1 and TH-GRN. This advantage arises because EH-GRN-1 generates an oversized entrapping pattern that simultaneously encircles both the target and obstacles. When obstacles are large or clustered, this results in inefficient patterns, as depicted in Figure 5.14b, c. Meanwhile, TH-GRN's rigid, circular-like pattern limits its adaptability to dynamic obstacle environments, further reducing its effectiveness, as shown in Figure 5.15b, c. Consequently, robots

Table 5.1 Two Statistical metrics are calculated for EH-GRN-1 (EH), TH-GRN (TH), and GRN-1 on OBS-1 and OBS-3.

Convergence Error	instance		GRN-1	EH [364]	TH [366]	t-test (p-value)	
						GRN-1 vs EH	GRN-1 vs TH
	OBS-1	mean	**1.3581E-01**	3.0819E-01	2.7143E-01	6.35E-41	7.33E-37
		std	3.1688E-03	7.2254E-03	7.1285E-03		
	OBS-3	mean	**8.7313E-02**	2.2014E-01	1.1120E-01	1.35E-18	8.34E-25
		std	2.1643E-02	3.0694E-02	8.4375E-03		
Distributed Variance	**instance**		**GRN-1**	**EH [364]**	**TH [366]**	**t-test (p-value)**	
						GRN-1 vs EH	**GRN-1 vs TH**
	OBS-1	mean	**8.7313E-02**	2.2014E-01	1.1120E-01	1.61E-19	4.25E-06
		std	2.1643E-02	3.0694E-02	8.4375E-03		
	OBS-3	mean	**3.4045E-02**	4.0491E-01	3.3760E-01	1.09E-36	3.11E-27
		std	4.2570E-03	2.3030E-02	4.0088E-02		

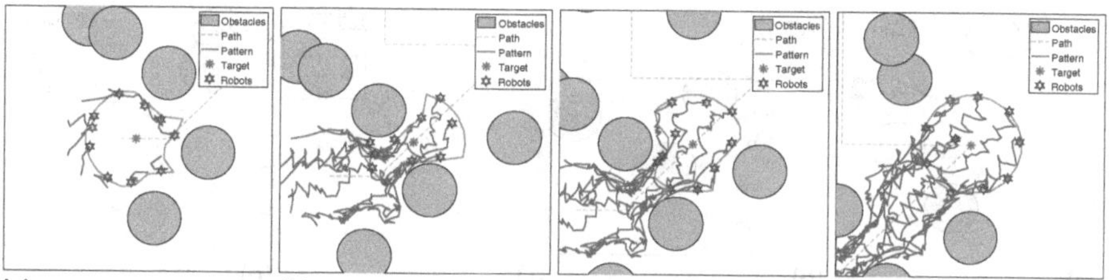

(a) (b) (c) (d)

Figure 5.12 In OBS-3, where numerous circular obstacles are randomly distributed, the GRN-1 generates an entrapping pattern that effectively entraps the passing target.

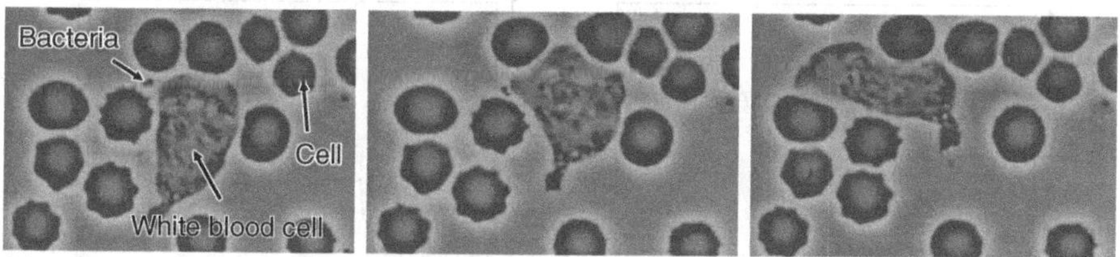

(a) (b) (c)

Figure 5.13 The white blood cell performs phagocytosis, wherein it surrounds and breaks down a bacterium during the process. [17] / Elsevier.

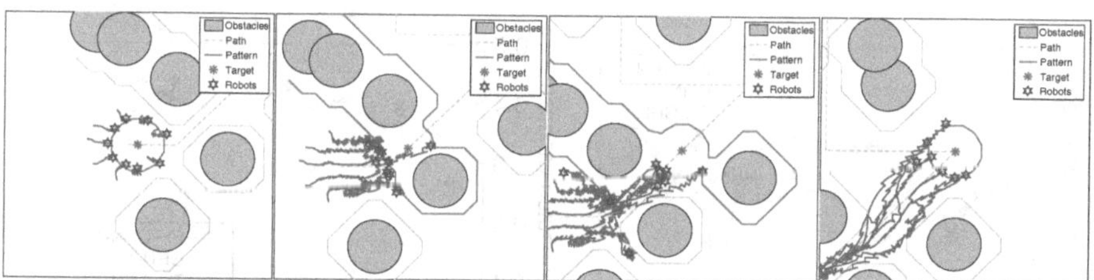

(a) (b) (c) (d)

Figure 5.14 In OBS-3, where numerous circular obstacles are randomly distributed, the EH-GRN-1 generates an entrapping pattern that entraps the passing target.

using EH-GRN-1 and TH-GRN achieve poorer convergence and distribution compared to GRN-1. Figure 5.16c, d presents results from 30 independent runs, showing that C_e and D_v values for GRN-1 are consistently lower than those for EH-GRN-1 and TH-GRN in every trial. Statistical analysis, supported by p-values in Table 5.1, confirms these differences are significant. These results demonstrate GRN-1's superior performance and its successful transfer to the complex obstacle-dense OBS-3 environment.

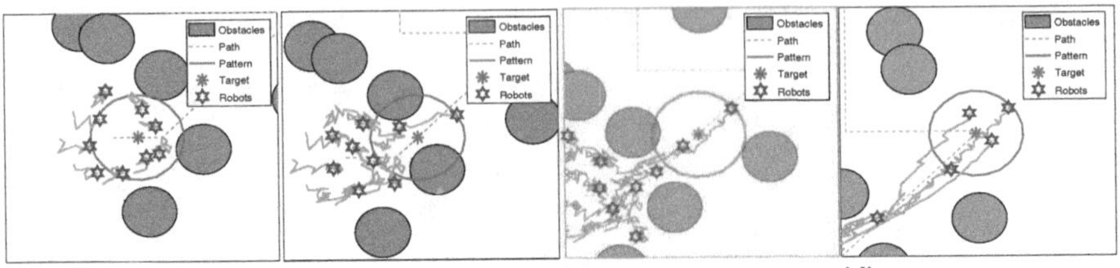

(a) (b) (c) (d)

Figure 5.15 In OBS-3, where numerous circular obstacles are randomly distributed, the TH-GRN generates an entrapping pattern that entraps the passing target.

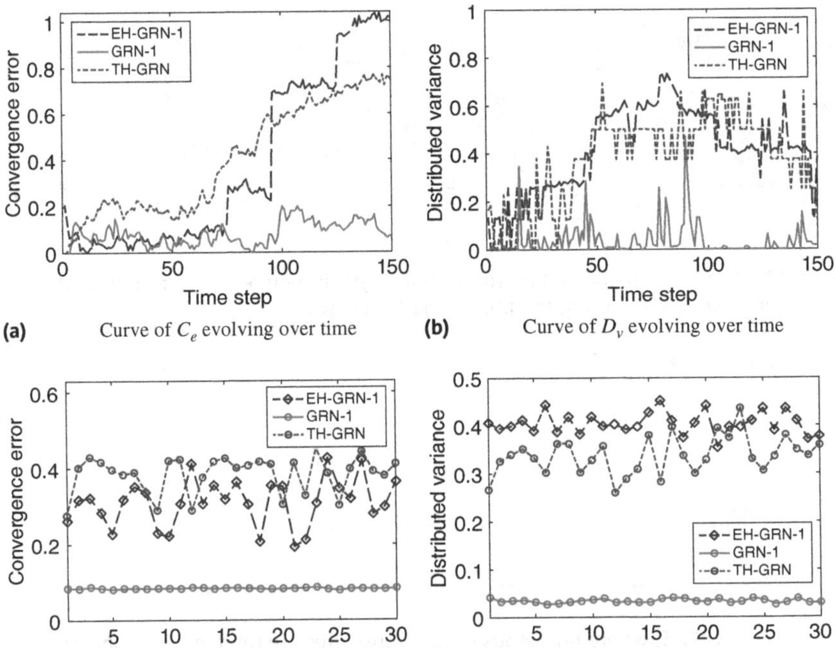

(a) Curve of C_e evolving over time
(b) Curve of D_v evolving over time
(c) For 30 independent runs, C_v trends
(d) For 30 independent runs, D_v trends

Figure 5.16 Statistical analysis of two evaluation metrics for OBS-3, achieved by EH-GRN-1, TH-GRN, and GRN-1.

5.2.3.3 Experiments Results Using E-puck

A. Entrapping multiple moving targets

The proposed method effectively generates dynamic entrapping patterns for multiple targets without predefined formations, adapting seamlessly to changes in target positions. For example, when clustered targets require a large encircling

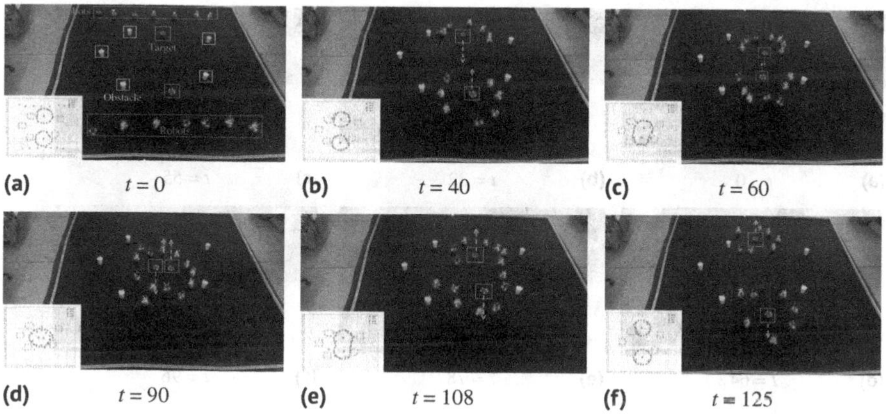

Figure 5.17 The proposed method generates an entrapping pattern that successfully captures two targets within an environment containing obstacles. [17] / Elsevier.

pattern that later splits into smaller formations to trap scattered targets escaping in different directions, the method performs this adjustment autonomously. A physical experiment with E-puck robots, shown in Figure 5.17, demonstrates this capability. Initially, two swarms of robots (in red lights) form separate circles around two targets (in blue lights). As the targets move closer together, the two circles merge into a single larger circle, trapping both targets. When the targets later escape in opposite directions and move far enough apart, the large circle divides into two smaller ones, each encircling a single target. Notably, these transitions occur without specific commands, resembling biological processes such as cell fusion and division.

B. Capturing targets effectively even when certain robots experience functional failures

The proposed method utilizes a fully distributed control scheme, ensuring the swarm remains operational even when some robots fail. As shown in Figure 5.18, the initial snapshots (Figure 5.18a, b) depict the swarm robots (in red) successfully encircling the target (in blue). When the target attempts to escape through obstacles (in white), the swarm adapts dynamically, maintaining the entrapping formation while following the target (Figure 5.18c). Later, two robots fail and become stationary (in yellow), as seen in Figure 5.18d. Despite these failures, the remaining robots effectively adjust their movements, bypassing both the obstacles and the non-functional robots. The swarm eventually re-establishes a secure entrapment around the target, as illustrated in Figure 5.18e, f. This highlights the robustness of the method and its ability to adapt seamlessly to unforeseen disruptions within the swarm.

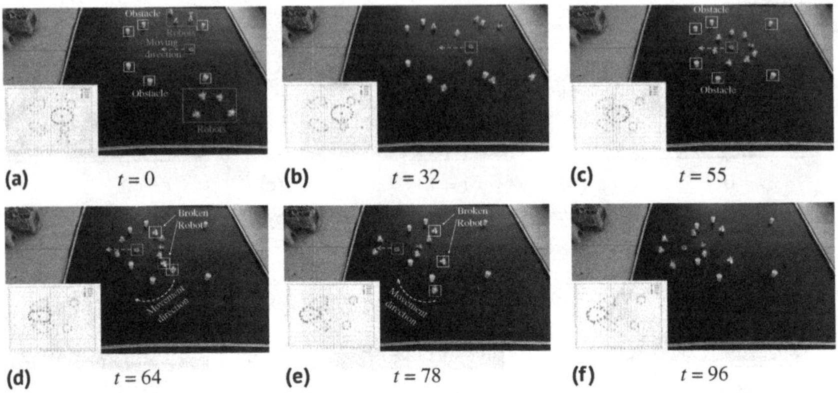

Figure 5.18 Achieving target entrapment successfully even in scenarios where some robots are unable to operate properly. [17] / Elsevier.

5.2.4 Conclusion

Designing GRN models for swarm robot self-organization relies heavily on human intuition and expertise. Additionally, models designed for specific scenarios often fail to generalize to new environments. To overcome these challenges, this research introduces an automated framework based on CMOGP. This framework utilizes CMOGP to construct GRN models by systematically reconfiguring a set of predefined network motifs, enabling efficient synthesis and optimization of the models. By systematically exploring evolved models within the PF, the framework provides valuable insights for designing GRN models capable of generating patterns for specific tasks. The experimental findings demonstrate that the GRN models, developed through the evolutionary process, are capable of effectively directing swarm robots to entrap targets, even in diverse and highly complex scenarios. These evolved models exhibit three essential features: (1) they dynamically adapt to environmental changes, generating irregular patterns that effectively avoid obstacles while entrapping targets; (2) they facilitate group merging and division, allowing for the simultaneous entrapment of multiple targets without predefined instructions; (3) the fully distributed control scheme ensures robustness, enabling the swarm to maintain functionality and complete the task even if some robots fail.

5.3 Swarm Control in Communication-denied Environments

The above study has confirmed the feasibility of the dynamic transformation mechanism of swarm pattern. Building on this foundation, the subsection has conducted research on unmanned aerial vehicle (UAV) swarm control in communication-denied environments. In practical scenarios, UAV swarms often encounter various uncertainties, including environmental noise, communication disruptions, strong electromagnetic interference, and even uncertainties in individual UAV behaviors [368]. These factors significantly increase the difficulty of swarm control. Moreover, many existing swarm control methods are based on assumptions of a known environment and reliable communication conditions. However, when faced with unknown environments and communication-denied conditions, these methods often fail [369]. To address these challenges, this study presents a UAV swarm control approach that relies solely on visual data in communication-denied environments. The proposed VG-Swarm builds upon the feasibility of the dynamic transformation mechanism of swarm patterns and further optimizes it with practical experiments. It successfully addresses the challenges of designing effective swarm control strategies in communication-denied environments, where other approaches or technological pathways have struggled. The innovation of our method lies in the fact that UAV swarms do not require communication. Instead, they rely solely on vision (without GPS signals or positional information) to achieve intelligent environmental perception. This approach enables tasks such as target search, stable tracking, and comprehensive encirclement. In this section, the study has innovatively proposed a vision-based UAV swarm control model, referred to as VG-Swarm [40], overcoming the challenge of controlling UAV swarms solely with visual input in communication-denied environments. Additionally, this innovative technology has numerous civilian applications, including urban counterterrorism, target search and rescue, and post-disaster recovery operations.

5.3.1 Method Architecture of VG-Swarm

The proposed framework, as shown in Figure 5.19, consists of three core components: (1) the visual perception module, which serves as the sensing layer of the system, responsible for collecting and analyzing environmental information, including two key functions: object recognition and spatial localization; (2) the V-GRN module, which utilizes visual information to generate the swarm formation of the UAV swarm; and (3) the UAV behavior design module, which generates specific behavior strategies for each UAV based on the outputs from the visual perception module and the V-GRN module.

5.3.1.1 Object Detection

Figure 5.20 provides a detailed illustration of how each UAV utilizes the visual perception module to construct a local coordinate system and achieve group formation. Specifically, in Figure 5.20a, each UAV establishes a right-handed local coordinate system based on its perception, with the forward direction defined as the Y-axis to standardize its movement and environmental sensing.

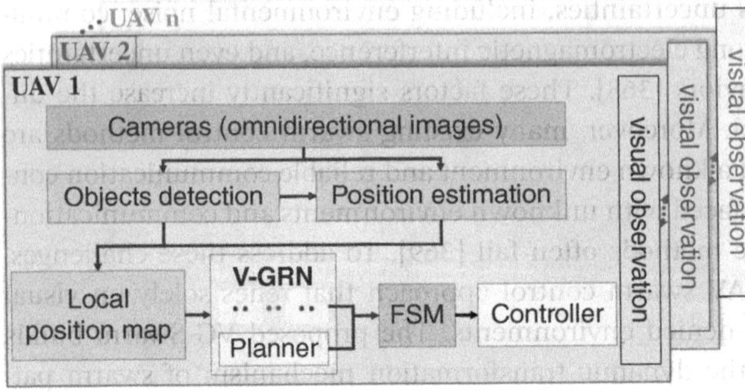

Figure 5.19 The architectural framework of the proposed swarm control method. For visual perception, omnidirectional images are captured using four cameras. The detected object positions are stored in the local position map. Using this data, V-GRN creates concentration fields within the perception range, which are then used to develop candidate entrapping patterns. Movement control is handled by the planner in V-GRN, while the finite state machine module oversees action outputs and transitions between behavior states.

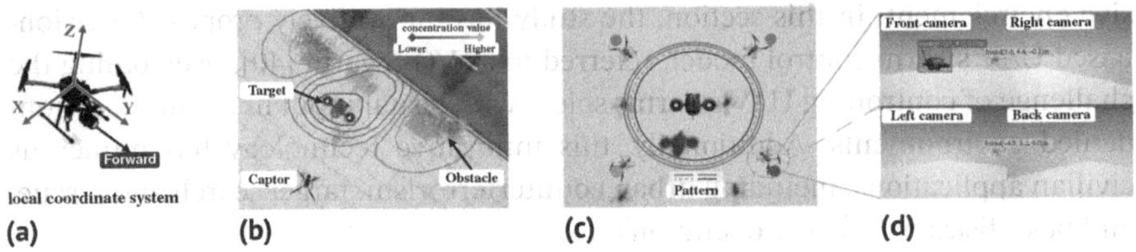

Figure 5.20 The proposed approach is illustrated through a schematic diagram that explains the coordinate system and pattern formation process. [40] / IEEE.

In Figure 5.20b, a captor UAV generates a concentration field in an environment with obstacles, visualized using multiple contour lines. In this concentration field, areas around the target indicate lower concentrations, representing regions that are farther away or less influenced, while areas around the obstacle signify higher concentrations, usually near targets or critical regions. Figure 5.20c further illustrates four independent entrapping patterns generated by four UAVs in an open environment. These patterns are designed to ideally overlap to achieve optimal target encirclement, thereby improving coordination efficiency and control accuracy. Finally, Figure 5.20d presents a detailed view of the environment as captured by the UAV located in the lower-right corner of Figure 5.20c through its visual perception module, providing an intuitive basis for analyzing the visual input and decision-making process of a single UAV. The key to achieving the above functionalities lies in the UAV's ability to possess excellent visual perception capabilities. Therefore, the following content will provide a detailed explanation of the visual perception method employed in this approach.

YOLO models are widely recognized for their effectiveness in detection tasks due to their high accuracy and fast inference speed. However, their direct application in drone-based scenarios poses significant challenges, primarily due to the typically small size of detected objects and the limited computational resources available on onboard devices. This study enhances the detection performance for tiny objects by introducing modifications to the YOLOv5s network architecture, tailored specifically for deployment on onboard devices [370].

This study observed that some prediction heads in the original YOLOv5s network rely on low-resolution, high-level feature maps, rendering them suboptimal for detecting small objects. To address this limitation, the study introduced an additional path that combines bottom-up and top-down processing and replaced one prediction head to process lower-level, higher-resolution feature maps, resulting in a 0.7% *mAP* improvement with a trade-off of a 9.5% increase in parameters. To further optimize efficiency, the number of convolutional kernels was reduced by 60%, retaining 40% of the original YOLOv5s network, to address the limited feature diversity in small object detection, which is attributed to the blurring of image details.

Despite the use of a lightweight model, real-time omnidirectional visual detection on embedded edge devices is constrained to 5.7 FPS. To address this limitation, the study implemented a set of targeted optimization measures:

1. The inference speed of the custom model was accelerated with the TensorRT engine, and data preparation for the detector was enhanced through multiprocessing, achieving a performance of 17.56 FPS.
2. Images captured from four cameras were stitched into a unified 1280×960 frame, allowing the detector to perform omnidirectional detection on a single frame, resulting in a performance of 45.46 FPS.

5.3.1.2 Object Position Estimation

This study improves online object detection by incorporating camera field of view (FOV) data with detection results to estimate the spatial positions of identified objects. Previous research by Haseeb et al. [371] demonstrated a linear correlation between the inverse of the bounding box (Bbox) size and the distance to the object. Building on this, our study identifies a power–law relationship between the object distance and the Bbox area, which is accurately represented by Equation 5.26.

$$D = \alpha \cdot s^n \tag{5.26}$$

where the object distance D is correlated with the pixel area of the Bbox s. The values of α and n are calculated through the least squares method using various data sets comprising s and D.

In addition, Griffin et al. [372] proposed estimating object distance (D) based on camera motion and the height of the Bbox. However, for small objects, the height of the Bbox changes only slightly, even with substantial camera movement, limiting the method's accuracy. To address this, the study demonstrates that D can be more effectively calculated using the Bbox area (s) combined with the camera's relative motion, as outlined in Equation 5.27.

$$D_i = \frac{C_{z_j} - C_{z_i}}{1 - \sqrt{\frac{s_i}{s_j}}} \tag{5.27}$$

where at i-th time step, the object distance (D_i) is calculated using the camera's relative movement toward the target, represented by ($C_{z_j} - C_{z_i}$). s_i is the Bbox area of i-th time step and s_j is the Bbox area of j-th time step. The indices i and j can correspond to any timestamps, provided the distances from the observation points to the target are not identical.

This study recreated a real-world target in a simulated environment at a 1:1 scale, gathering multiple Bbox areas (s) and corresponding distances (D) calculated using Equation 5.27. The collected data were used to develop a distance estimation function (Equation 5.26), which was later applied directly to real-world experiments. Camera motion was tracked using the UWB system, while object detection provided the corresponding s values to complete the process. This approach allows a UAV to estimate object distances by applying s values to pre-established distance estimation models (each tailored to specific detection classes) even when the UAV remains stationary.

The object's relative spatial coordinates (x, y, z) are determined within the local coordinate frame using the camera's horizontal and vertical fields of view (FOV_h, FOV_v) and the estimated distance D, as shown in Figure 5.20a.

$$\begin{aligned} x &= \frac{2 \cdot p_x \cdot D \cdot \sin \frac{FOV_h}{2}}{W} \\ y &= \sqrt{D^2 - x^2} \\ z &= \frac{2 \cdot p_y \cdot D \cdot \sin \frac{FOV_v}{2}}{H} \end{aligned} \tag{5.28}$$

where p_x and p_y represent the pixel offsets of the Bbox center relative to the image center, while W and H denote the image's width and height in pixels, respectively.

5.3.1.3 Local Position Map Generation

The local position map plays a critical role in maintaining the relative spatial location of objects within the UAV's perception range. As detailed in Equation 5.29, the map stores several key attributes for each detected object, including its position coordinates $[x, y, z]$, its object type (OT), and a detection loss counter (LT).

$$object := [x, y, z, OT, LT] \tag{5.29}$$

The OT represents the classification of the object as determined by the visual perception system. The detection loss counter LT, on the other hand, tracks the number of consecutive frames in which an object is not detected. If an object remains undetected for 50 consecutive frames, it is removed from the local position map, ensuring the map remains accurate and free from outdated information. Given that modern visual perception systems operate at high frequencies, the spatial position of a single object typically changes minimally over short intervals. This allows the local position map to employ self-motion compensation to adjust for the UAV's movement. By doing so, it effectively addresses temporary detection failures, ensuring that transient lapses in the object detector do not compromise the consistency and reliability of the local position map.

5.3.1.4 Vision-based Gene Regulatory Network

In contrast to conventional GRN models [364, 373], the proposed V-GRN model (Figure 5.21) operates without a communication component or access to global information. Instead, it relies solely on inputs (p_1, p_2, and p_3) derived from

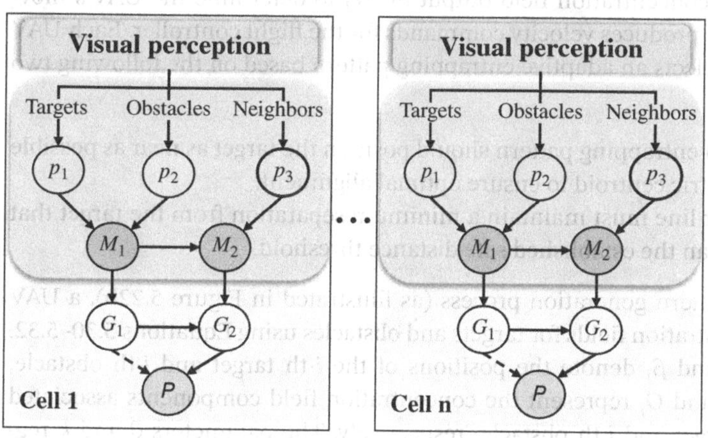

Figure 5.21 Proposed V-GRN model.

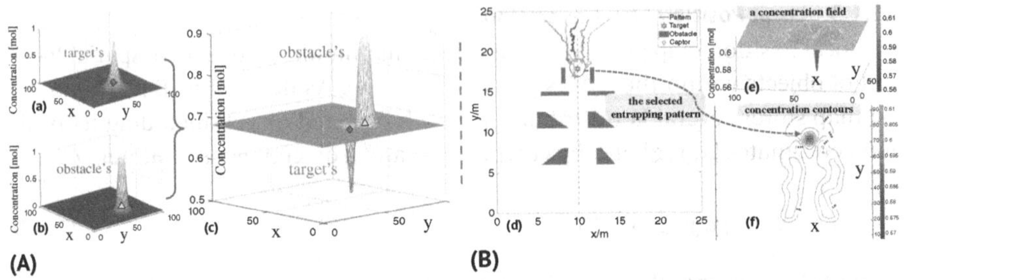

Figure 5.22 A schematic diagram of concentration field generation and swarm entrapping patterns using V-GRN. In Part A, subfigure (a) illustrates the concentration field generated by a target, while (b) depicts the concentration field created by an obstacle or a neighboring UAV; (c) combines these individual fields, presenting the resulting fused concentration field. In part B, (d) illustrates a scenario where five robots dynamically work together to entrap a target within an obstacle-filled environment. In (e), a concentration field generated by a robot is depicted, representing the area within its perception range. Lastly, (f) displays several contour lines derived from the concentration field in (e), with the contour lines surrounding the target serving as potential entrapping pattern candidates.

local information. By defining a local coordinate system centered on the UAV, p_1, p_2, and p_3 are used to create concentration fields representing targets, obstacles, and neighboring agents, respectively (Figure 5.22). These generated concentration fields are progressively fused through two components: protein M_1 (illustrated in Figure 5.22a) and protein M_2. The output concentration field from M_1 is processed by G_1, which generates candidate entrapping patterns. These patterns are represented as the concentration contours entrapping a target. Meanwhile, the planner G_2 uses the concentration field output by M_2 to determine the UAV's moving direction and produces velocity commands for the flight controller. Each UAV independently selects an adaptive entrapping pattern based on the following two conditions:

1. The selected entrapping pattern should position the target as near as possible to its geometric centroid to ensure optimal alignment.
2. The contour line must maintain a minimum separation from the target that is greater than the established safe distance threshold.

During the pattern generation process (as illustrated in Figure 5.22b), a UAV produces concentration fields for targets and obstacles using Equations 5.30–5.32. Specifically, γ_i and β_j denote the positions of the i-th target and j-th obstacle, respectively. T_i and O_j represent the concentration field components associated with the i-th target and j-th obstacle, respectively. The parameters θ and k regulate the field generation, and ∇^2 is the Laplace operator, defined as the second derivative of T_i and O_j. A fused concentration field (M) is calculated using

Equations 5.33 and 5.34 with the flag parameter set to 1, effectively ignoring neighbor concentration fields (as illustrated in Figure 5.22e). This fused field is then utilized to generate candidate entrapping patterns, which correspond to the concentration contours around the target (as illustrated in Figure 5.22f) and align with G_1 in Figure 5.21. It is important to note that the first derivative terms in Equations 5.30–5.32 are zero because this study excludes the spatiotemporal dynamics of protein diffusion. The equations governing this process are as follows:

$$\frac{dT_i}{dt} = \nabla^2 T_i + \gamma_i - T_i \tag{5.30}$$

$$\frac{dO_j}{dt} = \nabla^2 O_j + \beta_j - O_j \tag{5.31}$$

$$\frac{dN_m}{dt} = \nabla^2 N_m + \eta_m - N_m \tag{5.32}$$

$$\begin{aligned} M = {} & sig(1 - (\sum_{i=1}^{n_t} T_i)^2, \theta, k) + sig((\sum_{j=1}^{n_o} O_j)^2, \theta, k) \\ & +(1 - Flag) \cdot sig((\sum_{m=1}^{n_n} N_m)^2, \theta, k) \end{aligned} \tag{5.33}$$

$$sig(x, \theta, k) = \frac{1}{1 + e^{-k(x-\theta)}} \tag{5.34}$$

To calculate the moving orientation, a UAV generates the concentration field for all neighboring robots using Equations 5.32 and 5.34. Here, η_m represents the position of the m-th neighbor and N_m denotes the concentration field component created by that neighbor. The fused concentration field (M_2) is obtained by combining the output of M_1 with the neighbor concentration field through Equation 5.33, setting $Flag = 0$. Importantly, neighboring robots are treated as obstacles when determining the moving direction. When generating entrapping patterns, however, the concentration field of neighbors is ignored. As a result, the candidate entrapping patterns are influenced solely by targets and obstacles, without any contribution from the neighbors' fields. In the fused concentration field created by M_2, the optimal moving direction is identified using a sampling-based method. Specifically, the UAV evaluates points on five concentric circles, starting with a radius of 0.5 m from the center of its rigid body. Four additional circles are sampled, each with an incrementally larger radius, increasing by 0.6 m outward. To ensure a more globally optimal movement, the UAV evaluates 180 possible directions, spaced at intervals of 2°, and selects the direction with the minimum total concentration values across all five circles.

Planner G_2 generates velocity control signals for the UAV's flight controller based on the selected pattern and target direction. The velocity vector G_2, defined in Equation 5.35, specifies the relative motion components in each direction, focusing on directional guidance rather than exact magnitudes.

$$\vec{v} = [k \cdot m_s \cdot \sin\theta, k \cdot m_s \cdot \cos\theta, k \cdot \Delta z] \tag{5.35}$$

where θ represents the angle between the UAV's forward direction and the target, while k serves as a scaling factor. The parameter m_s determines the step length for dynamic motion and gradually decreases to zero as the UAV reaches the designated pattern. Additionally, Δz denotes the height difference relative to the target. To control the UAV's dynamic velocity, a sigmoid function is employed for smooth and efficient regulation.

5.3.1.5 Behavior Design of Swarm UAVs

This study developed a uniform finite state machine model (Figure 5.23) for each UAV to operate in complex environments, focusing on the entrapment task. The model includes searching and entrapping states, with the latter covering behaviors like approaching, staying away from targets, and maintaining encirclement. The conditions for activation between states are also defined.

Each captor UAV begins in the "Initialization" state, transitioning to the "Searching Targets" state to locate targets. In this state, the UAV moves randomly at a speed of 0.4 m/s, navigating toward areas of lower concentration within the fused concentration field. It continues searching until a target is detected. Upon detecting a target, the UAV leverages the V-GRN model to generate and select an appropriate entrapping pattern. It then follows the velocity command vector $\vec{v}$ to

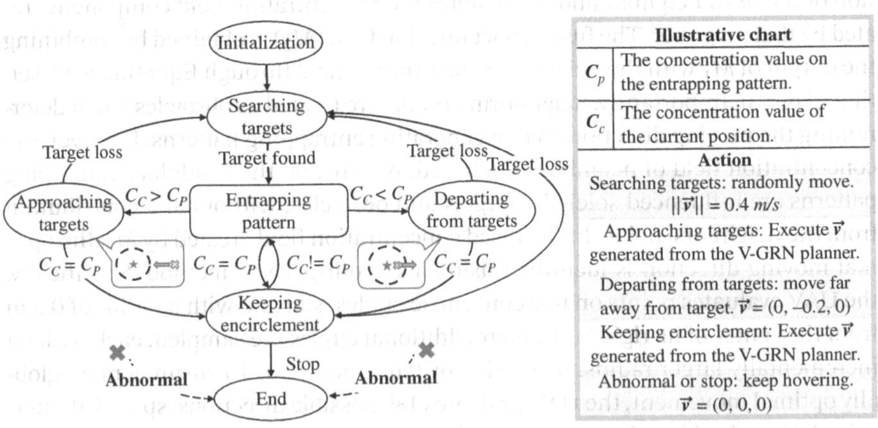

Figure 5.23 The finite state machine model used in this study.

either approach or encircle the target. The UAV monitors its progress by comparing the current concentration value C_C with the target concentration value of the selected pattern C_P. Based on this comparison, it transitions to the appropriate next state, such as approaching the target, disengaging, or continuing with the entrapping process. If the target becomes stationary, the UAV maintains its current entrapping pattern to ensure stable engagement. This systematic process allows for dynamic and adaptive responses to the target's behavior.

5.3.2 Experimental Setup

5.3.2.1 Hardware Setup

1. Simulation Experiment Setup: To better simulate real-world environments, this study uses AirSim [374] and Unreal Engine 4 platforms for simulation experiments. A custom-designed quadcopter model is employed in the simulation scenario, built to a 1:1 scale based on the actual dimensions of the DJI Matrice 200, as shown in Figure 5.24 (left). This customized quadcopter is equipped with cameras mounted on its front, rear, left, and right sides, enabling comprehensive environmental perception. In the simulations, the quadcopter serves as the captor, while a dual-rotor drone acts as the target for testing and validation.

2. Real-world Experiment Setup: The H350 quadcopter is used as the captor, while the DJI MATRICE 200 serves as the target. The captor is equipped with four wide-angle RGB cameras (two connected via CSI and two via USB) providing a 120° horizontal FOV and a 90° vertical FOV, ensuring full omnidirectional visibility. The cameras capture frames at a resolution of 640 × 480 at 20 Hz, which are stitched together to create a single panoramic image. Additionally, each captor is outfitted with a Jetson Xavier NX with SSD storage and a DJI N3 autopilot for processing and control (Figure 5.24, right).

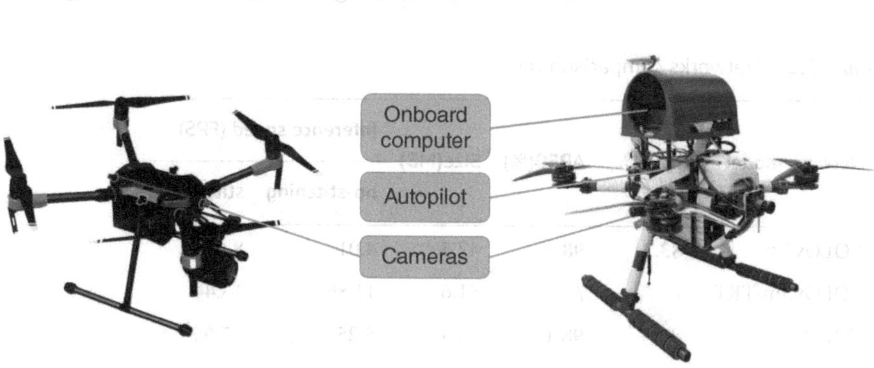

Figure 5.24 The UAVs used in this study and their configurations. [40] / IEEE.

5.3.2.2 Object Detection and Position Estimation

The customized network model consists of 269 layers, including 1.3 million parameters (17% of YOLOv5s), and has a compact size of 3.01 MB. The network is not pretrained and directly trains the detector model using both simulation and real-world datasets for a total of 200 epochs. The synthetic dataset includes 1849 images with more than 5300 Bboxes, while the real-world dataset contains 5567 images with more than 5700 Bboxes.

For the detection tasks, the settings are configured with an IoU threshold of 0.5, a confidence threshold of 0.6, and an inference resolution of 1280 for the combined images. The network model is optimized using a TensorRT engine with 16-bit floating-point precision to enhance performance. Table 5.2 compares the performance of YOLOv5m, YOLOv5s, and the proposed approach on the Jetson Xavier NX, where "TRT" indicates the TensorRT-optimized version of each model.

5.3.2.3 Target Entrapping Experiments

To demonstrate the effectiveness of the proposed V-GRN method in guiding a swarm of UAVs to form adaptive entrapping formations within dynamic environments, this research conducts a variety of simulation experiments using the AirSim simulator. These experiments encompass several distinct scenarios, such as open environments, constrained narrow roadways, and settings with randomly placed obstacles, designed to evaluate the method's performance under diverse conditions.

In the experimental setup designed to achieve target encirclement, the captors are initially deployed at positions trailing behind the target. These captors utilize onboard visual sensors to autonomously detect and determine the target's location. Once the target has been successfully identified, the captors shift their operational mode from searching to actively executing an encirclement strategy. This

Table 5.2 Networks comparison results.

Network model	mAP(%)	AP50(%)	Size(MB)	Inference speed (FPS)	
				no-stitching	stitching
YOLOv5m	**83.2**	**98.5**	42.5	4.01	8.57
YOLOv5m-TRT	/	/	51.6	11.86	13.44
YOLOv5s	80.8	98.1	14.4	5.25	17.65
YOLOv5s-TRT	/	/	18.7	13.83	31.25
Ours	80.8	97.6	**3.01**	5.71	24.17
Ours-TRT	/	/	7.89	**17.56**	**45.46**

strategy, rooted in the principles of the concentration field, is inherently adaptive, dynamically adjusting to environmental variations in real-time. Simulation studies revealed that the captors can achieve a maximum speed of 5 m/s under idealized conditions. However, in practical implementations involving physical experiments, their speed is limited to 1 m/s. This limitation is due to a substantial processing delay, exceeding 500 ms, primarily caused by the computational demands of processing the raw image data captured by the CSI cameras. To evaluate the performance of the proposed method in various experimental scenarios, we selected average entrapment distance error, average speed, and success rate (SR) as the evaluation metrics. The specific definitions and descriptions of these metrics can be found in Ref. [40].

5.3.3 Experimental Results

5.3.3.1 Target Detection

This research presents the results of the model comparison in Table 5.2, highlighting the advantages of the used prediction head. Despite reducing the number of convolution kernels by half, the detector retains the same mean average precision as YOLOv5s. Furthermore, when using the TRT mode with stitched images, the detector exhibits a significant speed improvement, operating 45.5% faster. In real-world experiments, the detector trained for 200 epochs achieved an impressive average precision of 97.6% (AP_{50}) and 80.78% (AP) on a hold-out validation dataset of 650 images, with the confidence threshold set at $p^{conf} = 0.1\%$.

The accuracy of monocular visual relative localization is predominantly influenced by the precision of the detected Bboxes. However, the detector generates slightly varying Bbox outputs when observing the target from different perspectives, introducing inconsistencies in localization. Consequently, the monocular position estimation system achieves a positioning accuracy with an error rate of less than 11% for distances up to 10 m, and less than 13% for distances up to 15 m.

5.3.3.2 Swarm Entrapment

The proposed algorithm was rigorously evaluated in the UE4 simulation under three scenarios: open environments, constrained narrow roadways, and settings with randomly placed obstacles. A summary of the experimental results is presented below:

1. In the open environment: After identifying the target, the captors operate at variable speeds ranging from 0 m/s to 5 m/s to carry out the entrapment. For instance, as illustrated in Figure 5.25, the captors successfully surrounded a stationary target located over 10 m away in less than 10 s. Following this, the target began to move at a slow pace, prompting the captors to maintain their

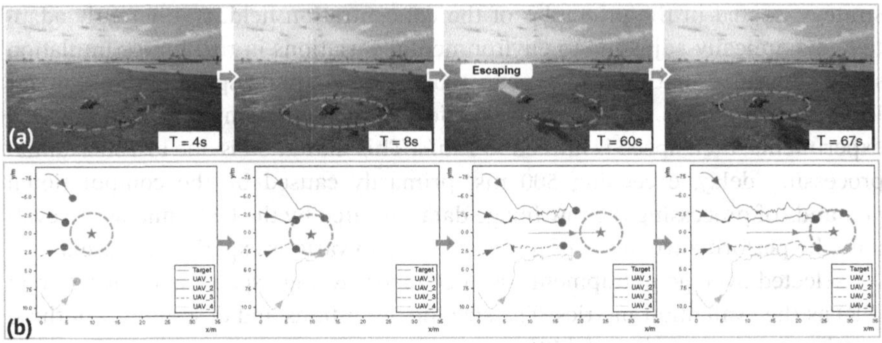

Figure 5.25 The complete process of a UAV swarm entrapping a moving target is illustrated: (a) represents the experimental procedure; (b) displays the flight paths of all UAVs, which were tracked using GPS positional data within the simulation environment. [40] / IEEE.

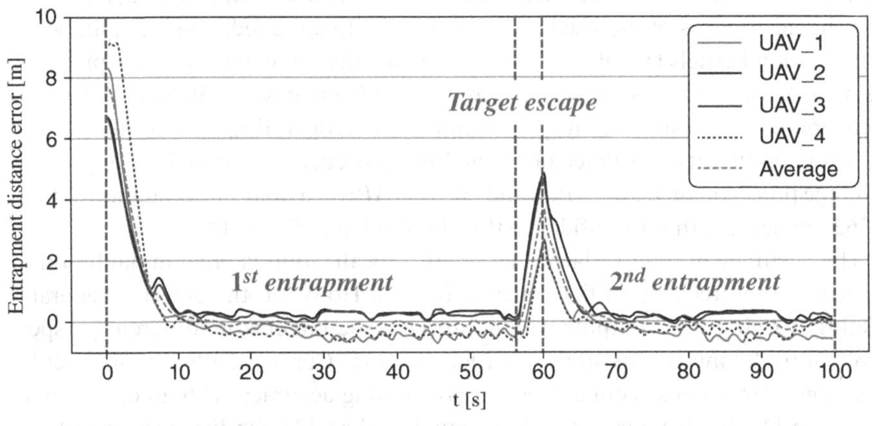

Figure 5.26 The average entrapment distance error of UAVs.

encirclement formation. After a short period, the target attempted to break free from the encirclement by accelerating to a high speed. In response, the captors quickly pursued and managed to re-establish the entrapment in a brief timeframe. The entrapment distance error recorded throughout the experiment, depicted in Figure 5.26, highlights the capability of the proposed method to effectively generate dynamic entrapment behaviors. Additionally, this study evaluated the SRs of entrapment across more than 20 experiments, considering varying average distances and time durations, as summarized in Table 5.3.

2. In the constrained narrow roadways: The captors initially navigate through parallel obstacle lanes while maintaining an encirclement formation, as illustrated in Figure 5.27a. The presence of obstacles causes the encirclement pattern

Table 5.3 Experimental result of simulation.

	Success rate (%)			
Initial average distance (m)	**6 s**	**10 s**	**14 s**	**Average speed (m/s)**
6	0	93	100	0.76
10	80	100	100	2.01
14	73	100	100	2.53

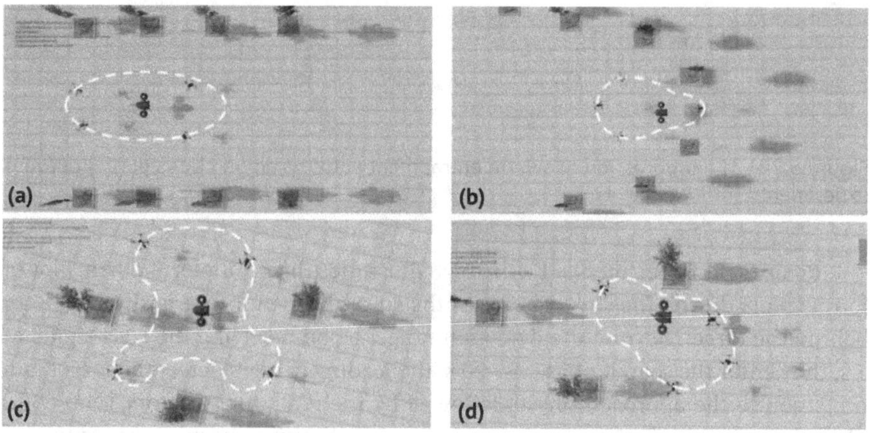

Figure 5.27 The swarm of UAVs generates distinct entrapping patterns tailored to various obstacle scenarios. [40] / IEEE.

to shift from a circular to an elliptical shape. Guided by real-time input from the V-GRN-generated concentration field, the captors adjust their formation to sustain effective target encirclement. As the captors and the target enter conical obstacle lanes (Figure 5.27b), the concentration contours around the target transform into a droplet-like shape. In response, the captors dynamically adapt their formation to maintain the entrapment, ensuring continued pursuit and containment of the target.

3. In the Random Obstacle Scenario: As the captors and the target move into an environment with obstacles distributed randomly throughout the scene (Figure 5.27c, d), the proposed V-GRN system continues to generate adaptive entrapment strategies. This allows the captors to dynamically adjust their formations, transitioning seamlessly between different encirclement patterns to maintain effective pursuit and containment of the target.

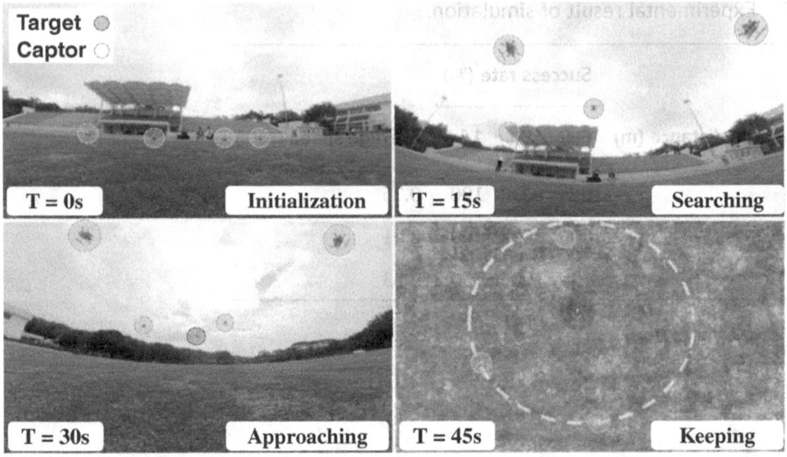

Figure 5.28 The progression of swarm entrapment patterns among UAVs during outdoor experiments. [40] / IEEE.

4. Real-world Experiments: Figure 5.28 illustrates the positions and behavior of four captors during the experiment. At the start, the captors are positioned 15 m behind the target, beyond the detection range. Upon receiving the start signal at 5 s, they enter the search phase. By 20 s, the leading captors detect the target and transition to the approach phase, followed by the remaining captors, who complete the transition by 30 s. The swarm then synchronizes its movements to successfully execute the entrapment task. The entire process is completed within an average time of 50 s, without any collisions. Furthermore, this study expands the number of UAVs to 10, and as shown in Figure 5.29, it successfully achieves target entrapment.

To evaluate the scalability of the method, the experiment was repeated using a larger swarm of 10 UAVs. The results, shown in Figure 5.29, demonstrate the system's ability to handle increased swarm sizes effectively. Additional details and footage of these experiments can be found in the online video: https://youtube/R1SD1YlC-dc

5.3.4 Conclusion

This study presents the V-GRN model, which incorporates an omnidirectional visual perception algorithm to enable multiple UAVs to execute target entrapment and form adaptive swarm formations in dynamic environments. The method is designed to function without inter-agent communication or reliance on external positioning systems, ensuring reliability in GNSS-denied environments or under unstable communication conditions. While GPS was used in this study for ground

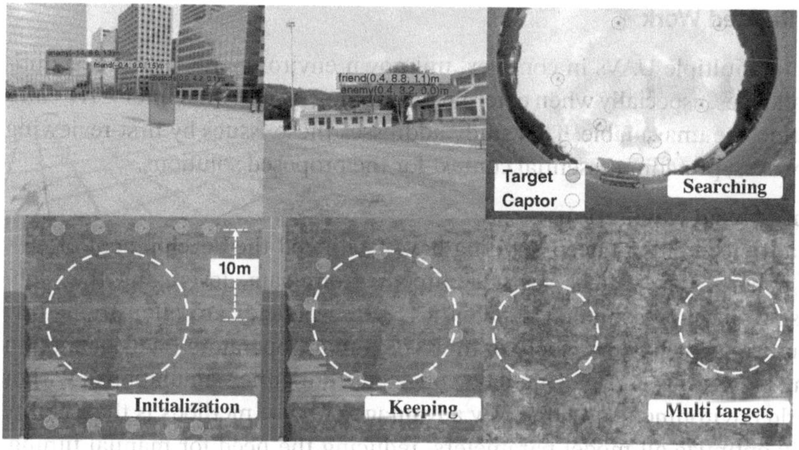

Figure 5.29 Ten UAVs entrap two dynamic targets. [40] / IEEE.

speed measurement and hovering, the system can be adapted to use visual-inertial odometry or optical flow modules to further enhance its resilience. Both simulations and real-world experiments confirm the robustness and effectiveness of the proposed approach, showing its capability to handle partial agent failures while successfully completing the entrapment task.

5.4 Vision-based Distributed Multi-UAV Collision Avoidance

This study presents a novel collision-avoidance method that leverages DRL in a decentralized framework [375]. By utilizing onboard visual sensors and inertial measurement data, the proposed method enables multiple UAVs to efficiently navigate and plan their flight paths. Unlike traditional methods, it eliminates the need for external positioning systems, pre-built maps, or additional computational resources. To address the inefficiencies of reinforcement learning models that rely on pixel-based input, the proposed method incorporates an autoencoder. This innovation allows for the simultaneous learning of latent state representations and the development of an effective collision-avoidance policy. Experimental results demonstrate that the proposed method outperforms reinforcement learning models based on convolutional neural networks (CNNs), offering greater robustness and stability. Additionally, the learned strategy successfully generates safe, collision-free trajectories for multi-UAV systems. Even in complex scenarios where UAV density increases significantly, the system consistently delivers reliable navigation performance, highlighting its scalability and adaptability in challenging environments.

5.4.1 Related Work

Navigating multiple UAVs in complex, unknown environments presents significant challenges, especially when external positioning systems and inter-UAV communication are unavailable. This study addresses these issues by first reviewing related works, providing essential context for the proposed solutions.

Learning-based Navigation:
Recent advancements in deep learning have facilitated the development of navigation policies that can be learned directly from data, leveraging the technology's strong representation capabilities. These policies are often trained by imitating expert behaviors, such as those of human operators [376] or motion planning algorithms [377], using data collected from either simulations [378] or real-world environments [379]. A key advantage of these methods is their ability to jointly optimize all model parameters, reducing the need for manual tuning. However, challenges remain, particularly in designing appropriate loss functions for training. Additionally, expert trajectories used for imitation are often suboptimal, which can hinder the convergence of these models to robust and generalizable solutions. DRL-based methods offer an alternative approach. For instance, Faust et al. [380] incorporate global information from traditional planners to train collision-avoidance policies, while Leiva et al. [381] use two-dimensional point clouds from range sensors to achieve map-less navigation. Ma et al. [382] propose a collision-avoidance framework for UAVs in three-dimensional environments that combines saliency map learning with reinforcement learning. These DRL methods are framed as Markov Decision Processes (MDPs) and rely on sensor inputs to directly learn navigation strategies. They excel in environments without pre-built maps and are less dependent on sensor accuracy, making them highly adaptable and efficient in dynamic settings.

Decentralized Multi-robot Collision Avoidance:
Decentralized multi-robot navigation has recently seen significant advancements, with many approaches drawing inspiration from the optimal reciprocal collision avoidance framework [383]. This framework relies on precise detection of the shape, position, and velocity of surrounding robots and obstacles to ensure collision avoidance [384], [385]. To mitigate this reliance on perfect sensing, Godoy et al. [386] introduced communication to share velocity information among robots. However, this method has limitations, including the need for precise parameter tuning and reduced robustness in environments with unreliable communication. In response, DRL-based collision-avoidance systems have emerged as an alternative [387], [388]. These systems use sensors to estimate nearby agents and obstacles for path planning. However, they often require complex state-estimation pipelines. Notable works, such as [389] and its extension [390], have developed fully decentralized collision-avoidance policies based on raw sensor inputs like two-dimensional laser scans and inertial measurements. While effective

in two-dimensional workspaces for UGVs, these methods are unsuitable for UAVs due to the higher degrees of freedom involved in UAV navigation. Bastien et al. [391] demonstrated that vision-based models can achieve organized swarm behavior without relying on spatial representations or explicit collision-avoidance mechanisms. Using vision-based input, this study applies DRL to enable decentralized multi-UAV collision avoidance in three-dimensional environments, ensuring scalability and robustness.

5.4.2 Proposed Method

This subsection provides a comprehensive explanation of the proposed method. It begins by outlining the DRL framework specifically designed for multi-UAV systems. Following this, it delves into the details of the neural network architecture used to implement the control policy, highlighting its design and functionality.

5.4.2.1 Problem Formulation

In this study, each UAV relies solely on its onboard sensors for collision detection and avoidance. The collision avoidance task for multi-UAV systems is formulated as a Partially Observable Markov Decision Process (POMDP), characterized by a 6-tuple $(\mathcal{S}, \mathcal{A}, \mathcal{T}, \mathcal{R}, \Omega, \mathcal{O})$. In this model, $\mathcal{S}$ represents the set of environment states, which are only partially observable to the UAVs ($s \in \mathcal{S}$). $\mathcal{A}$ denotes the set of possible actions available to the UAVs ($a \in \mathcal{A}$). The transition function $\mathcal{T}$ describes the likelihood of transitioning from one state to another, while the reward function $\mathcal{R}$ assigns a numerical reward to each state–action pair ($\mathcal{S} \times \mathcal{A} \rightarrow \mathbb{R}$). The observations available to the UAVs are expressed by the set $\Omega(o \in \Omega)$, and $\mathcal{O}$ provides the probabilistic distribution of observations conditioned on the current state ($o \sim \mathcal{O}(s)$).

In this study, we consider a scenario where a set of N UAVs operate within a shared environment. At each timestep t, the i-th UAV ($1 \leq i \leq N$) receives an observation denoted as $\boldsymbol{oi}^t$, which it uses to calculate and execute an action $\boldsymbol{ai}^t$. This process ensures that all UAVs move collision-free from their current positions $\boldsymbol{pi}^t$ toward their respective goal positions $\boldsymbol{gi}^t$. Throughout the navigation process, each UAV relies solely on its own observations and has no access to the states of other UAVs. The observation vector of each UAV consists of three components: $\boldsymbol{o}^t = [\boldsymbol{o}^t z, \boldsymbol{o}^t g, \boldsymbol{o}^t v]$. Here, $\boldsymbol{o}^t z$ represents images of the surrounding environment captured by the vision sensor, $\boldsymbol{o}^t g$ denotes the UAV's relative position to its goal, and $\boldsymbol{o}^t v$ indicates its current velocity.

Based on the partial observation $\boldsymbol{o}^t$, each UAV independently computes its action $\boldsymbol{a}^t$, which is sampled from a shared policy π:

$$\boldsymbol{a}^t \backsim \pi_\theta(\cdot|\boldsymbol{o}^t) \tag{5.36}$$

where θ is the parameter vector of the policy. The action $\boldsymbol{a}^t$ specifies the UAV's velocity command, guiding it toward its goal while avoiding collisions. This velocity remains effective for a fixed time horizon Δt, after which the UAV receives an updated observation $\boldsymbol{o}^{t+1}$ and repeats the process.

5.4.2.2 Reinforcement Learning Setting

In this fully decentralized framework, each UAV operates independently, relying exclusively on its own observations to make real-time decisions. Unlike centralized approaches, the state transition model $\mathcal{T}$, which represents the kinematics and dynamics of the UAVs, is not explicitly used. Instead, UAVs refine their decision-making policies by optimizing the reward function, allowing them to adapt to dynamic environments and navigate collision-free. This methodology ensures that UAVs can function autonomously and flexibly without requiring external coordination or global state information. The following sections provide an in-depth analysis of this approach, detailing the framework's design, its implementation, and how it achieves robust and efficient collision avoidance in multi-UAV systems.

1. Observation Space: The observation $\boldsymbol{o}_z^t$ consists of depth images captured at time step t by the forward-facing camera mounted on the UAV. These images provide information about the distance from the UAV's viewpoint to the surfaces of objects in the scene. To enhance the extraction of both spatial and temporal features, this study combines three consecutive depth frames into a single stacked input. The observation $\boldsymbol{o}_g^t$ represents the goal position in a self-centered coordinate system, as the UAV is controlled using body frame commands: forward velocity ($\boldsymbol{v}_x^{cmd}$), climb rate ($\boldsymbol{v}_z^{cmd}$), and angular velocity for steering ($\boldsymbol{v}_\omega^{cmd}$). Additionally, $\boldsymbol{o}_v^t$ describes the UAV's current velocity state, expressed as $\boldsymbol{o}_v^t = [\boldsymbol{v}_x^t, \boldsymbol{v}_z^t, \boldsymbol{v}_\omega^t]$, encompassing forward speed, vertical speed, and rotational speed.

2. Action Space: To ensure that the movement of the UAV is both smooth and controllable, the action space is defined as a range of constrained velocities within a continuous space. The action $a = [\boldsymbol{v}_x^{cmd}, \boldsymbol{v}_z^{cmd}, \boldsymbol{v}_\omega^{cmd}]$, generated by the policy network $\pi(s)$, includes two linear velocities (forward and climb) and one angular velocity (steering). The forward velocity $\boldsymbol{v}_x^{cmd}$ is restricted to the range (0.0, 2.0) m/s, the climb velocity $\boldsymbol{v}_z^{cmd}$ is limited to (−0.5, 0.5) m/s, and the steering velocity $\boldsymbol{v}_\omega^{cmd}$ is constrained to (−0.5, 0.5) rad/s. It is important to note that backward movement ($\boldsymbol{v}_x^{cmd} < 0$ m/s) is not permitted. This restriction is due to the UAV's reliance on a front-facing camera, which cannot capture the environment behind the UAV, making it impossible to detect obstacles or prevent collisions when moving backward.

3. Reward Function: The reward function is designed to maximize both path efficiency and safety for all UAVs. It guides the UAVs to learn an optimal policy

that balances goal-reaching and collision avoidance. The overall reward at each timestep is expressed as:

$$\boldsymbol{r}^t = \boldsymbol{r}^t_{\text{goal}} + \boldsymbol{r}^t_{\text{avoid}} \tag{5.37}$$

The component $\boldsymbol{r}^t_{\text{goal}}$ encourages UAVs to efficiently reach their assigned goals. If a UAV comes within 0.5 m of its target position at the current timestep, it receives a fixed reward, $\boldsymbol{r}_{\text{arrival}}$. Otherwise, the reward is determined by the reduction in the distance to the goal from the previous timestep to the current timestep, scaled by a weighting factor ω_{goal} to promote efficient path planning. The goal-reaching reward is defined as:

$$\boldsymbol{r}^t_{\text{goal}} = \begin{cases} \boldsymbol{r}_{\text{arrival}} & \text{if } \| \boldsymbol{p}^t_i - \boldsymbol{g}^t_i \| < 0.5 \\ \omega_{\text{goal}} \cdot (\| \boldsymbol{p}^{t-1}_i - \boldsymbol{g}^t_i \| - \| \boldsymbol{p}^t_i - \boldsymbol{g}^t_i \|) & \text{otherwise} \end{cases} \tag{5.38}$$

where $\boldsymbol{pi}^t$ is the UAV's current position, $\boldsymbol{gi}^t$ is its goal position, and ω_{goal} is a hyper-parameter to encourage more direct and efficient movement.

The term $\boldsymbol{r}^t_{\text{avoid}}$ penalizes UAVs for unsafe behavior, encouraging them to maintain safe distances from obstacles and other UAVs. Each UAV is assigned a collision radius R. If a UAV collides with another UAV (i.e., the distance between them is less than $2R$), it incurs a fixed penalty $\boldsymbol{r}_{\text{collision}}$. Additionally, if the shortest distance to any object detected in the depth image, denoted as d^t_{min}, falls below the predetermined safe distance $\boldsymbol{d}_{\text{safe}}$, a penalty is imposed. The magnitude of this penalty is proportional to the difference between $\boldsymbol{d}_{\text{safe}}$ and d^t_{min}, ensuring that closer proximity to obstacles results in a higher penalty. The collision-avoidance reward is given by:

$$\boldsymbol{r}^t_{\text{avoid}} = \begin{cases} \boldsymbol{r}_{\text{collision}} & \text{if } \| \boldsymbol{p}^t_i - \boldsymbol{p}^t_j \| < 2R \\ \omega_{\text{avoid}} \cdot max(\boldsymbol{d}_{\text{safe}} - \boldsymbol{d}^t_{\text{min}}, 0) & \text{otherwise} \end{cases} \tag{5.39}$$

where ω_{avoid} is a penalty weight to discourage risky proximity to obstacles or other UAVs.

During training, the following parameter values are used:

a. $\boldsymbol{r}_{\text{arrival}} = 50$: Reward for successfully reaching the goal.
b. $\boldsymbol{r}_{\text{collision}} = -10$: Penalty for a collision.
c. $\omega_{\text{goal}} = 3$: Weight to encourage efficient paths.
d. $\omega_{\text{avoid}} = -0.05$: Weight to penalize unsafe proximity.
e. $\boldsymbol{d}_{\text{safe}} = 5$: Safe distance threshold in meters.

These settings strike a balance between incentivizing goal-reaching efficiency and ensuring safe operation in multi-UAV environments.

5.4.2.3 Network Architecture

Enabling agents to perform control tasks using model-free reinforcement learning with high-dimensional image pixels is a complex and challenging task. To address this, this study introduces a neural network architecture based on **SAC + AE** [392], a robust and efficient framework for learning policies from high-dimensional visual data. This architecture, designed for the proposed multi-UAV navigation policy π_θ, is illustrated in Figure 5.30. A deterministic autoencoder is employed to extract both visual and temporal features from three consecutive depth images $\boldsymbol{o}_z^t$. The encoder is composed of four convolutional layers followed by a fully connected layer, which maps the depth image inputs into a 50-dimensional latent representation. The first convolutional layer applies 32 two-dimensional filters with a kernel size of 3 and a stride of 2, followed by ReLU activation. The remaining convolutional layers also use 32 filters with a kernel size of 3 and a stride of 1, each followed by ReLU activations. The fully connected layer generates a 50-dimensional latent representation, which is further stabilized using layer normalization. To reconstruct the original image from this latent representation, the decoder employs a structure mirroring the encoder. It consists of a fully connected layer followed by four deconvolutional layers, configured identically to the encoder, ensuring the effective reconstruction of the visual input. The latent representation is combined with two additional observations, $\boldsymbol{o}_g^t$ (representing the goal position) and $\boldsymbol{o}_v^t$ (representing the current velocity), to form the input for both the actor and critic networks. The actor network is a three-layer multilayer perceptron with a hidden layer size of 1024. It produces two outputs, a three-dimensional mean vector and a covariance matrix, which together parameterize a Gaussian distribution used to determine the three actions of the UAV. Similarly, the critic

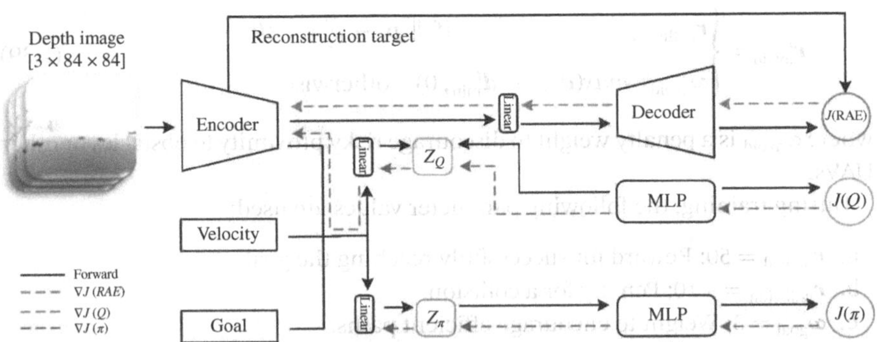

Figure 5.30 The neural network for collision avoidance takes depth images, current velocity, and the relative goal position as inputs. To enable dimensionality reduction of the input images and support policy learning, the approach employs a regularized autoencoder.

network, which evaluates the Q-function, is also designed as a three-layer multilayer perceptron with 1024 units in each hidden layer. To enhance the stability and effectiveness of the learned policy, the study ensures that gradients from the actor network do not propagate back to the convolutional encoder, preserving the integrity of the extracted visual features. The extracted latent representation is concatenated with two additional observation components, og^t (the goal position) and ov^t (the current velocity), forming the input for the actor and critic networks. The actor network is a 3-layer multilayer perceptron with a hidden layer size of 1024. It outputs a three-dimensional mean vector and covariance matrix that parameterize a Gaussian action distribution, corresponding to the three control actions. The critic network, representing the Q-function, is similarly constructed with a 3-layer multilayer perceptron and 1024 hidden units. To enhance the performance of the learned policy, the study ensures that the gradients from the actor network do not update the convolutional encoder, preserving the stability and reliability of the extracted visual features.

5.4.2.4 Multi-UAV Policy Training

This study implements a collision-free navigation strategy for multi-UAV systems using an extended **SAC + AE** framework [392]. The detailed algorithm for this approach is outlined in Ref. [375]. As shown in Figure 5.31, the method adopts a centralized learning framework during training and a distributed execution model

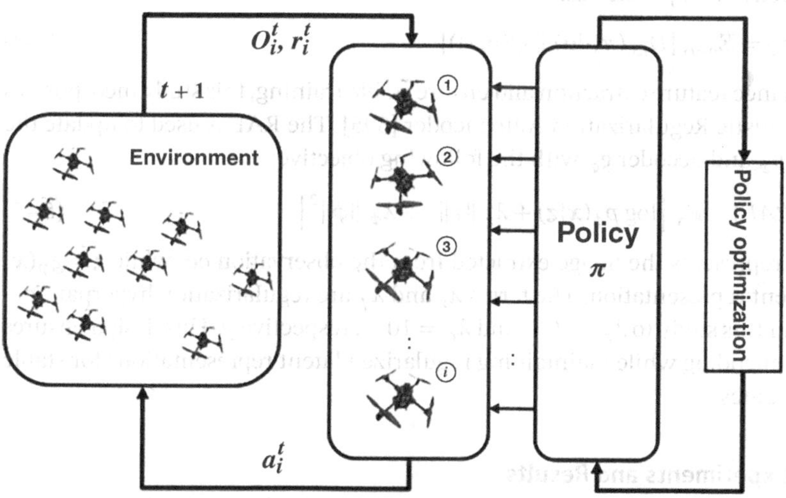

Figure 5.31 The overall framework of the proposed method. Within an environment of N UAVs, the proposed method allows each UAV to perceive its state through observations and rewards, denoted as $\boldsymbol{o}_i^t$ and $\boldsymbol{r}_i^t$, at every timestep. Actions $\boldsymbol{a}_i^t$ are then executed based on a shared policy π.

during operation [393]. During training, the policy is learned from trajectories collected across all UAVs in the system. At each timestep, UAV i observes its state $\mathbf{o}_i^t$ and executes an action $\boldsymbol{a}_i^t$ based on the shared policy π_θ. The training process is divided into two key phases. In the first phase, UAVs execute the current policy and store their trajectories $(\mathbf{o}_i^t, \boldsymbol{a}_i^t, r_i^t, \mathbf{o}_i^{t'}, d_i^t)$ in a replay buffer $\mathcal{B}$. In the second phase, batches of trajectories $(\mathbf{o}, \boldsymbol{a}, r, \mathbf{o}', d)$ are sampled from the replay buffer $\mathcal{B}$ to update both the policy and the autoencoder. This iterative process ensures efficient learning and policy optimization.

The updated policy is optimized using the following objective functions. In the policy evaluation phase, the goal is to fit the Q-function [394], which is achieved by minimizing the following loss:

$$J(Q) = \mathbb{E}_{(\mathbf{o},\boldsymbol{a},r,\mathbf{o}')\sim\mathcal{B}}\left[(Q(\mathbf{o}, \boldsymbol{a}) - r - \gamma\overline{V}(\mathbf{o}'))^2\right] \tag{5.40}$$

where $\overline{V}(\mathbf{o})$ represents the expected value function, defined as:

$$\overline{V}(\mathbf{o}) = \mathbb{E}_{\boldsymbol{a}\sim\pi_\theta}\left[(\overline{Q}(\mathbf{o}, \boldsymbol{a}) - r - \alpha \log \pi(\boldsymbol{a}|\mathbf{o})\right] \tag{5.41}$$

In this formulation, $\overline{Q}$ refers to the target Q-function, which is parameterized using a weight vector updated through an exponentially moving average of the Q-function weights.

In the policy improvement phase, the objective is to minimize the KL divergence between the policy and the Boltzmann distribution induced by the Q-function. This objective is expressed as:

$$J(\pi) = \mathbb{E}_{\mathbf{o}\sim\mathcal{B}}\left[D_{KL}(\pi(\cdot|\mathbf{o}) \parallel \mathcal{Q}(\mathbf{o}, \cdot))\right] \tag{5.42}$$

To enhance feature extraction and ensure stable training, this study incorporates a deterministic Regularization Autoencoder [395]. The RAE is used to update the encoder p_ϕ and decoder g_φ with the following objective:

$$J(RAE) = \mathbb{E}_{\boldsymbol{x}}\left[\log p_\phi(\boldsymbol{x}|\mathbf{z}) + \lambda_z \left\|\mathbf{z}\right\|^2 + \lambda_\phi \left\|\phi\right\|^2\right] \tag{5.43}$$

where $\boldsymbol{x}$ represents the image extracted from the observation $o\mathbf{o}$, while $\mathbf{z} = g_\varphi(\boldsymbol{x})$ is the latent representation. The terms λ_z and λ_ϕ are regularization hyperparameters, set in this study to $\lambda_z = 10^{-6}$ and $\lambda_\phi = 10^{-7}$, respectively. This design ensures efficient encoding while maintaining regularized latent representations for stable policy updates.

5.4.3 Experiments and Results

5.4.3.1 Simulation Environment and Experiments Setup

To develop a robust collision avoidance policy for multi-UAV systems, the proposed algorithm was implemented using ROS Noetic and evaluated in a multi-UAV scenario within the AirSim simulator [374] (Figure 5.32).

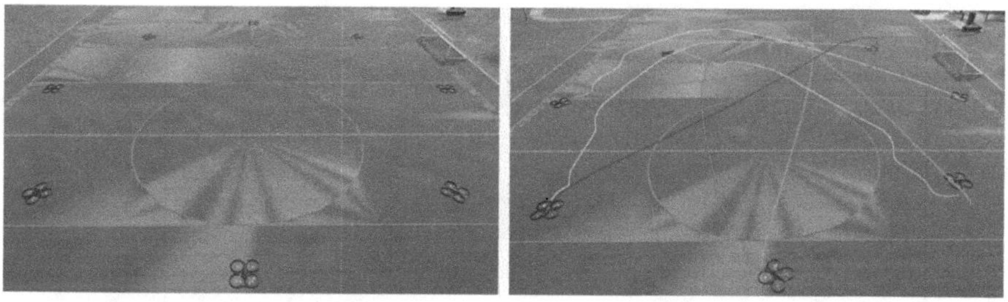

Figure 5.32 Left: The AirSim simulator, used as the testing platform for this study. Right: UAV trajectories in the Circle scenario, illustrating the performance of the proposed learned policy. [375] / IEEE.

At the start of each episode, the UAVs are assigned random initial positions and goal points within a three-dimensional space. This randomization enables more effective exploration of the high-dimensional observation space, improving the robustness and efficiency of the learning policy.

This study evaluates the proposed method using two distinct types of testing scenarios:

1. Random Scenarios: In these scenarios, the initial and goal positions of each UAV are randomly generated within a defined area. Depending on the size of the area, the scene is categorized into three types based on UAV density: a Large Scene has a low density of approximately 0.04 UAVs per cubic meter, while a Medium Scene has a moderate density of approximately 0.06 UAVs per cubic meter. Lastly, a Small Scene features a high density, with approximately 0.1 UAVs per cubic meter.
2. Cycle Scenarios: In this scenario, all UAVs are uniformly positioned along a circle at a specific altitude. Each UAV is tasked with reaching its goal located on the opposite side of the circle.

It is important to note that the proposed policy is trained exclusively on medium-sized random scenarios.

To comprehensively evaluate and compare the performance of the proposed method, this study introduces four key metrics: SR, Success weighted by Path Length (SPL), Extra Distance, and Average Speed. The detailed definitions of these metrics, as well as the hyperparameter settings for the proposed method, can also be found in Ref. [375].

5.4.3.2 Experiment Results on Various Scenarios

This study evaluates the performance of the proposed policy by comparing it with two other DRL-based methods. The first method (RL+CNN) employs a CNN

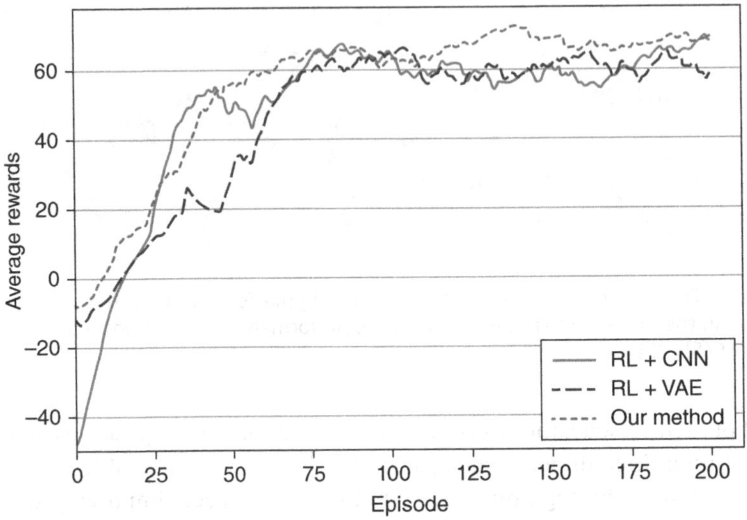

Figure 5.33 The average training rewards are compared among the methods: RL+CNN (solid line), RL+VAE (dashed line), and the proposed approach (dotted line).

alongside reinforcement learning [396]. The second method (RL+VAE) integrates a variational autoencoder (VAE) for image compression with a policy-based DRL model [397].

In this study, different methods were trained using medium-sized random scenarios, with the corresponding learning curves shown in Figure 5.33. Although all approaches successfully converged during training, the proposed method outperformed others by attaining significantly higher reward values, showcasing its effectiveness.

Table 5.4 presents the performance evaluation of different methods across various sizes of random scenarios after 1000 episodes, with the best results highlighted in **bold**. The proposed method demonstrates exceptional performance consistency across different scene sizes. Specifically, as the scene size increases, metrics such as SR, SPL, and average speed show a steady improvement, while the extra distance traveled decreases. In all three types of scenarios, the proposed method achieves a higher SR compared to other methods, even though it incurs a slightly longer extra distance for enhanced collision avoidance. For the simplest scenario (large scene), the proposed method outperforms RL+CNN and RL+VAE by 6.6% and 5.2%, respectively. In the most challenging scenario (small scene), it achieves even greater competitiveness, exceeding RL+CNN and RL+VAE by 13.1% and 7.7%, respectively. Remarkably, the proposed method not only achieves the highest SR and fastest average speed in the large scene but also reaches a peak SR of approximately 90% in the small scene. These results indicate that the proposed approach is

Table 5.4 The performance metrics, expressed as mean and standard deviation, are assessed across scenarios of different scene sizes for each method.

Scene size	Method	Success rate	SPL	Extra distance	Average speed
Small	Our method	**0.895**	0.715	2.035/2.270	0.914/0.130
	RL+CNN	0.764	0.637	1.505/1.154	0.715/0.070
	RL+VAE	0.818	**0.745**	**0.686/0.494**	**0.967/0.099**
Medium	Our method	**0.934**	**0.791**	1.784/1.774	0.991/0.104
	RL+CNN	0.829	0.717	1.499/1.080	0.771/0.065
	RL+VAE	0.851	0.785	**0.777/0.607**	**1.038/0.090**
Large	Our method	**0.952**	**0.852**	1.334/1.202	**1.066/0.094**
	RL+CNN	0.886	0.774	1.663/1.133	0.773/0.072
	RL+VAE	0.900	0.833	**0.886/0.536**	1.063/0.082

highly suitable for real-world multi-UAV group tasks. Additionally, both the proposed method and RL+VAE outperform RL+CNN, underscoring the effectiveness of incorporating an autoencoder for policy learning.

This study evaluated various methods within a circular scenario (12-m radius, involving 8 UAVs). RL+CNN was excluded from subsequent experiments due to its poor performance during testing. The trajectories generated by the proposed method and RL+VAE were visualized, with their performance metrics shown in Figure 5.34. While the proposed method demonstrated superior navigation performance across most metrics, it showed higher extra distance compared to RL+VAE. This is because the proposed method requires UAVs to adjust their paths more frequently to avoid collisions. Future work will aim to optimize the proposed method by reducing extra distance while maintaining high SR.

Scalability and robustness are critical for the effective operation of multi-UAV systems. To evaluate these factors, this study examines a circular scenario where the radius is fixed at 12 m while the number of UAVs is gradually increased, followed by a detailed quantitative analysis. As illustrated in Figure 5.35, although the performance of the proposed method experiences a slight decline as the number of UAVs grows, it maintains consistently high success rates (SR, close to 100%) and SPL (above 90%). To avoid collisions, the extra distance increases, and the average speed decreases slightly, demonstrating the system's ability to adapt to growing complexity. In contrast, RL+VAE exhibits only minimal changes in extra distance and average speed; however, its SR and SPL show significant declines as the number of UAVs increases, especially when the UAV count exceeds 12. These findings

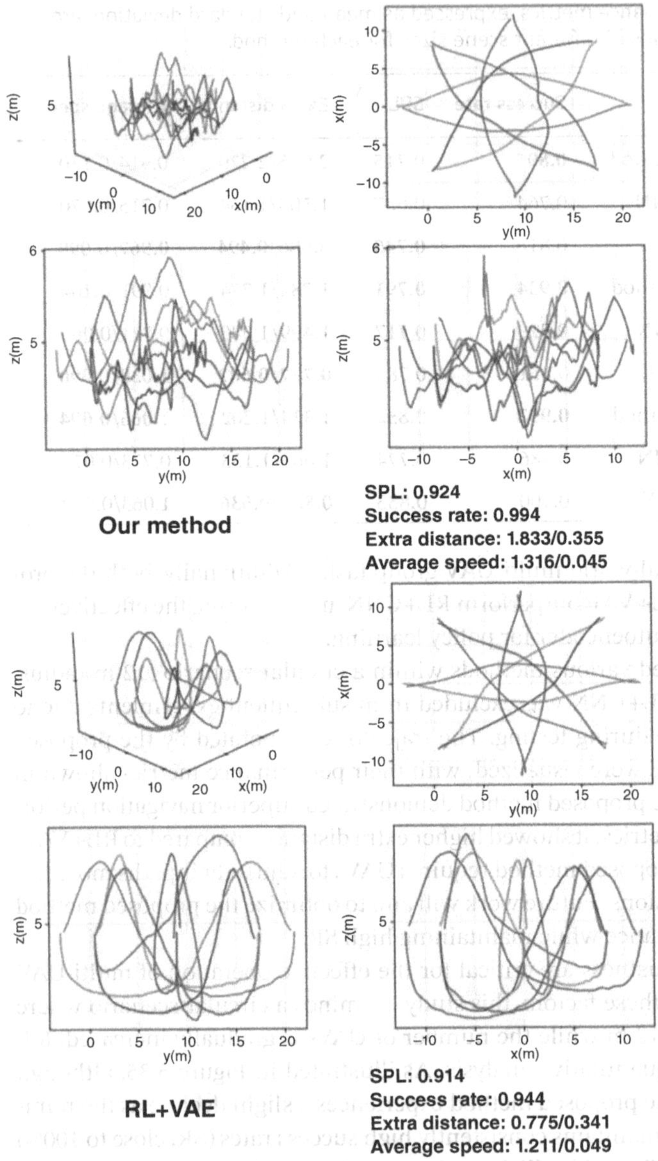

Figure 5.34 The trajectories of the proposed method and RL+VAE are illustrated in a circular scenario with a 12-m radius involving eight UAVs. To enhance visualization, both perspective and three-view diagrams are presented, with each UAV's trajectory distinguished by a unique color.

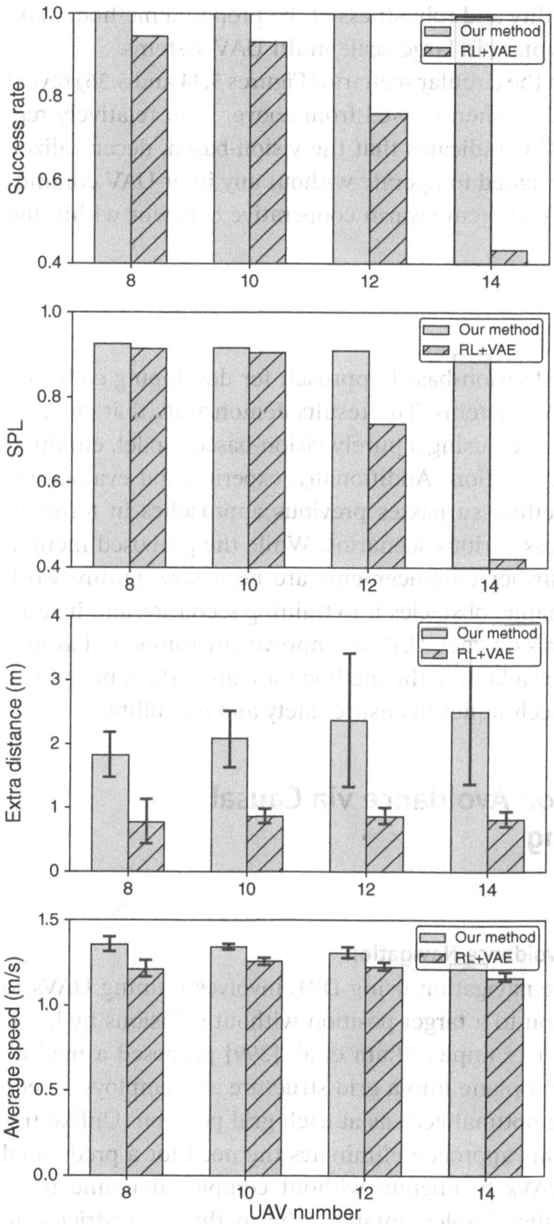

Figure 5.35 Four performance metrics for both the proposed method and RL+VAE are evaluated in a circular scenario with a fixed radius of 12 m.

underscore the superior scalability and robustness of the proposed method, making it highly suitable for deployment in large-scale multi-UAV systems.

Furthermore, observations in the circular scenario (Figures 5.34 and 5.36) reveal that the trajectories of the UAVs, when viewed from above, form relatively regular and organized patterns. This indicates that the vision-based, decentralized collision-avoidance policy – designed to operate without any inter-UAV communication – successfully fosters emergent swarm cooperative behavior within the system.

5.4.4 Conclusion

This study presents an efficient vision-based approach for developing collision-avoidance policies in multi-UAV systems. The results demonstrate that effective collision avoidance can be achieved using a purely vision-based model, eliminating the need for spatial representation. Additionally, experimental evaluations confirm that the proposed method surpasses previous approaches in terms of robustness and scalability across various scenarios. While the proposed method shows significant potential, further enhancements are necessary. Future work will incorporate static and dynamic obstacles into training scenarios and investigate the use of multiple cameras on each UAV to improve environmental awareness [398]. A key priority will be adapting the method for real-world applications, which will require additional techniques to ensure safety and reliability.

5.5 Multi-UAV Collision Avoidance via Causal Representation Learning

5.5.1 Related Work

5.5.1.1 DRL-based Collision Avoidance Navigation

Multi-UAV collision avoidance navigation using DRL involves training UAVs to navigate from an initial position to a target position without collisions by leveraging actions and rewards. For example, Pham et al. [399] proposed a method that discretizes the UAV's flight plane into a grid structure and employs a deep Q-network [400] to determine optimal actions at each grid position. Unlike traditional methods [401–403], this approach eliminates the need for a predefined dynamics model, enabling UAVs to operate without complex dynamic modeling [404]. While this simplifies implementation, the method is restricted to obstacle-free environments and does not accommodate high-dimensional inputs like images, limiting its applicability to more complex scenarios. Walvekar et al. [405] addressed the challenge of UAV collision avoidance in a three-dimensional environment using a deep Q-network. Their approach utilized first-person view images captured by the UAV as inputs to the DQN, allowing the system to make

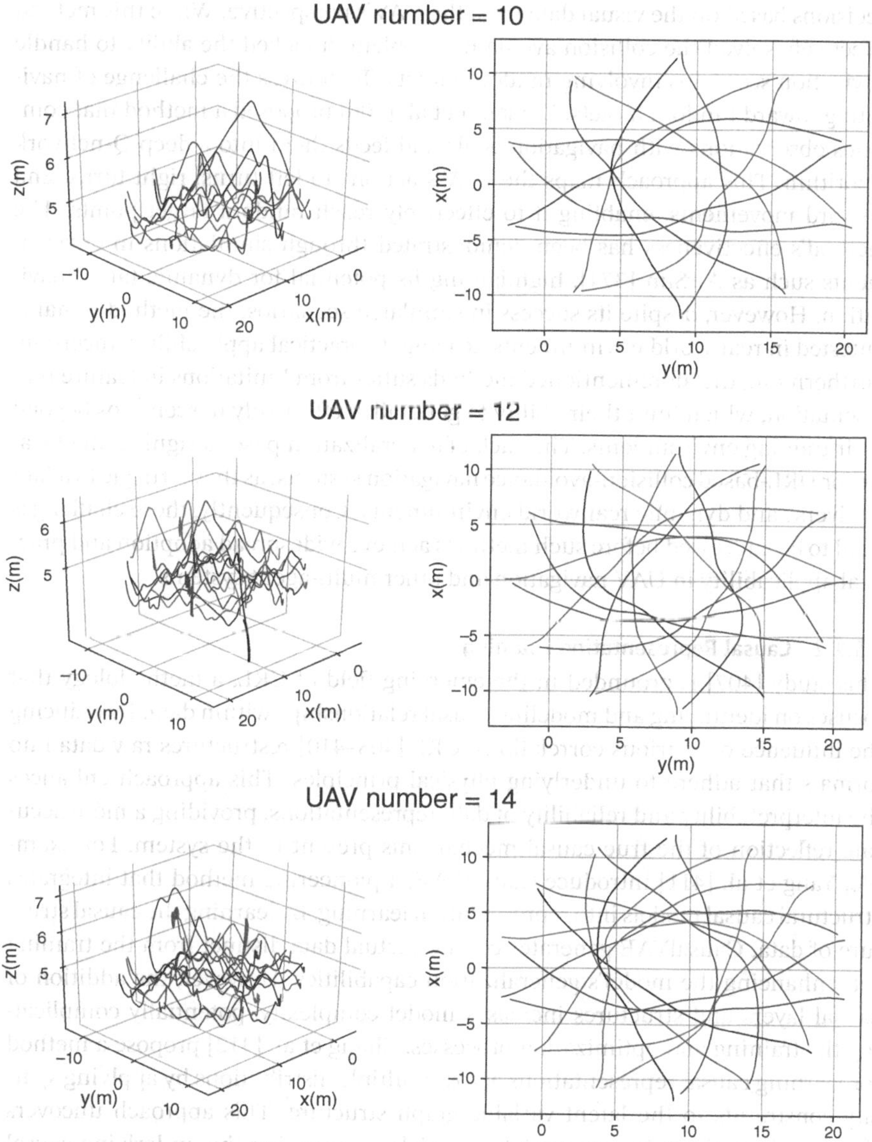

Figure 5.36 The trajectories of the proposed method in a circular scenario (radius = 12 m) with 10, 12, and 14 UAVs are visualized through both perspective and bird's-eye views. Each UAV's trajectory is distinguished by a unique color for clarity.

decisions based on the visual data from the UAV's perspective. While this method effectively solved the collision avoidance problem, it lacked the ability to handle navigation scenarios involving random targets. To address the challenge of navigating toward random targets, Kersandt et al. [406] proposed a method that combines observations with navigation goals and feeds them into a deep Q-network algorithm. This approach maps the UAV's actions to left turns, right turns, and forward movements, enabling it to effectively reach random target points. The method's effectiveness has been demonstrated through simulations in environments such as AirSim [374], highlighting its potential for dynamic target navigation. However, despite its success in simulated scenarios, the method remains untested in real-world environments, leaving its practical applicability uncertain. Furthermore, the aforementioned methods suffer from limitations in feature representation, which limit their ability to generalize effectively to scenarios beyond their training environments. This lack of generalization poses a significant obstacle for DRL-based collision-avoidance navigation systems, as they struggle to adapt to diverse and dynamic real-world environments. Consequently, these challenges need to be addressed before such methods achieve widespread adoption and practical applicability in UAV navigation and other multi-agent systems.

5.5.1.2 Causal Representation Learning

This study [407] is grounded in the emerging field of CRL, a methodology that focuses on identifying and modeling causal relationships within data. By reducing the influence of spurious correlations, CRL [408–410] restructures raw data into formats that adhere to underlying physical principles. This approach enhances the interpretability and reliability of data representations, providing a more accurate reflection of the true causal mechanisms present in the system. For example, Yang et al. [411] introduce CausalVAE, a pioneering method that integrates structural causal models into representation learning. By learning the causal structure of data, CausalVAE generates counterfactual data distinct from the training set, enhancing the model's generalization capabilities. However, the addition of causal layers and structures increases model complexity, potentially complicating the training and optimization processes. Zhang et al. [412] propose a method for learning causal representations across multiple distributions by applying sparsity constraints to the latent variable graph structure. This approach uncovers the causal relationships among latent variables, revealing the underlying causal structure of the data, which aids in understanding system behavior and supporting decision-making. However, while the method effectively identifies causal relationships, interpreting these connections often requires domain expertise, particularly in complex systems with numerous latent variables. In addition,

Zhao et al. [413] introduce DCCL, a general framework designed to tackle the sparsity of supervisory signals and the long-tail problem in CRL. By leveraging contrastive learning grounded in causal mechanisms and causal graphs, DCCL learns disentangled causal embeddings, significantly improving the precision and robustness of model predictions. While the framework effectively addresses challenges posed by sparse data, its performance remains partially dependent on the quality and diversity of the training dataset. Building on previous research, this work conducts causal intervention on the shape of obstacles to extract invariant causal representations inspired by the above studies. This method can effectively reduce the effect of spurious correlations on the generalization ability of DRL.

5.5.2 Proposed Method

5.5.2.1 Problem Formulation

DRL [414–416] focuses on enabling an intelligent agent to interact with its environment in a way that maximizes cumulative rewards. The agent develops strategies based on the current environmental state, following an iterative process with Markov properties. As a result, DRL can be modeled as a POMDP, represented by a 6-tuple $(\mathcal{S}, \mathcal{A}, \mathcal{P}, \mathcal{R}, \Omega, \mathcal{O})$. Here, $\mathcal{S}$ is the set of agent states, Ω represents the observation space ($o \in \Omega$), and $\mathcal{A}$ is the set of possible actions. The state transition probability is denoted by $\mathcal{P}$, with $P^a_{ss'} = P[S_{t+1} = S' — S_t = s, A_t = a]$, while $\mathcal{O}$ describes the observational distribution ($o = \mathcal{O}(s)$). The primary objective is to determine a policy that maps states to optimal control actions.

1. Observation Space

The observation space of the UAV is represented as $O = [I, V, G]$, where I consists of three consecutive depth images, V represents the current speed of the UAV, and G indicates the Euclidean distance between the UAV's current position and the target point.

2. Action Space

To ensure smooth and continuous movement, this study defines the UAV's action space as $A = [v_x, v_y, v_z]$, where v_x represents forward velocity, v_y denotes turning velocity, and v_z specifies climbing velocity.

3. Reward Function

Collision avoidance navigation aims to guide UAVs to their target positions in complex environments without collisions. This process involves two subtasks: navigation (r_g) and obstacle avoidance (r_c). To reinforce desired behavior, the UAVs receive positive rewards upon safely reaching the target and negative feedback as punishment in the event of a collision.

$$r = r_g + r_c \tag{5.44}$$

$$r_g = \begin{cases} r_{\text{arrival}} & \text{if } d_t < 0.5 \\ \alpha_{\text{goal}} \cdot (d_t - d_{t-1}) & \text{otherwise} \end{cases} \tag{5.45}$$

where d_t represents the distance between the UAV's position and the target position at time t, r_{arrival} denotes the reward for UAVs reaching their target positions, and α_{goal} serves as the reward weight.

$$r_c = \begin{cases} r_{\text{collision}} & \text{if crash} \\ \alpha_{\text{avoid}} \cdot max(d_{\text{safe}} - d_{\text{min}}, 0) & \text{otherwise} \end{cases} \tag{5.46}$$

where $r_{\text{collision}}$ represents the penalty assigned for collisions, while α_{avoid} denotes the weight of the penalty. The parameter d_{safe} defines the safe distance that UAVs must maintain to avoid collisions, and d_{min} refers to the minimum distance detected in a depth image.

5.5.2.2 Architectural Overview

Building on previous work [375], this approach, shown in Figure 5.37, includes two main parts: representation learning and policy learning. First, a causal representation structure based on an auto-encoder learns features that diminish the influence of obstacle shapes. Then, these learned representations, combined with the current velocity and target position, are fed into a policy network. Using the

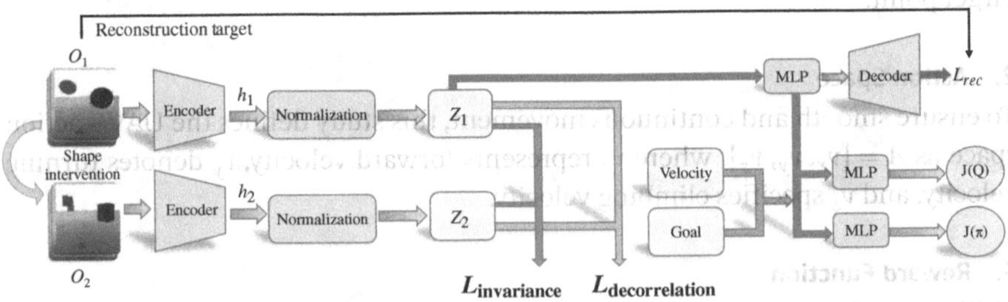

Figure 5.37 The illustration of the proposed method. To extract invariant causal representation, we intervene the shape of obstacles and maximize the mutual information through the $\mathcal{L}_{\text{invariance}}$. In addition, we introduce the $\mathcal{L}_{\text{decorrelation}}$ to ensure that different dimensions capture different information.

SAC algorithm [417], the policy network outputs three-dimensional velocity commands that guide the UAVs toward their targets while avoiding obstacles.

In the policy improvement phase, the actor network is updated by maximizing the following objective $J(\pi)$:

$$J(\pi) = \mathbb{E}_{o\sim\mathcal{B}}\left[D_{KL}(\pi(a|o) \parallel Q(o,a))\right] \tag{5.47}$$

where $\pi(a|o)$ denotes the probability of selecting action a given observation o, and $Q(o, a)$ represents the target action distribution for that observation.

To improve the critic network, this study minimizes the following loss function $J(Q)$:

$$J(Q) = \mathbb{E}_{(o,a,r,o')\sim\mathcal{B}}\left[\left(Q(o,a) - r - \gamma\bar{V}(o')\right)^2\right] \tag{5.48}$$

where $\bar{V}(o')$ denotes the target value function, γ is the discount factor, and r is the received reward.

This study employs a reconstruction loss function $\mathcal{L}(rec)$ to update the encoder–decoder pair ($z = g_x$) with the following objective:

$$\mathcal{L}(rec) = \mathbb{E}_x\left[\log p_\phi(x|z) + \lambda_z \left\|z\right\|^2 + \lambda_\phi \left\|\phi\right\|^2\right] \tag{5.49}$$

where x is the depth image extracted from observations, and $\lambda_z|z|^2$ and $\lambda_\phi|\phi|^2$ serve as regularization terms for the latent variable z and the model parameters ϕ, respectively. Both λ_z and λ_ϕ are hyperparameters.

5.5.2.3 Extract Invariant Causal Representation

This subsection explores the learning of causal representations from images to develop robust policies that remain effective across different training domains. The approach focuses on identifying and utilizing the key factors that drive successful actions. Prior research [421, 422] has verified that CRL can uncover invariant mechanisms within data from multiple domains and identify hidden causal variables. Building on this foundation, the study employs causal interventions [423, 424] to vary the shapes of obstacles while keeping other environmental features constant. Notably, this operation can be achieved using AirSim's image rendering technology [425]. To disentangle these causal representations, the study employs an auto-encoder [426, 427], as shown in Figure 5.37. Depth cameras capture two types of observations: $\mathcal{O}_1$, containing images of UAVs and spherical obstacles, and $\mathcal{O}_2$, showing UAVs and cubic obstacles. These observations are processed through an encoder to extract representations (such as h_1 and h_2), which then undergo instance-dimensional normalization. This normalization ensures that each feature has a mean of zero and a standard deviation of $\frac{1}{\sqrt{N}}$, as defined by the formula:

$$z = \frac{h_i - \mu(h_i)}{\sigma(h_i) * \sqrt{N}} \tag{5.50}$$

The resulting normalized representations, $z1$ and $z2$, are further refined to maximize mutual information by minimizing their differences. This is achieved through the invariance term:

$$\mathcal{L}_{\text{invariance}} = ||z_1 - z_2||^2 \tag{5.51}$$

In addition, previous research [428] has shown that different dimensions within representations often share overlapping information. To address this and ensure each dimension captures unique content, this study introduces a decorrelation term:

$$\mathcal{L}_{\text{decorrelation}} = \mathcal{F}(z_1, z_1^T, I) + \mathcal{F}(z_2, z_2^T, I) \tag{5.52}$$

where $F(\cdot, \cdot)$ represents the Frobenius norm, and I is the identity matrix. This term minimizes the off-diagonal elements of the autocorrelation matrix, reducing redundancy between dimensions. To eliminate correlation terms, we attempt to set the off-diagonal elements of each represented autocorrelation matrix to zero. This approach avoids trivial solutions that produce identical output vectors, thereby preventing scenarios where all inputs and outputs are the same.

5.5.3 Experiment Results and Analysis

5.5.3.1 Simulation Software and Experiments Setup

AirSim [374] is a reliable, high-fidelity simulation platform commonly used for testing UAV algorithms because of its accuracy and adaptability. In this study, the platform provides the foundation for developing and evaluating UAV collision avoidance algorithms. The simulations are conducted on a computer with the Ubuntu 20.04 operating system and a single NVIDIA RTX 4090 GPU, ensuring efficient computation and realistic performance.

5.5.3.2 Performance Metrics and Experiment Scenarios

This study assesses the performance of the proposed method using four key metrics: SR, SPL, Extra Distance, and Average Speed:

1. SR: The proportion of UAVs that reach their target position within a set time limit without collisions.
2. SPL: A measure that combines SR with path efficiency, prioritizing shorter and more optimal routes.

Figure 5.38 Simulation scenarios for evaluating the generalization ability. [407] / IEEE.

3. Extra Distance: The additional distance traveled by UAVs compared to the direct straight-line distance between their starting point and target.
4. Average Speed: The overall average speed of all UAVs during the task.

As shown in Figure 5.38, this study assesses the proposed method's generalization capability by testing it on obstacles of various shapes, including cubes, spheres, triangles, cylinders, pentahedrons, and cuboids. The simulations are carried out in a fixed 16×16×4 area with eight UAVs and four static obstacles. To introduce variability, the initial and target positions of the UAVs, as well as the obstacle locations, are randomly generated.

5.5.3.3 Performance Comparison

1. Evaluation Obstacle Shapes

This study compares the proposed method with the state-of-the-art approach, SAC+RAE, across different obstacle scenarios. As shown in Table 5.5, the proposed method achieves a higher SR and SPL, demonstrating its improved ability to generalize to unseen obstacles. Additionally, it achieves a faster average speed, enabling UAVs to reach their targets in less time. However, the method incurs a longer extra distance compared to SAC+RAE, indicating that UAVs adjust their paths more frequently to avoid collisions, which increases the overall distance traveled.

2. More Evaluation Baselines

To further evaluate the generalization capability of the proposed method, this study compares it with three evaluation baselines: (1) AutoAugment [418], (2) DrAC [419], and (3) SE [420]. The evaluation is conducted in a playground environment with four new cuboid obstacles. The results in Table 5.6 show that the proposed CRL method achieves a notable improvement in navigation SR and average speed compared to previous approaches. This demonstrates its ability to enhance generalization performance, enabling UAVs to navigate efficiently and reliably in unseen environments.

Table 5.5 Performance (as mean/std) comparison under different obstacle shape.

Obstacle shape	Seen/Unseen	Method	Success rate (%)	SPL (%)	Extra distance (m)	Average speed (m/s)
Cube	Seen	SAC+RAE	67.6	58.3	**1.483/1.436**	0.771/0.158
		Our method	**74.4** (↑ 6.8)	**62.5** (↑ 4.2)	1.542/1.519	**0.806/0.125**
Sphere	Seen	SAC+RAE	68.3	58.4	**1.680/1.709**	0.783/0.160
		Our method	**73.4** (↑ 5.1)	**62.2** (↑ 3.8)	1.735/1.718	**0.814/0.132**
Triangle	Unseen	SAC+RAE	72.5	62.1	**1.535/1.622**	0.775/0.164
		Our method	**76.3** (↑ 3.8)	**64.4** (↑ 2.3)	1.815/1.802	**0.809/0.135**
Cylinder	Unseen	SAC+RAE	67.6	58.6	**1.646/1.621**	0.783/0.155
		Our method	**72.1** (↑ 4.5)	**61.7** (↑ 3.1)	1.853/1.810	**0.810/0.138**
Pentahedron	Unseen	SAC+RAE	67.3	56.7	**1.575/1.546**	0.784/0.153
		Our method	**73.8** (↑ 6.5)	**61.8** (↑ 5.1)	1.651/1.623	**0.813/0.130**
Cuboid	Unseen	SAC+RAE	68.2	59.3	**1.635/1.578**	0.799/0.131
		Our method	**74.4** (↑ 6.2)	**63.0** (↑ 3.7)	1.851/1.820	**0.803/0.136**

Table 5.6 Performance comparison of various methods in the testing scenario with four cuboid obstacles.

Method	Success Rate (%)	SPL (%)	Extra Distance (m)	Average Speed (m/s)
SAC+RAE	68.2	59.3	1.635/1.578	0.799/0.1311
+ AutoAugment [418]	68.8	61.2	**1.505/2.881**	0.398/0.131
+ DrAC [419]	69.8	61.5	1.538/2.693	0.379/0.122
+ SE [420]	70.1	61.6	1.564/2.648	0.377/0.122
+ Our method	**74.4**	**62.5**	1.542/1.519	**0.860/0.125**

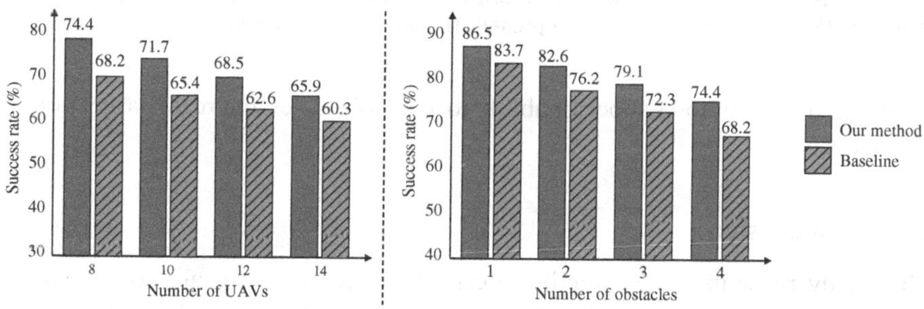

Figure 5.39 Scalability analysis experiments. Performance comparison between our proposed CRL and the baseline model under conditions with different numbers of UAVs and obstacles. The SR is used as the evaluation metric. Best viewed in color.

3. Scalability

To evaluate the scalability of the proposed method, experiments were conducted in environments with varying numbers of UAVs and obstacles, including four new cuboid obstacles. As shown in Figure 5.39, the proposed method consistently outperformed the SAC+RAE approach across all scenarios. These results highlight its robustness and ability to scale effectively, maintaining reliable performance even as environmental complexity increases.

4. Visualization

To clearly demonstrate the collision avoidance performance of the proposed method, this study visualizes UAV trajectories through both perspective and three-view drawings. As shown in Figure 5.40, the proposed method achieves more efficient path planning and collision avoidance compared to the SAC+RAE approach. The visualized results highlight smoother and more optimized navigation paths,

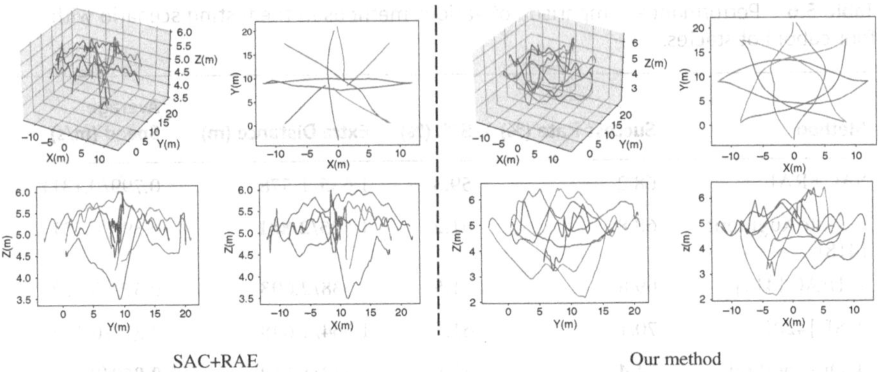

Figure 5.40 Visualization of UAV trajectories in perspective drawing and three-view drawing. We use different colors to represent trajectories of different UAVs.

reinforcing the method's superior ability to avoid obstacles and reach target points effectively.

5.5.4 Conclusion

This study presents a CRL method designed to identify causal variables from images, improving the generalization ability of DRL. The approach involves intervening on obstacle shapes and maximizing mutual information using an invariance term. To ensure that each dimension of the representation captures unique information, a decorrelation term is introduced. Experimental results confirm that the proposed method outperforms state-of-the-art approaches in generalization across various testing scenarios. While the proposed method effectively addresses generalization issues related to unknown obstacles, it shows limited adaptability to background variations. Additionally, due to hardware constraints, this study focuses solely on simulated environments, and the learned strategies remain untested in real-world settings. Future work will explore more advanced disentanglement techniques and causal intervention strategies. Efforts will also focus on applying CRL to real-world experiments and expanding its use to other DRL tasks for broader applicability.

6

Conclusions and Future Directions

6.1 Conclusions

The modular design automation of intelligent robot systems, known as MODENA, constructs a unified modular graph model for the morphology, controller, and vision system of the robot within a digital twin architecture. It achieves design automation with the assistance of various AI technologies. This approach can enhance design efficiency and performance, avoid the trial-and-error procedures of traditional design methods, and stimulate innovative design.

This book demonstrates that the integration of morphology and controller design automation represents a trend. By introducing the background, foundation, framework, and application scenarios of MODENA, it shows that this can be accomplished by incorporating the concept of biological evolution and constructing search spaces, designing search strategies, and evaluating indicators. These methods include those based on evolutionary computation, learning, and a combination of evolution and learning.

6.2 Research Prospects

While MODENA has made significant progress over the past two decades, several challenges remain unsolved, and new areas of application are emerging. The following sections will highlight opportunities for future research, with a focus on advancing theoretical foundations:

1. Multi-view Unified Modeling

Creating a unified model that combines the morphology, controller, and vision systems of intelligent robots is essential for improving the efficiency of automation design. Currently, these components are designed using separate tools and methods, each tailored to specific needs [429, 430]. For example, BGs are often used for

Modular Design Automation of Intelligent Robot Systems, First Edition. Zhun Fan et al.

mechanical systems, finite state machines for swarm controllers, and deep neural networks for vision systems. While effective individually, these varied approaches make it challenging to address the interconnections between modules. Developing an integrated, multi-view modeling tool that captures these relationships will be a key step toward achieving more streamlined and cohesive robot design automation.

2. Surrogate-assisted Optimization

Optimizing intelligent robots requires balancing multiple competing objectives while satisfying various constraints. These tasks are often computationally demanding and rely on external simulation tools, making the process complex and time-consuming. To address this, CMOEAs have become a widely adopted approach in MODENA research.

Conventional methods treat all infeasible regions equally, though only those near the UPF influence the final solution. Recognizing and leveraging this distinction is crucial for improving algorithm performance, particularly in balancing convergence, diversity, and feasibility. Additionally, integrating adaptive surrogate models that combine global and local approximations for objectives and constraints offers a promising pathway to further enhance CMOEAs and streamline optimization processes.

3. Knowledge Extraction and Reuse

The design automation of intelligent robots relies on two types of knowledge: explicit and implicit. Explicit knowledge, often called human knowledge, is easily interpretable by experts. In contrast, implicit knowledge, or machine knowledge, is not directly understandable by humans but can be stored, processed, and potentially interpreted in the future.

To extract explicit knowledge, symbolic regression, a method based on genetic programming (GP), is often applied. By leveraging prior domain knowledge to define functions and variables, this approach can automatically identify patterns within data. Similarly, causal reasoning, an emerging research field, provides another method for generating explainable knowledge. This knowledge can improve problem definitions and guide search processes in CMOEAs, further enhancing the efficiency of intelligent robot optimization.

4. Digital Twins and MODENA

The design of intelligent robots in MODENA relies heavily on simulations and experiments in both virtual and real-world environments. Digital twin technology provides an effective solution by creating precise virtual replicas of physical systems, significantly accelerating the design process. Unlike traditional trial-and-error methods, this approach maps robots into a virtual space across four key dimensions: geometry, contact dynamics, behavior, and rules.

With digital twins, the robot's morphology, controller, and vision systems can be easily adjusted, enabling the generation of countless design candidates. Machine learning and evolutionary computing further streamline the process by optimizing these candidates efficiently. Additionally, data produced during the design phase can be analyzed to uncover valuable insights and rules, supporting future design enhancements. Fully harnessing the potential of digital twin technology remains a vital area for advancing MODENA.

5. Domain Adaptation and Generalization

The limited ability of models to generalize, especially in vision systems, remains a significant challenge. Addressing this issue requires prioritizing model robustness when designing NAS methods. A promising solution involves integrating domain adaptation [431] and generalization evaluation [432] into the design process, enabling models to transfer knowledge from training scenarios to unseen testing environments.

One effective approach is introducing an evaluation metric that measures a model's generalization ability during candidate selection. This helps identify models that achieve a balance between high performance and robustness, ensuring they perform reliably in diverse and noisy conditions. Recent studies [433–435] have shown the potential of this strategy in improving model generalization, making it an essential direction for further exploration.

6. Active and Continual Learning

Another solution to the poor generalization of models is online learning in testing scenarios, allowing models to adapt quickly to new environments by adjusting their parameters. However, effective online learning must address two key challenges.

The first challenge is identifying what to learn in a given environment. Active vision and learning techniques [436, 437] help models focus on relevant targets and improve decision-making. These methods have been successfully applied to tasks such as robot exploration [438–440], UAV swarm localization, and similar applications [441, 442].

The second challenge is overcoming catastrophic forgetting, where learning new information causes the loss of previously acquired knowledge. This issue leads to overfitting and limits the model's ability to generalize. Continual learning [443] addresses this by enabling models to integrate new knowledge while preserving prior experiences. Recent studies [444–446] have incorporated continual learning into NAS, enhancing the adaptability and generalization of models in dynamic and evolving environments.

References

1 Mehdi Tarkian. *Design automation for multidisciplinary optimization: A high level CAD template approach*. PhD thesis, Linköping University Electronic Press, 2012.

2 Hod Lipson and Jordan B Pollack. Automatic design and manufacture of robotic lifeforms. *Nature*, 406(6799):974–978, 2000.

3 Robert Kwiatkowski and Hod Lipson. Task-agnostic self-modeling machines. *Science Robotics*, 4(26):eaau9354, 2019.

4 Michael Schmidt and Hod Lipson. Distilling free-form natural laws from experimental data. *Science*, 324(5923):81–85, 2009.

5 Zhun Fan. *Design automation of mechatronic systems using evolutionary computation and bond graph*. Michigan State University, 2004.

6 Saeed Behbahani and Clarence W de Silva. System-based and concurrent design of a smart mechatronic system using the concept of mechatronic design quotient (MDQ). *IEEE/ASME Transactions on Mechatronics*, 13(1):14–21, 2008.

7 Jean-François Dupuis, Zhun Fan, and Erik D Goodman. Evolutionary design of both topologies and parameters of a hybrid dynamical system. *IEEE Transactions on Evolutionary Computation*, 16(3):391–405, 2011.

8 Zhun Fan, Gui-Jie Zhu, Wen-Ji Li, Yu-Gen You, Xiao-Ming Li, Pei-Han Lin, and Bin Xin. Applications of evolutionary computation in the design automation of complex mechatronic system: A survey. *Acta Automatica Sinica*, 47(7):1495–1515, 2021.

9 S Ivvan Valdez, Salvador Botello-Aceves, Héctor M Becerra, and Eusebio E Hernández. Comparison between a concurrent and a sequential optimization methodology for serial manipulators using metaheuristics. *IEEE Transactions on Industrial Informatics*, 14(7):3155–3165, 2018.

Modular Design Automation of Intelligent Robot Systems, First Edition. Zhun Fan et al.

10 Sehoon Ha, Stelian Coros, Alexander Alspach, Joohyung Kim, and Katsu Yamane. Computational co-optimization of design parameters and motion trajectories for robotic systems. *The International Journal of Robotics Research*, 37(13–14):1521–1536, 2018.

11 Lelai Zhou and Shaoping Bai. A new approach to design of a lightweight anthropomorphic arm for service applications. *Journal of Mechanisms and Robotics*, 7(3):031001, 2015.

12 Prashant K Jamwal, Shahid Hussain, and Sheng Q Xie. Three-stage design analysis and multicriteria optimization of a parallel ankle rehabilitation robot using genetic algorithm. *IEEE Transactions on Automation Science and Engineering*, 12(4):1433–1446, 2014.

13 Zhun Fan, Yugen You, Xinye Cai, Haodong Zheng, Guijie Zhu, Wenji Li, Akhil Garg, Kalyanmoy Deb, and Erik Goodman. Analysis and multi-objective optimization of a kind of teaching manipulator. *Swarm and Evolutionary Computation*, 50:100554, 2019.

14 Zhun Fan, Wenji Li, Xinye Cai, Hui Li, Caimin Wei, Qingfu Zhang, Kalyanmoy Deb, and Erik Goodman. Push and pull search for solving constrained multi-objective optimization problems. *Swarm and Evolutionary Computation*, 44:665–679, 2019.

15 Guijie Zhu, Jiacheng Liu, Zhun Fan, Duan Yuan, Peili Ma, Meihua Wang, Weihua Sheng, and Kelvin CP Wang. A lightweight encoder–decoder network for automatic pavement crack detection. *Computer-Aided Civil and Infrastructure Engineering*, 39(12):1743–1765, 2024.

16 Agrim Gupta, Silvio Savarese, Surya Ganguli, and Li Fei-Fei. Embodied intelligence via learning and evolution. *Nature Communications*, 12(1):5721, 2021.

17 Zhun Fan, Zhaojun Wang, Wenji Li, Xiaomin Zhu, Bingliang Hu, An-Min Zou, Weidong Bao, Minqiang Gu, Zhifeng Hao, and Yaochu Jin. Automated pattern generation for swarm robots using constrained multi-objective genetic programming. *Swarm and Evolutionary Computation*, 81:101337, 2023.

18 Wenji Li, Zhaojun Wang, Ruitao Mai, Pengxiang Ren, Qinchang Zhang, Yutao Zhou, Ning Xu, JiaFan Zhuang, Bin Xin, Liang Gao, et al. Modular design automation of the morphologies, controllers, and vision systems for intelligent robots: A survey. *Visual Intelligence*, 1(1):2, 2023.

19 Frederick Hayes-Roth. The knowledge-based expert system: A tutorial. *Computer*, 17(09):11–28, 1984.

20 Michael F Reynolds, Alejandro J Cortese, Qingkun Liu, Zhangqi Zheng, Wei Wang, Samantha L Norris, Sunwoo Lee, Marc Z Miskin, Alyosha C Molnar, Itai Cohen, et al. Microscopic robots with onboard digital control. *Science Robotics*, 7(70):eabq2296, 2022.

21 Aude Billard and Danica Kragic. Trends and challenges in robot manipulation. *Science*, 364(6446):eaat8414, 2019.

22 Steven Macenski, Tully Foote, Brian Gerkey, Chris Lalancette, and William Woodall. Robot operating system 2: Design, architecture, and uses in the wild. *Science Robotics*, 7(66):eabm6074, 2022.

23 Mohammadali Honarpardaz, Johan Ölvander, and Mehdi Tarkian. Fast finger design automation for industrial robots. *Robotics and Autonomous Systems*, 113:120–131, 2019.

24 Hod Lipson. Evolutionary robotics and open-ended design automation. In *Biomimetics*, pages 147–174. CRC Press, 2005.

25 Timon Hoebert, Wilfried Lepuschitz, Markus Vincze, and Munir Merdan. Knowledge-driven framework for industrial robotic systems. *Journal of Intelligent Manufacturing*, 34(2):771–788, 2023.

26 Francisco Ramos, Andrés S Vázquez, Raúl Fernández, and Alberto Olivares Alarcos. Ontology based design, control and programming of modular robots. *Integrated Computer-Aided Engineering*, 25(2):173–192, 2018.

27 Michael Short and Kevin Burn. A generic controller architecture for intelligent robotic systems. *Robotics and Computer-Integrated Manufacturing*, 27(2): 292–305, 2011.

28 Xavier Armangué Quintana. *Modelling stereoscopic vision systems for robotic applications*. Universitat de Girona, 2003.

29 Askhat Diveev and Elena Sofronova. Automation of synthesized optimal control problem solution for mobile robot by genetic programming. In *Intelligent Systems and Applications: Proceedings of the 2019 Intelligent Systems Conference (IntelliSys) Volume 2*, pages 1054–1072. Springer, 2020.

30 Reem J Alattas, Sarosh Patel, and Tarek M Sobh. Evolutionary modular robotics: Survey and analysis. *Journal of Intelligent & Robotic Systems*, 95:815–828, 2019.

31 Harry A Pierson and Michael S Gashler. Deep learning in robotics: A review of recent research. *Advanced Robotics*, 31(16):821–835, 2017.

32 Bharat Singh, Rajesh Kumar, and Vinay Pratap Singh. Reinforcement learning in robotic applications: A comprehensive survey. *Artificial Intelligence Review*, 55(2):945–990, 2022.

33 Thomas Hellström. The relevance of causation in robotics: A review, categorization, and analysis. *Paladyn, Journal of Behavioral Robotics*, 12(1):238–255, 2021.

34 Victor Zykov, Efstathios Mytilinaios, Bryant Adams, and Hod Lipson. Self-reproducing machines. *Nature*, 435(7039):163–164, 2005.

35 Lorenzo Garattoni and Mauro Birattari. Autonomous task sequencing in a robot swarm. *Science Robotics*, 3(20):eaat0430, 2018.

36 Zhun Fan. *Mechatronic design automation: An emerging research and recent advances*. Nova Science Publishers, 2010.

37 Zhun Fan, Jiachuan Wang, and Erik Goodman. Exploring open-ended design space of mechatronic systems. *International Journal of Advanced Robotic Systems*, 1(4):24, 2004.

38 Grace W Lindsay. Convolutional neural networks as a model of the visual system: Past, present, and future. *Journal of Cognitive Neuroscience*, 33(10):2017–2031, 2021.

39 Yanmin Qian, Zhengyang Chen, and Shuai Wang. Audio-visual deep neural network for robust person verification. *IEEE/ACM Transactions on Audio, Speech, and Language Processing*, 29:1079–1092, 2021.

40 Huanlin Li, Yuwei Cai, Juncao Hong, Peng Xu, Hui Cheng, Xiaomin Zhu, Bingliang Hu, Zhifeng Hao, and Zhun Fan. VG-Swarm: A vision-based gene regulation network for UAVs swarm behavior emergence. *IEEE Robotics and Automation Letters*, 8(3):1175–1182, 2023.

41 Jie Li and Ying Tan. A probabilistic finite state machine based strategy for multi-target search using swarm robotics. *Applied Soft Computing*, 77:467–483, 2019.

42 David B Fogel. Practical advantages of evolutionary computation. *Applications of Soft Computing*, 3165:14–22, 1997.

43 Kalyanmoy Deb. Multi-objective optimisation using evolutionary algorithms: An introduction. In *Multi-Objective Evolutionary Optimisation for Product Design and Manufacturing*, pages 3–34. Springer, 2011.

44 Arash Asrari, Saeed Lotfifard, and Mohammad S Payam. Pareto dominance-based multiobjective optimization method for distribution network reconfiguration. *IEEE Transactions on Smart Grid*, 7(3):1401–1410, 2015.

45 Kalyanmoy Deb, Amrit Pratap, Sameer Agarwal, and TAMT Meyarivan. A fast and elitist multiobjective genetic algorithm: NSGA-II. *IEEE Transactions on Evolutionary Computation*, 6(2):182–197, 2002.

46 David W Corne, Nick R Jerram, Joshua Damian Knowles, and Martin J Oates. PESA-II: Region-based selection in evolutionary multiobjective optimization. In *Proceedings of the 3rd Annual Conference on Genetic and Evolutionary Computation*, pages 283–290, 2001.

47 Eckart Zitzler, Marco Laumanns, and Lothar Thiele. SPEA2: Improving the strength pareto evolutionary algorithm. *TIK Report*, 103:1–21, 2001.

48 Anupam Trivedi, Dipti Srinivasan, Krishnendu Sanyal, and Abhiroop Ghosh. A survey of multiobjective evolutionary algorithms based on decomposition. *IEEE Transactions on Evolutionary Computation*, 21(3):440–462, 2016.

49 Aimin Zhou, Qingfu Zhang, and Guixu Zhang. A multiobjective evolutionary algorithm based on decomposition and probability model. In *2012 IEEE Congress on Evolutionary Computation*, pages 1–8. IEEE, 2012.

50 Tadahiko Murata, Hisao Ishibuchi, and Mitsuo Gen. Specification of genetic search directions in cellular multi-objective genetic algorithms. In *Evolutionary Multi-Criterion Optimization: First International Conference, EMO 2001 Zurich, Switzerland, March 7–9, 2001 Proceedings 1*, pages 82–95. Springer, 2001.

51 Hai-Lin Liu, Fangqing Gu, and Qingfu Zhang. Decomposition of a multiobjective optimization problem into a number of simple multiobjective subproblems. *IEEE Transactions on Evolutionary Computation*, 18(3):450–455, 2013.

52 Ye Tian, Ran Cheng, Xingyi Zhang, Fan Cheng, and Yaochu Jin. An indicator-based multiobjective evolutionary algorithm with reference point adaptation for better versatility. *IEEE Transactions on Evolutionary Computation*, 22(4):609–622, 2017.

53 Eckart Zitzler and Simon Künzli. Indicator-based selection in multiobjective search. In *Parallel Problem Solving from Nature - PPSN VIII*, pages 832–842. Springer, 2004.

54 Michael Emmerich, Nicola Beume, and Boris Naujoks. An EMO algorithm using the hypervolume measure as selection criterion. In *International Conference on Evolutionary Multi-Criterion Optimization*, pages 62–76. Springer, 2005.

55 Johannes Bader and Eckart Zitzler. HypE: An algorithm for fast hypervolume-based many-objective optimization. *Evolutionary Computation*, 19(1):45–76, 2011.

56 Raquel Hernández Gómez and Carlos A Coello Coello. Improved metaheuristic based on the r2 indicator for many-objective optimization. In *Proceedings of the 2015 Annual Conference on Genetic and Evolutionary Computation*, pages 679–686, 2015.

57 Jing Liang, Xuanxuan Ban, Kunjie Yu, Boyang Qu, Kangjia Qiao, Caitong Yue, Ke Chen, and Kay Chen Tan. A survey on evolutionary constrained multiobjective optimization. *IEEE Transactions on Evolutionary Computation*, 27(2):201–221, 2022.

58 Carlos A Coello Coello. Theoretical and numerical constraint-handling techniques used with evolutionary algorithms: A survey of the state of the art. *Computer methods in Applied Mechanics and Engineering*, 191(11–12): 1245–1287, 2002.

59 Abdollah Homaifar, Charlene X Qi, and Steven H Lai. Constrained optimization via genetic algorithms. *Simulation*, 62(4):242–253, 1994.

60 Jeffrey A Joines and Christopher R Houck. On the use of non-stationary penalty functions to solve nonlinear constrained optimization problems with GA's. In *Proceedings of the first IEEE Conference on Evolutionary Computation. IEEE World Congress on Computational Intelligence*, pages 579–584. IEEE, 1994.

61 David W Coit, Alice E Smith, and David M Tate. Adaptive penalty methods for genetic optimization of constrained combinatorial problems. *INFORMS Journal on Computing*, 8(2):173–182, 1996.

62 Atidel Ben Hadj-Alouane and James C Bean. A genetic algorithm for the multiple-choice integer program. *Operations Research*, 45(1):92–101, 1997.

63 Yonas Gebre Woldesenbet, Gary G Yen, and Biruk G Tessema. Constraint handling in multiobjective evolutionary optimization. *IEEE Transactions on Evolutionary Computation*, 13(3):514–525, 2009.

64 Tetsuyuki Takahama and Setsuko Sakai. Constrained optimization by ε constrained particle swarm optimizer with ε-level control. In *Soft Computing as*

Transdisciplinary Science and Technology: Proceedings of the fourth IEEE International Workshop WSTST'05, pages 1019–1029. Springer, 2005.

65 Thomas Philip Runarsson and Xin Yao. Stochastic ranking for constrained evolutionary optimization. *IEEE Transactions on Evolutionary Computation*, 4(3):284–294, 2000.

66 Bingdong Li, Jinlong Li, Ke Tang, and Xin Yao. Many-objective evolutionary algorithms: A survey. *ACM Computing Surveys (CSUR)*, 48(1):1–35, 2015.

67 Efrén Mezura-Montes and Carlos A Coello Coello. Constraint-handling in nature-inspired numerical optimization: Past, present and future. *Swarm and Evolutionary Computation*, 1(4):173–194, 2011.

68 Ye Tian, Tao Zhang, Jianhua Xiao, Xingyi Zhang, and Yaochu Jin. A coevolutionary framework for constrained multiobjective optimization problems. *IEEE Transactions on Evolutionary Computation*, 25(1):102–116, 2020.

69 Tinkle Chugh, Yaochu Jin, Kaisa Miettinen, Jussi Hakanen, and Karthik Sindhya. A surrogate-assisted reference vector guided evolutionary algorithm for computationally expensive many-objective optimization. *IEEE Transactions on Evolutionary Computation*, 22(1):129–142, 2016.

70 Ke Li and Renzhi Chen. Batched data-driven evolutionary multiobjective optimization based on manifold interpolation. *IEEE Transactions on Evolutionary Computation*, 27(1):126–140, 2022.

71 Guangyong Sun, Guangyao Li, Zhihui Gong, Guanqiang He, and Qing Li. Radial basis functional model for multi-objective sheet metal forming optimization. *Engineering Optimization*, 43(12):1351–1366, 2011.

72 Ilya Loshchilov, Marc Schoenauer, and Michele Sebag. Dominance-based pareto-surrogate for multi-objective optimization. In *Asia-Pacific Conference on Simulated Evolution and Learning*, pages 230–239. Springer, 2010.

73 Taimoor Akhtar and Christine A Shoemaker. Multi objective optimization of computationally expensive multi-modal functions with rbf surrogates and multi-rule selection. *Journal of Global Optimization*, 64:17–32, 2016.

74 Ruitao Mai, Yifeng Qiu, Wenji Li, Zhaojun Wang, Biao Xu, Wei Chen, Jiafan Zhuang, Yun Li, and Zhun Fan. Expensive constrained multi-objective evolutionary algorithm with pareto set learning. In *Proceedings of the Genetic and Evolutionary Computation Conference Companion*, pages 351–354, 2024.

75 Wenji Li, Ruitao Mai, Pengxiang Ren, Zhaojun Wang, Qinchang Zhang, Ning Xu, Biao Xu, Zhun Fan, and Zhifeng Hao. A surrogate-ensemble assisted coevolutionary algorithm for expensive constrained multi-objective optimization problems. In *2023 IEEE Congress on Evolutionary Computation (CEC)*, pages 1–7. IEEE, 2023.

76 Wenji Li, Ruitao Mai, Zhaojun Wang, Yifeng Qiu, Biao Xu, Zhifeng Hao, and Zhun Fan. Surrogate-assisted push and pull search for expensive constrained multi-objective optimization problems. *Swarm and Evolutionary Computation*, 91:101728, 2024.

77 Eckart Zitzler and Lothar Thiele. Multiobjective evolutionary algorithms: A comparative case study and the strength Pareto approach. *IEEE Transactions on Evolutionary Computation*, 3(4):257–271, 1999.

78 Peter AN Bosman and Dirk Thierens. The balance between proximity and diversity in multiobjective evolutionary algorithms. *IEEE Transactions on Evolutionary Computation*, 7(2):174–188, 2003.

79 Laung-Terng Wang, Yao-Wen Chang, and Kwang-Ting Tim Cheng. *Electronic design automation: synthesis, verification, and test*. Morgan Kaufmann, 2009.

80 Kyung-Joong Kim and Sung-Bae Cho. Automated synthesis of multiple analog circuits using evolutionary computation for redundancy-based fault-tolerance. *Applied Soft Computing*, 12(4):1309–1321, 2012.

81 Mehdi Anjomshoa, Ali Mahani, and Salahedin Sadeghifard. A new automated design and optimization method of CMOS logic circuits based on modified imperialistic competitive algorithm. *Applied Soft Computing*, 21:423–432, 2014.

82 Zdenek Vasicek and Lukas Sekanina. Evolutionary approach to approximate digital circuits design. *IEEE Transactions on Evolutionary Computation*, 19(3):432–444, 2014.

83 Mohammad Shokouhifar and Ali Jalali. An evolutionary-based methodology for symbolic simplification of analog circuits using genetic algorithm and simulated annealing. *Expert Systems with Applications*, 42(3):1189–1201, 2015.

84 Shashi Poddar, Parag Narkhede, Vipan Kumar, and Amod Kumar. PSO aided adaptive complementary filter for attitude estimation. *Journal of Intelligent & Robotic Systems*, 87:531–543, 2017.

85 Bo Liu, Hao Yang, and Michael J Lancaster. Global optimization of microwave filters based on a surrogate model-assisted evolutionary algorithm. *IEEE Transactions on Microwave Theory and Techniques*, 65(6):1976–1985, 2017.

86 Jun Zhang, Hnery Shu-Hung Chung, Wai-Lun Lo, SY Hui, and Angus Wu. Implementation of a decoupled optimization technique for design of switching regulators using genetic algorithms. *IEEE Transactions on Power Electronics*, 16(6):752–763, 2001.

87 Federico Castejón and Enrique J Carmona. Automatic design of analog electronic circuits using grammatical evolution. *Applied Soft Computing*, 62:1003–1018, 2018.

88 Žiga Rojec, Árpád Bűrmen, and Iztok Fajfar. Analog circuit topology synthesis by means of evolutionary computation. *Engineering Applications of Artificial Intelligence*, 80:48–65, 2019.

89 John R Koza, Forrest H Bennett, David, Andre, Martin A Keane, and Frank Dunlap. Automated synthesis of analog electrical circuits by means of genetic programming. *IEEE Transactions on Evolutionary Computation*, 1(2):109–128, 1997.

90 Claudio Mattiussi and Dario Floreano. Analog genetic encoding for the evolution of circuits and networks. *IEEE Transactions on Evolutionary Computation*, 11(5):596–607, 2007.

91 Cevher Ak and Ali Yildiz. A novel closed-form expression obtained by using differential evolution algorithm to calculate pull-in voltage of MEMS cantilever. *Journal of Microelectromechanical Systems*, 27(3):392–397, 2018.

92 Mehran Rahmani. MEMS gyroscope control using a novel compound robust control. *ISA Transactions*, 72:37–43, 2018.

93 Valentina Zega, Attilio Frangi, Andrea Guercilena, and Gabriele Gattere. Analysis of frequency stability and thermoelastic effects for slotted tuning fork MEMS resonators. *Sensors*, 18(7):2157, 2018.

94 Mobayode Olusola Akinsolu, Bo Liu, Vic Grout, Pavlos I Lazaridis, Maria Evelina Mognaschi, and Paolo Di Barba. A parallel surrogate model assisted evolutionary algorithm for electromagnetic design optimization. *IEEE Transactions on Emerging Topics in Computational Intelligence*, 3(2):93–105, 2019.

95 Mohammad Reza Solouk, Mohammad Hassan Shojaeefard, and Masoud Dahmardeh. Parametric topology optimization of a MEMS gyroscope for automotive applications. *Mechanical Systems and Signal Processing*, 128:389–404, 2019.

96 Seyedfakhreddin Nabavi and Lihong Zhang. Design and optimization of piezoelectric MEMS vibration energy harvesters based on genetic algorithm. *IEEE Sensors Journal*, 17(22):7372–7382, 2017.

97 Seyedfakhreddin Nabavi and Lihong Zhang. Frequency tuning and efficiency improvement of piezoelectric MEMS vibration energy harvesters. *Journal of Microelectromechanical Systems*, 28(1):77–87, 2018.

98 Zhun Fan, Jinchao Liu, Torben Sorensen, and Pan Wang. Improved differential evolution based on stochastic ranking for robust layout synthesis of MEMS components. *IEEE Transactions on Industrial Electronics*, 56(4):937–948, 2008.

99 Gary K Fedder and Qi Jing. A hierarchical circuit-level design methodology for microelectromechanical systems. *IEEE Transactions on Circuits and Systems II: Analog and Digital Signal Processing*, 46(10):1309–1315, 1999.

100 Tamal Mukherjee and Gary K Fedder. Hierarchical mixed-domain circuit simulation, synthesis and extraction methodology for MEMS. *Journal of VLSI Signal Processing Systems for Signal, Image and Video Technology*, 21:233–249, 1999.

101 Zhun Fan, Jiachuan Wang, Sofiane Achiche, Erik Goodman, and Ronald Rosenberg. Structured synthesis of MEMS using evolutionary approaches. *Applied Soft Computing*, 8(1):579–589, 2008.

102 Zhun Fan, Erik D Goodman, Jiachuan Wang, Ronald Rosenberg, Kisung Seo, and Jianjun Hu. Hierarchical evolutionary synthesis of MEMS. In *Proceedings of the 2004 Congress on Evolutionary Computation (IEEE Cat. No. 04TH8753)*, volume 2, pages 2320–2327. IEEE, 2004.

103 Ying Zhang. *MEMS design synthesis based on hybrid evolutionary computation*. PhD thesis, University of California, Berkeley, 2006.

104 Kamal Youcef-Toumi. Modeling, design, and control integration: A necessary step in mechatronics. *IEEE/ASME Transactions on Mechatronics*, 1(1):29–38, 1996.

105 Pushpendra Kumar, Ismail Bensekrane, Blaise Conrard, Armand Toguyeni, and Rochdi Merzouki. Functionability analysis of redundant mechatronic systems in bond graph framework. *IEEE/ASME Transactions on Mechatronics*, 24(6):2465–2476, 2019.

106 Lijie Zhang, Fei Guo, Yongquan Li, and Wenjuan Lu. Global dynamic modeling of electro-hydraulic 3-UPS/S parallel stabilized platform by bond graph. *Chinese Journal of Mechanical Engineering*, 29(6):1176–1185, 2016.

107 Barbara Lafarge, Sébastien Grondel, Christophe Delebarre, and Eric Cattan. A validated simulation of energy harvesting with piezoelectric cantilever beams on a vehicle suspension using Bond Graph approach. *Mechatronics*, 53:202–214, 2018.

108 Eng Hock Tay, Woodie Flowers, and John Barrus. Automated generation and analysis of dynamic system designs. *Research in Engineering Design*, 10:15–29, 1998.

109 Kisung Seo, Zhun Fan, Jianjun Hu, Erik D Goodman, and Ronald C Rosenberg. Toward a unified and automated design methodology for multi-domain dynamic systems using bond graphs and genetic programming. *Mechatronics*, 13(8–9):851–885, 2003.

110 Zhun Fan, Kisung Seo, Jianjun Hu, Erik D Goodman, and Ronald C Rosenberg. A novel evolutionary engineering design approach for mixed-domain systems. *Engineering Optimization*, 36(2):127–147, 2004.

111 Jiachuan Wang, Zhun Fan, Janis P Terpenny, and Erik D Goodman. Knowledge interaction with genetic programming in mechatronic systems design using bond graphs. *IEEE Transactions on Systems, Man, and Cybernetics, Part C (Applications and Reviews)*, 35(2):172–182, 2005.

112 Yi Ye, Chen-Bo Yin, Yue Gong, and Jun-jing Zhou. Position control of nonlinear hydraulic system using an improved PSO based PID controller. *Mechanical Systems and Signal Processing*, 83:241–259, 2017.

113 Alejandro Rodíguez-Molina, Miguel Gabriel Villarreal-Cervantes, and Mario Aldape-Pérez. Adaptive control for the four-bar linkage mechanism based on differential evolution. In *2018 IEEE Congress on Evolutionary Computation (CEC)*, pages 1–7. IEEE, 2018.

114 Camilo Caraveo, Fevrier Valdez, and Oscar Castillo. Optimization of fuzzy controller design using a new bee colony algorithm with fuzzy dynamic parameter adaptation. *Applied Soft Computing*, 43:131–142, 2016.

115 Aliasghar Arab and Alireza Alfi. An adaptive gradient descent-based local search in memetic algorithm applied to optimal controller design. *Information Sciences*, 299:117–142, 2015.

116 Ibrahim Mustafa Mehedi, Ubaid M Al-Saggaf, Rachid Mansouri, and Maamar Bettayeb. Two degrees of freedom fractional controller design: Application to the ball and beam system. *Measurement*, 135:13–22, 2019.

117 Patricia Ochoa, Oscar Castillo, and José Soria. Optimization of fuzzy controller design using a differential evolution algorithm with dynamic parameter adaptation based on type-1 and interval type-2 fuzzy systems. *Soft Computing*, 24(1):193–214, 2020.

118 Laatra Yousfi, Amel Bouchemha, Mohcene Bechouat, and Abdelhani Boukrouche. Vector control of induction machine using PI controller optimized by genetic algorithms. In *2014 16th International Power Electronics and Motion Control Conference and Exposition*, pages 1272–1277. IEEE, 2014.

119 Liang Li, Yahui Zhang, Chao Yang, Xiaohong Jiao, Lipeng Zhang, and Jian Song. Hybrid genetic algorithm-based optimization of powertrain and control parameters of plug-in hybrid electric bus. *Journal of the Franklin Institute*, 352(3):776–801, 2015.

120 Amal Moharam, Mostafa A El-Hosseini, and Hesham A Ali. Design of optimal PID controller using hybrid differential evolution and particle swarm optimization with an aging leader and challengers. *Applied Soft Computing*, 38:727–737, 2016.

121 Fubin Wang, Jie Liu, Zhikun Chen, and Chunwang Jiao. Auto disturbances refection visual servoing control for excavator robot based on niching particle swarm optimization. *Jixie Gongcheng Xuebao (Chinese Journal of Mechanical Engineering)*, 48(1):32–38, 2012.

122 Dong-Li Zhang, Tang Ying-Gan, and Guan Xin-Ping. Optimum design of fractional order PID controller for an AVR system using an improved artificial bee colony algorithm. *Acta Automatica Sinica*, 40(5):973–979, 2014.

123 Weijia Zheng and Youguo Pi. Study of the fractional order proportional integral controller for the permanent magnet synchronous motor based on the differential evolution algorithm. *ISA transactions*, 63:387–393, 2016.

124 WeiJia Zheng, Ying Luo, XiaoHong Wang, YouGuo Pi, and YangQuan Chen. Fractional order piλdμ controller design for satisfying time and frequency domain specifications simultaneously. *ISA Transactions*, 68:212–222, 2017.

125 John R Koza, Martin A Keane, Jessen Yu, William Mydlowec, and Forrest H Bennett. Automatic synthesis of both the control law and parameters for a controller for a three-lag plant with five-second delay using genetic programming and simulation techniques. In *Proceedings of the 2000 American Control Conference. ACC (IEEE Cat. No. 00CH36334)*, volume 1, pages 453–459. IEEE, 2000.

126 John R Koza, Martin A Keane, Jessen Yu, Forrest H Bennett, and William Mydlowec. Automatic creation of human-competitive programs and controllers by means of genetic programming. *Genetic Programming and Evolvable Machines*, 1:121–164, 2000.

127 Martin A Keane, Jessen Yu, and John R Koza. Automatic synthesis of both the topology and tuning of a common parameterized controller for two families of plants using genetic programming. In *Proceedings of the 2nd Annual Conference on Genetic and Evolutionary Computation*, pages 496–504. Citeseer, 2000.

128 Martin A Keane, John R Koza, and Matthew J Streeter. Automatic synthesis using genetic programming of an improved general-purpose controller for industrially representative plants. In *Proceedings 2002 NASA/DoD Conference on Evolvable Hardware*, pages 113–122. IEEE, 2002.

129 Dennis Bruijnen, Iris Soute, René van de Molengraft, and Maarten Steinbuch. Exploration of automatic controller synthesis for motion systems using genetic programming. *IFAC Proceedings Volumes*, 37(14):229–234, 2004.

130 Krystian Łapa and Krzysztof Cpałka. Flexible fuzzy PID controller (FFPIDC) and a nature-inspired method for its construction. *IEEE Transactions on Industrial Informatics*, 14(3):1078–1088, 2017.

131 Zhan Jiaoyang, Xin Bin, and Chen Jie. Evolutionary design of controllers with optimized structure and its application in a maglev ball control system. In *2017 36th Chinese Control Conference (CCC)*, pages 2545–2550. IEEE, 2017.

132 Zhan Jiao-Yang. *The intelligent optimization design of disturbance rejection controllers with optimized structures.* PhD thesis, Beijing Institute of Technology, 2019.

133 Mitchell A Potter and Kenneth A De Jong. Cooperative coevolution: An architecture for evolving coadapted subcomponents. *Evolutionary Computation*, 8(1):1–29, 2000.

134 Rudolf Paul Wiegand, William C Liles, and Kenneth A De Jong. An empirical analysis of collaboration methods in cooperative coevolutionary algorithms. In *Proceedings of the Genetic and Evolutionary Computation Conference (GECCO)*, volume 2611, pages 1235–1245. Morgan Kaufmann San Francisco, 2001.

135 Jiachuan Wang, Zhun Fan, Janis P Terpenny, and Erik D Goodman. Cooperative body–brain coevolutionary synthesis of mechatronic systems. *AI EDAM*, 22(3):219–234, 2008.

136 Jean-François Dupuis, Zhun Fan, and Erik Goodman. Evolved finite state controller for hybrid system. In *Proceedings of the first ACM/SIGEVO Summit on Genetic and Evolutionary Computation*, pages 105–112, 2009.

137 Jean-Francois Dupuis and Zhun Fan. Evolved finite state controller for hybrid system in reduced search space. In *2009 IEEE/ASME International Conference on Advanced Intelligent Mechatronics*, pages 833–838. IEEE, 2009.

138 Jean-François Dupuis and Zhun Fan. Comparing an evolved finite state controller for hybrid system to a lookahead design. In *IEEE Congress on Evolutionary Computation*, pages 1–6. IEEE, 2010.

139 Jean-Francois Dupuis, Zhun Fan, and Erik Goodman. Evolutionary design of discrete controllers for hybrid mechatronic systems. *International Journal of Systems Science*, 46(2):303–316, 2015.

140 Mehran Rahmani, Ahmad Ghanbari, and Mir Mohammad Ettefagh. A novel adaptive neural network integral sliding-mode control of a biped robot using bat algorithm. *Journal of Vibration and Control*, 24(10):2045–2060, 2018.

141 Marco Dorigo, Mauro Birattari, and Manuele Brambilla. Swarm robotics. *Scholarpedia*, 9(1):1463, 2014.

142 Guanqiang Gao, Yi Mei, Bin Xin, Ya-Hui Jia, and Will N Browne. Automated coordination strategy design using genetic programming for dynamic multipoint dynamic aggregation. *IEEE Transactions on Cybernetics*, 52(12):13521–13535, 2021.

143 Sanza Kazadi. Model independence in swarm robotics. *International Journal of Intelligent Computing and Cybernetics*, 2(4):672–694, 2009.

144 Spring Berman, Vijay Kumar, and Radhika Nagpal. Design of control policies for spatially inhomogeneous robot swarms with application to commercial pollination. In *2011 IEEE international conference on robotics and automation*, pages 378–385. IEEE, 2011.

145 Manuele Brambilla, Carlo Pinciroli, Mauro Birattari, and Marco Dorigo. Property-driven design for swarm robotics. In *AAMAS*, volume 2012, pages 1–8. 2012.

146 Gianpiero Francesca, Manuele Brambilla, Arne Brutschy, Vito Trianni, and Mauro Birattari. Automode: A novel approach to the automatic design of control software for robot swarms. *Swarm Intelligence*, 8:89–112, 2014.

147 Gianpiero Francesca, Manuele Brambilla, Arne Brutschy, Lorenzo Garattoni, Roman Miletitch, Gaëtan Podevijn, Andreagiovanni Reina, Touraj Soleymani, Mattia Salvaro, Carlo Pinciroli, et al. Automode-Chocolate: Automatic design of control software for robot swarms. *Swarm Intelligence*, 9:125–152, 2015.

148 Marco Dorigo, Guy Theraulaz, and Vito Trianni. Swarm robotics: Past, present, and future [point of view]. *Proceedings of the IEEE*, 109(7):1152–1165, 2021.

149 Ernesto Nunes, Marie Manner, Hakim Mitiche, and Maria Gini. A taxonomy for task allocation problems with temporal and ordering constraints. *Robotics and Autonomous Systems*, 90:55–70, 2017.

150 Meng Wu, Xiaomin Zhu, Li Ma, Ji Wang, Weidong Bao, Wenji Li, and Zhun Fan. Torch: Strategy evolution in swarm robots using heterogeneous–homogeneous coevolution method. *Journal of Industrial Information Integration*, 25:100239, 2022.

151 Gábor Vásárhelyi, Csaba Virágh, Gergő Somorjai, Tamás Nepusz, Agoston E Eiben, and Tamás Vicsek. Optimized flocking of autonomous drones in confined environments. *Science Robotics*, 3(20):eaat3536, 2018.

152 Rolf Pfeifer, Max Lungarella, and Fumiya Iida. Self-organization, embodiment, and biologically inspired robotics. *Science*, 318(5853):1088–1093, 2007.

153 Dylan Shah, Bilige Yang, Sam Kriegman, Michael Levin, Josh Bongard, and Rebecca Kramer-Bottiglio. Shape changing robots: Bioinspiration, simulation, and physical realization. *Advanced Materials*, 33(19):2002882, 2021.

154 Karine Miras, Eliseo Ferrante, and Agoston E Eiben. Environmental influences on evolvable robots. *PloS One*, 15(5):e0233848, 2020.

155 Gongjin Lan, Milan Jelisavcic, Diederik M Roijers, Evert Haasdijk, and Agoston E Eiben. Directed locomotion for modular robots with evolvable morphologies. In *Parallel Problem Solving from Nature–PPSN XV: 15th International Conference*, Coimbra, Portugal, September 8–12, 2018, *Proceedings, Part I 15*, pages 476–487. Springer, 2018.

156 Aguston E Eiben, Nicolas Bredeche, Mark Hoogendoorn, Jürgen Stradner, Jon Timmis, Andy Tyrrell, and Alan Winfield. The triangle of life: Evolving robots in real-time and real-space. In *European Conference on Artificial Life (ECAL-2013)*, pages 1–8, 2013.

157 Agoston E Eiben and Jim Smith. From evolutionary computation to the evolution of things. *Nature*, 521(7553):476–482, 2015.

158 Agoston E Eiben, Serge Kernbach, and Evert Haasdijk. Embodied artificial evolution: Artificial evolutionary systems in the 21st century. *Evolutionary Intelligence*, 5:261–272, 2012.

159 Josh C Bongard. Evolutionary robotics. *Communications of the ACM*, 56(8):74–83, 2013.

160 Daniel Marbach and Auke Jan Ijspeert. Co-evolution of configuration and control for homogenous modular robots. In *Proceedings of the Eighth Conference on Intelligent Autonomous Systems (IAS8)*, pages 712–719. IOS Press, 2004.

161 Charles Schaff. Neural approaches to co-optimization in robotics. *arXiv preprint arXiv:2209.00579*, 2022.

162 Lisa Meeden and Deepak Kumar. Trends in evolutionary robotics. In *Soft Computing for Intelligent Robotic Systems*, pages 215–233. Springer, 1998.

163 Allan Zhao, Jie Xu, Mina Konaković-Luković, Josephine Hughes, Andrew Spielberg, Daniela Rus, and Wojciech Matusik. Robogrammar: Graph grammar for terrain-optimized robot design. *ACM Transactions on Graphics (TOG)*, 39(6):1–16, 2020.

164 Jie Xu, Andrew Spielberg, Allan Zhao, Daniela Rus, and Wojciech Matusik. Multi-objective graph heuristic search for terrestrial robot design. In *2021 IEEE International Conference on Robotics and Automation (ICRA)*, pages 9863–9869. IEEE, 2021.

165 Aslan Miriyev and Mirko Kovač. Skills for physical artificial intelligence. *Nature Machine Intelligence*, 2(11):658–660, 2020.

166 Jiafu Wan, Shenglong Tang, Di Li, Muhammad Imran, Chunhua Zhang, Chengliang Liu, and Zhibo Pang. Reconfigurable smart factory for drug packing in healthcare industry 4.0. *IEEE transactions on industrial informatics*, 15(1):507–516, 2018.

167 Qianli Ma, Haiyuan Li, and Gregory S Chirikjian. New probabilistic approaches to the AX= XB hand-eye calibration without correspondence. In *2016 IEEE International Conference on Robotics and Automation (ICRA)*, pages 4365–4371. IEEE, 2016.

168 Chaoqun Niu, Qing Zhu, Yaonan Wang, Xianen Zhou, and Wenting Shen. Real time counting system of glass bottle based on multi objects tracking. In *2021 China Automation Congress (CAC)*, pages 5402–5407. IEEE, 2021.

169 Bin Yan, Houwen Peng, Kan Wu, Dong Wang, Jianlong Fu, and Huchuan Lu. Lighttrack: Finding lightweight neural networks for object tracking via one-shot architecture search. In *Proceedings of the IEEE/CVF Conference on Computer Vision and Pattern Recognition*, pages 15180–15189, 2021.

170 Thanaporn Viriyasaranon and Jang-Hwan Choi. Object detectors involving a NAS-gate convolutional module and capsule attention module. *Scientific Reports*, 12(1):3916, 2022.

171 Ning Wang, Yang Gao, Hao Chen, Peng Wang, Zhi Tian, Chunhua Shen, and Yanning Zhang. NAS-FCOS: Fast neural architecture search for object detection. In *Proceedings of the IEEE/CVF Conference on Computer Vision and Pattern Recognition*, pages 11943–11951, 2020.

172 Yukang Chen, Tong Yang, Xiangyu Zhang, Gaofeng Meng, Xinyu Xiao, and Jian Sun. Detnas: Backbone search for object detection. *Advances in Neural Information Processing Systems*, 32:6642–6652, 2019.

173 Lewei Yao, Hang Xu, Wei Zhang, Xiaodan Liang, and Zhenguo Li. SM-NAS: Structural-to-modular neural architecture search for object detection. In *Proceedings of the AAAI Conference on Artificial Intelligence*, volume 34, pages 12661–12668, 2020.

174 Hui Zhang, Liuchen Wu, Yurong Chen, Ruibo Chen, Senlin Kong, Yaonan Wang, Jianwen Hu, and Jonathan Wu. Attention-guided multitask convolutional neural network for power line parts detection. *IEEE Transactions on Instrumentation and Measurement*, 71:1–13, 2022.

175 Zongwei Zhou, Md Mahfuzur Rahman Siddiquee, Nima Tajbakhsh, and Jianming Liang. UNet++: A nested U-Net architecture for medical image segmentation. In *Deep Learning in Medical Image Analysis and Multimodal Learning for Clinical Decision Support: 4th International Workshop, DLMIA 2018, and 8th International Workshop, ML-CDS 2018, Held in Conjunction with*

MICCAI 2018, Granada, Spain, September 20, 2018, *Proceedings 4*, pages 3–11. Springer, 2018.

176 Fabian Isensee, Paul F Jaeger, Simon AA Kohl, Jens Petersen, and Klaus H Maier-Hein. nnU-Net: A self-configuring method for deep learning-based biomedical image segmentation. *Nature Methods*, 18(2):203–211, 2021.

177 Chenxi Liu, Liang-Chieh Chen, Florian Schroff, Hartwig Adam, Wei Hua, Alan L Yuille, and Li Fei-Fei. Auto-deeplab: Hierarchical neural architecture search for semantic image segmentation. In *Proceedings of the IEEE/CVF Conference on Computer Vision and Pattern Recognition*, pages 82–92, 2019.

178 Vladimir Nekrasov, Hao Chen, Chunhua Shen, and Ian Reid. Fast neural architecture search of compact semantic segmentation models via auxiliary cells. In *Proceedings of the IEEE/CVF Conference on Computer Vision and Pattern Recognition*, pages 9126–9135, 2019.

179 Jiahong Wei, Guijie Zhu, Zhun Fan, Jinchao Liu, Yibiao Rong, Jiajie Mo, Wenji Li, and Xinjian Chen. Genetic U-Net: automatically designed deep networks for retinal vessel segmentation using a genetic algorithm. *IEEE Transactions on Medical Imaging*, 41(2):292–307, 2021.

180 Feng Zhang, Xiatian Zhu, and Mao Ye. Fast human pose estimation. In *Proceedings of the IEEE/CVF Conference on Computer Vision and Pattern Recognition*, pages 3517–3526, 2019.

181 Anton Milan. MOT16: A benchmark for multi-object tracking. *arXiv preprint arXiv:1603.00831*, 2016.

182 Zhengxia Zou, Keyan Chen, Zhenwei Shi, Yuhong Guo, and Jieping Ye. Object detection in 20 years: A survey. *Proceedings of the IEEE*, 111(3):257–276, 2023.

183 Han Cai, Ligeng Zhu, and Song Han. ProxylessNAS: Direct neural architecture search on target task and hardware. *arXiv preprint arXiv:1812.00332*, 2018.

184 Bichen Wu, Xiaoliang Dai, Peizhao Zhang, Yanghan Wang, Fei Sun, Yiming Wu, Yuandong Tian, Peter Vajda, Yangqing Jia, and Kurt Keutzer. FBNet: Hardware-aware efficient convnet design via differentiable neural architecture search. In *Proceedings of the IEEE/CVF Conference on Computer Vision and Pattern Recognition*, pages 10734–10742, 2019.

185 Javier García López, Antonio Agudo, and Francesc Moreno-Noguer. E-DNAS: Differentiable neural architecture search for embedded systems. In *2020 25th International Conference on Pattern Recognition (ICPR)*, pages 4704–4711. IEEE, 2021.

186 Xiangzhong Luo, Di Liu, Hao Kong, Shuo Huai, Hui Chen, and Weichen Liu. LightNAS: On lightweight and scalable neural architecture search for embedded platforms. *IEEE Transactions on Computer-Aided Design of Integrated Circuits and Systems*, 42(6):1784–1797, 2022.

187 Amber Cassimon, Simon Vanneste, Stig Bosmans, Siegfried Mercelis, and Peter Hellinckx. Designing resource-constrained neural networks using neural architecture search targeting embedded devices. *Internet of Things*, 12:100234, 2020.

188 Alvin Wan, Xiaoliang Dai, Peizhao Zhang, Zijian He, Yuandong Tian, Saining Xie, Bichen Wu, Matthew Yu, Tao Xu, Kan Chen, et al. FBNetV2: Differentiable neural architecture search for spatial and channel dimensions. In *Proceedings of the IEEE/CVF Conference on Computer Vision and Pattern Recognition*, pages 12965–12974, 2020.

189 Yihui He, Ji Lin, Zhijian Liu, Hanrui Wang, Li-Jia Li, and Song Han. AMC: AutoML for model compression and acceleration on mobile devices. In *Proceedings of the European Conference on Computer Vision (ECCV)*, pages 784–800, 2018.

190 TP Lillicrap. Continuous control with deep reinforcement learning. *arXiv preprint arXiv:1509.02971*, 2015.

191 Manas Gupta, Siddharth Aravindan, Aleksandra Kalisz, Vijay Chandrasekhar, and Lin Jie. Learning to prune deep neural networks via reinforcement learning. *arXiv preprint arXiv:2007.04756*, 2020.

192 Sixing Yu, Arya Mazaheri, and Ali Jannesari. Topology-aware network pruning using multi-stage graph embedding and reinforcement learning. In *International Conference on Machine Learning*, pages 25656–25667. PMLR, 2022.

193 Zi Wang and Chengcheng Li. Channel pruning via lookahead search guided reinforcement learning. In *Proceedings of the IEEE/CVF Winter Conference on Applications of Computer Vision*, pages 2029–2040, 2022.

194 Miao Yin, Yang Sui, Siyu Liao, and Bo Yuan. Towards efficient tensor decomposition-based DNN model compression with optimization framework. In *Proceedings of the IEEE/CVF Conference on Computer Vision and Pattern Recognition*, pages 10674–10683, 2021.

195 Babak Rokh, Ali Azarpeyvand, and Alireza Khanteymoori. A comprehensive survey on model quantization for deep neural networks in image classification. *ACM Transactions on Intelligent Systems and Technology*, 14(6):1–50, 2023.

196 Pengguang Chen, Shu Liu, Hengshuang Zhao, and Jiaya Jia. Distilling knowledge via knowledge review. In *Proceedings of the IEEE/CVF Conference on Computer Vision and Pattern Recognition*, pages 5008–5017, 2021.

197 Jian Cheng, Pei-song Wang, Gang Li, Qing-hao Hu, and Han-qing Lu. Recent advances in efficient computation of deep convolutional neural networks. *Frontiers of Information Technology & Electronic Engineering*, 19:64–77, 2018.

198 Seema A Bhalgaonkar, Mousami V Munot, and Devidas Gopal Anuse. Pruning for compression of visual pattern recognition networks: A survey from deep neural networks perspective. *Pattern Recognition and Data Analysis with Applications*, pages 675–687, 2022.

199 Yu Cheng, Duo Wang, Pan Zhou, and Tao Zhang. Model compression and acceleration for deep neural networks: The principles, progress, and challenges. *IEEE Signal Processing Magazine*, 35(1):126–136, 2018.

200 Ching-Hao Wang, Kang-Yang Huang, Yi Yao, Jun-Cheng Chen, Hong-Han Shuai, and Wen-Huang Cheng. Lightweight deep learning: An overview. *IEEE Consumer Electronics Magazine*, 13(4):51–64, 2022.

201 Chenghao Zhang, Kun Tian, Bin Fan, Gaofeng Meng, Zhaoxiang Zhang, and Chunhong Pan. Continual stereo matching of continuous driving scenes with growing architecture. In *Proceedings of the IEEE/CVF Conference on Computer Vision and Pattern Recognition*, pages 18901–18910, 2022.

202 Xiaofang Wang, Xuehan Xiong, Maxim Neumann, AJ Piergiovanni, Michael S Ryoo, Anelia Angelova, Kris M Kitani, and Wei Hua. AttentionNAS: Spatiotemporal attention cell search for video classification. In *Computer Vision–ECCV 2020: 16th European Conference*, Glasgow, UK, August 23–28, 2020, Proceedings, Part VIII 16, pages 449–465. Springer, 2020.

203 Lumin Xu, Yingda Guan, Sheng Jin, Wentao Liu, Chen Qian, Ping Luo, Wanli Ouyang, and Xiaogang Wang. ViPNAS: Efficient video pose estimation via neural architecture search. In *Proceedings of the IEEE/CVF Conference on Computer Vision and Pattern Recognition*, pages 16072–16081, 2021.

204 Michael S Ryoo, AJ Piergiovanni, Mingxing Tan, and Anelia Angelova. AssembleNet: Searching for multi-stream neural connectivity in video architectures. *arXiv preprint arXiv:1905.13209*, 2019.

205 AJ Piergiovanni, Anelia Angelova, and Michael S Ryoo. Tiny video networks. *Applied AI Letters*, 3(1):e38, 2022.

206 Wei Peng, Xiaopeng Hong, and Guoying Zhao. Video action recognition via neural architecture searching. In *2019 IEEE International Conference on Image Processing (ICIP)*, pages 11–15. IEEE, 2019.

207 Xuelian Cheng, Yiran Zhong, Mehrtash Harandi, Yuchao Dai, Xiaojun Chang, Hongdong Li, Tom Drummond, and Zongyuan Ge. Hierarchical neural architecture search for deep stereo matching. *Advances in Neural Information Processing Systems*, 33:22158–22169, 2020.

208 Tonmoy Saikia, Yassine Marrakchi, Arber Zela, Frank Hutter, and Thomas Brox. AutoDispNet: Improving disparity estimation with autoML. In *Proceedings of the IEEE/CVF International Conference on Computer Vision*, pages 1812–1823, 2019.

209 Lam Huynh, Phong Nguyen, Jiří Matas, Esa Rahtu, and Janne Heikkilä. Lightweight monocular depth with a novel neural architecture search method. In *Proceedings of the IEEE/CVF Winter Conference on Applications of Computer Vision*, pages 3643–3653, 2022.

210 Shaoli Liu, Chengjian Zheng, Kaidi Lu, Si Gao, Ning Wang, Bofei Wang, Diankai Zhang, Xiaofeng Zhang, and Tianyu Xu. EVSRNet: Efficient video super-resolution with neural architecture search. In *Proceedings of the IEEE/CVF Conference on Computer Vision and Pattern Recognition*, pages 2480–2485, 2021.

211 AJ Piergiovanni, Anelia Angelova, and Michael S Ryoo. Tiny video networks. *Applied AI Letters*, 3(1):e38, 2022.

212 Qiang Wang, Shaohuai Shi, Kaiyong Zhao, and Xiaowen Chu. EASNet: Searching elastic and accurate network architecture for stereo matching. In *European Conference on Computer Vision*, pages 437–453. Springer, 2022.

213 Bowen Baker, Otkrist Gupta, Nikhil Naik, and Ramesh Raskar. Designing neural network architectures using reinforcement learning. *arXiv preprint arXiv:1611.02167*, 2016.

214 Barret Zoph. Neural architecture search with reinforcement learning. *arXiv preprint arXiv:1611.01578*, 2016.

215 Barret Zoph, Vijay Vasudevan, Jonathon Shlens, and Quoc V Le. Learning transferable architectures for scalable image recognition. In *Proceedings of the IEEE Conference on Computer Vision and Pattern Recognition*, pages 8697–8710, 2018.

216 Zhao Zhong, Junjie Yan, Wei Wu, Jing Shao, and Cheng-Lin Liu. Practical block-wise neural network architecture generation. In *Proceedings of the IEEE Conference on Computer Vision and Pattern Recognition*, pages 2423–2432, 2018.

217 Hanxiao Liu, Karen Simonyan, and Yiming Yang. Darts: Differentiable architecture search. *arXiv preprint arXiv:1806.09055*, 2018.

218 Xin Chen, Lingxi Xie, Jun Wu, and Qi Tian. Progressive differentiable architecture search: Bridging the depth gap between search and evaluation. In *Proceedings of the IEEE/CVF International Conference on Computer Vision*, pages 1294–1303, 2019.

219 Kenneth O Stanley and Risto Miikkulainen. Evolving neural networks through augmenting topologies. *Evolutionary Computation*, 10(2):99–127, 2002.

220 Risto Miikkulainen, Jason Liang, Elliot Meyerson, Aditya Rawal, Dan Fink, Olivier Francon, Bala Raju, Hormoz Shahrzad, Arshak Navruzyan, Nigel Duffy, et al. Evolving deep neural networks. In *Artificial Intelligence in the Age of Neural Networks and Brain Computing*, pages 269–287. Elsevier, 2024.

221 Zhichao Lu, Ian Whalen, Vishnu Boddeti, Yashesh Dhebar, Kalyanmoy Deb, Erik Goodman, and Wolfgang Banzhaf. NSGA-Net: neural architecture search using multi-objective genetic algorithm. In *Proceedings of the Genetic and Evolutionary Computation Conference*, pages 419–427, 2019.

222 Yanan Sun, Gary G Yen, and Zhang Yi. IGD indicator-based evolutionary algorithm for many-objective optimization problems. *IEEE Transactions on Evolutionary Computation*, 23(2):173–187, 2018.

223 Ashraf Darwish, Aboul Ella Hassanien, and Swagatam Das. A survey of swarm and evolutionary computing approaches for deep learning. *Artificial Intelligence Review*, 53(3):1767–1812, 2020.

224 Peter Spirtes, Clark Glymour, and Richard Scheines. *Causation, prediction, and search*. MIT press, 2001.

225 Chang Gong, Chuzhe Zhang, Di Yao, Jingping Bi, Wenbin Li, and YongJun Xu. Causal discovery from temporal data: An overview and new perspectives. *ACM Computing Surveys*, 57(4):1–38, 2023.

226 Hector Geffner, Rina Dechter, and Joseph Y Halpern. *Probabilistic and Causal Inference: The Works of Judea Pearl*. ACM, 2022.

227 Peter L Spirtes, Christopher Meek, and Thomas S Richardson. Causal inference in the presence of latent variables and selection bias. *arXiv preprint arXiv:1302.4983*, 2013.

228 David Maxwell Chickering. Optimal structure identification with greedy search. *Journal of Machine Learning Research*, 3(Nov):507–554, 2002.

229 Daphane Koller. *Probabilistic graphical models: Principles and techniques*. MIT press, 2009.

230 Kenneth A Bollen. *Structural equations with latent variables*, volume 210. John Wiley & Sons, 1989.

231 Doug Fisher and Hans-J Lenz. *Learning from data: Artificial intelligence and statistics V*, volume 5. Springer Science & Business Media, 1996.

232 Jie Qiao, Ruichu Cai, Kun Zhang, Zhenjie Zhang, and Zhifeng Hao. Causal discovery with confounding cascade nonlinear additive noise models. *ACM Transactions on Intelligent Systems and Technology (TIST)*, 12(6):1–28, 2021.

233 Kun Zhang, Zhikun Wang, Jiji Zhang, and Bernhard Schölkopf. On estimation of functional causal models: General results and application to the post-nonlinear causal model. *ACM Transactions on Intelligent Systems and Technology (TIST)*, 7(2):1–22, 2015.

234 Shohei Shimizu, Patrik O Hoyer, Aapo Hyvärinen, Antti Kerminen, and Michael Jordan. A linear non-gaussian acyclic model for causal discovery. *Journal of Machine Learning Research*, 7:2003–2030, 2006.

235 Jonas Peters, Joris M Mooij, Dominik Janzing, and Bernhard Schölkopf. Causal discovery with continuous additive noise models. *Journal of Machine Learning Research*, 15:2009–2053, 2014.

236 CWJ Granger. Investigating causal relations by econometric models and cross-spectral methods. In *Essays in Econometrics: Collected Papers of Clive WJ Granger*, pages 31–47, 2001.

237 Ali Shojaie and Emily B Fox. Granger causality: A review and recent advances. *Annual Review of Statistics and Its Application*, 9(1):289–319, 2022.

238 Anil K Seth, Adam B Barrett, and Lionel Barnett. Granger causality analysis in neuroscience and neuroimaging. *Journal of Neuroscience*, 35(8):3293–3297, 2015.

239 Ioannis Kontoyiannis and Maria Skoularidou. Estimating the directed information and testing for causality. *IEEE Transactions on Information Theory*, 62(11):6053–6067, 2016.

240 Abolfazl Safikhani and Ali Shojaie. Joint structural break detection and parameter estimation in high-dimensional nonstationary var models. *Journal of the American Statistical Association*, 117(537):251–264, 2022.

241 Alex Tank, Ian Covert, Nicholas Foti, Ali Shojaie, and Emily B Fox. Neural granger causality. *IEEE Transactions on Pattern Analysis and Machine Intelligence*, 44(8):4267–4279, 2021.

242 Alex Tank, Xiudi Li, Emily B Fox, and Ali Shojaie. The convex mixture distribution: Granger causality for categorical time series. *SIAM Journal on Mathematics of Data Science*, 3(1):83–112, 2021.

243 Zizhen Deng, Xiaolong Zheng, Hu Tian, and Daniel Dajun Zeng. Deep causal learning: Representation, discovery and inference. *arXiv preprint arXiv:2211.03374*, 2022.

244 Matthew J Vowels, Necati Cihan Camgoz, and Richard Bowden. D'ya like dags? A survey on structure learning and causal discovery. *ACM Computing Surveys*, 55(4):1–36, 2022.

245 Xun Zheng, Bryon Aragam, Pradeep K Ravikumar, and Eric P Xing. DAGs with NO TEARS: Continuous optimization for structure learning. In *32nd Conference on Neural Information Processing Systems (NeurIPS 2018)*, pages 1–12, 2018.

246 Mohammadali Honarpardaz, Mehdi Tarkian, Johan Ölvander, and Xiaolong Feng. Finger design automation for industrial robot grippers: A review. *Robotics and Autonomous Systems*, 87:104–119, 2017.

247 Richard S Sutton and Andrew G Barto. *Reinforcement learning: An introduction*. MIT press, 2018.

248 Frank Veenstra, Andres Faina, Sebastian Risi, and Kasper Stoy. Evolution and morphogenesis of simulated modular robots: A comparison between a direct and generative encoding. In *Applications of Evolutionary Computation: 20th European Conference, EvoApplications 2017*, Amsterdam, The Netherlands, April 19–21, 2017, *Proceedings, Part I 20*, pages 870–885. Springer, 2017.

249 Yann LeCun, Yoshua Bengio, and Geoffrey Hinton. Deep learning. *Nature*, 521(7553):436–444, 2015.

250 Kaisa Miettinen. *Nonlinear multiobjective optimization*. Springer Science & Business Media, 1999.

251 Qingfu Zhang and Hui Li. MOEA/D: A multiobjective evolutionary algorithm based on decomposition. *IEEE Transactions on Evolutionary Computation*, 11(6):712–731, 2007.

252 Efrén Mezura-Montes, Jesús Velázquez-Reyes, and CA Coello Coello. Modified differential evolution for constrained optimization. In *2006 IEEE International Conference on Evolutionary Computation*, pages 25–32. IEEE, 2006.

253 Kalyanmoy Deb. An efficient constraint handling method for genetic algorithms. *Computer Methods in Applied Mechanics and Engineering*, 186(2–4):311–338, 2000.

254 Zhixiang Yang, Xinye Cai, and Zhun Fan. Epsilon constrained method for constrained multiobjective optimization problems: Some preliminary results. In *Proceedings of the Companion Publication of the 2014 Annual Conference on Genetic and Evolutionary Computation*, pages 1181–1186, 2014.

255 Md Asafuddoula, Tapabrata Ray, Ruhul Sarker, and Khairul Alam. An adaptive constraint handling approach embedded MOEA/D. In *2012 IEEE Congress on Evolutionary Computation*, pages 1–8. IEEE, 2012.

256 Muhammad Asif Jan and Rashida Adeeb Khanum. A study of two penalty-parameterless constraint handling techniques in the framework of MOEA/D. *Applied Soft Computing*, 13(1):128–148, 2013.

257 Zhun Fan, Wenji Li, Xinye Cai, Han Huang, Yi Fang, Yugen You, Jiajie Mo, Caimin Wei, and Erik Goodman. An improved epsilon constraint-handling method in MOEA/D for CMOPs with large infeasible regions. *Soft Computing*, 23:12491–12510, 2019.

258 Zhun Fan, Zhaojun Wang, Wenji Li, Yutong Yuan, Yugen You, Zhi Yang, Fuzan Sun, and Jie Ruan. Push and pull search embedded in an M2M framework for solving constrained multi-objective optimization problems. *Swarm and Evolutionary Computation*, 54:100651, 2020.

259 Hai-Lin Liu, Chaoda Peng, Fangqing Gu, and Jiechang Wen. A constrained multi-objective evolutionary algorithm based on boundary search and archive. *International Journal of Pattern Recognition and Artificial Intelligence*, 30(01):1659002, 2016.

260 Bo Yang Qu and Ponnuthurai Nagarathnam Suganthan. Constrained multi-objective optimization algorithm with an ensemble of constraint handling methods. *Engineering Optimization*, 43(4):403–416, 2011.

261 Chaoda Peng, Hai-Lin Liu, and Fangqing Gu. An evolutionary algorithm with directed weights for constrained multi-objective optimization. *Applied Soft Computing*, 60:613–622, 2017.

262 Yongkuan Yang, Jianchang Liu, Shubin Tan, and Honghai Wang. A multi-objective differential evolutionary algorithm for constrained multi-objective optimization problems with low feasible ratio. *Applied Soft Computing*, 80:42–56, 2019.

263 Zhun Fan, Wenji Li, Xinye Cai, Hui Li, Caimin Wei, Qingfu Zhang, Kalyanmoy Deb, and Erik Goodman. Difficulty adjustable and scalable constrained multiobjective test problem toolkit. *Evolutionary Computation*, 28(3):339–378, 2020.

264 Zhi-Zhong Liu, Bing-Chuan Wang, and Ke Tang. Handling constrained multiobjective optimization problems via bidirectional coevolution. *IEEE Transactions on Cybernetics*, 52(10):10163–10176, 2021.

265 Matthias Seeger. Gaussian processes for machine learning. *International Journal of Neural Systems*, 14(02):69–106, 2004.

266 Yajie Zhang, Hao Jiang, Ye Tian, Haiping Ma, and Xingyi Zhang. Multigranularity surrogate modeling for evolutionary multiobjective optimization with expensive constraints. *IEEE Transactions on Neural Networks and Learning Systems*, 35(3):2956–2968, 2024.

267 Zhenshou Song, Handing Wang, Bing Xue, Mengjie Zhang, and Yaochu Jin. Balancing objective optimization and constraint satisfaction in expensive constrained evolutionary multi-objective optimization. *IEEE Transactions on Evolutionary Computation*, 28(5):1286–1300, 2024.

268 Kalyanmoy Deb, Rayan Hussein, Proteek Chandan Roy, and Gregorio Toscano-Pulido. A taxonomy for metamodeling frameworks for evolutionary multiobjective optimization. *IEEE Transactions on Evolutionary Computation*, 23(1):104–116, 2018.

269 Rayan Hussein and Kalyanmoy Deb. A generative kriging surrogate model for constrained and unconstrained multi-objective optimization. In *Proceedings of the Genetic and Evolutionary Computation Conference 2016*, pages 573–580, 2016.

270 Haofeng Wu, Qingda Chen, Yaochu Jin, Jinliang Ding, and Tianyou Chai. A surrogate-assisted expensive constrained multi-objective optimization algorithm based on adaptive switching of acquisition functions. *IEEE Transactions on Emerging Topics in Computational Intelligence*, 8(2):2050–2064, 2024.

271 Alaleh Ahmadianshalchi, Syrine Belakaria, and Janardhan Rao Doppa. Preference-aware constrained multi-objective Bayesian optimization. In *Proceedings of the 7th Joint International Conference on Data Science & Management of Data (11th ACM IKDD CODS and 29th COMAD)*, pages 182–191, 2024.

272 Zhongwei Ma and Yong Wang. Evolutionary constrained multiobjective optimization: Test suite construction and performance comparisons. *IEEE Transactions on Evolutionary Computation*, 23(6):972–986, 2019.

273 Kalyanmoy Deb, Mohamed Abouhawwash, and Haitham Seada. A computationally fast convergence measure and implementation for single-, multiple-, and many-objective optimization. *IEEE Transactions on Emerging Topics in Computational Intelligence*, 1(4):280–293, 2017.

274 Yoshua Bengio, Patrice Simard, and Paolo Frasconi. Learning long-term dependencies with gradient descent is difficult. *IEEE Transactions on Neural Networks*, 5(2):157–166, 1994.

275 Zhengyang Wang, Na Zou, Dinggang Shen, and Shuiwang Ji. Non-local u-nets for biomedical image segmentation. In *Proceedings of the AAAI Conference on Artificial Intelligence*, volume 34, pages 6315–6322, 2020.

276 Xavier Glorot and Yoshua Bengio. Understanding the difficulty of training deep feedforward neural networks. In *Proceedings of the Thirteenth International Conference on Artificial Intelligence and Statistics*, pages 249–256. JMLR Workshop and Conference Proceedings, 2010.

277 Md Zahangir Alom, Mahmudul Hasan, Chris Yakopcic, Tarek M Taha, and Vijayan K Asari. Recurrent residual convolutional neural network based on u-net (r2u-net) for medical image segmentation. *arXiv preprint arXiv:1802.06955*, 2018.

278 Zaiwang Gu, Jun Cheng, Huazhu Fu, Kang Zhou, Huaying Hao, Yitian Zhao, Tianyang Zhang, Shenghua Gao, and Jiang Liu. Ce-net: Context encoder network for 2D medical image segmentation. *IEEE Transactions on Medical Imaging*, 38(10):2281–2292, 2019.

279 Steven Guan, Amir A Khan, Siddhartha Sikdar, and Parag V Chitnis. Fully dense UNet for 2-D sparse photoacoustic tomography artifact removal. *IEEE Journal of Biomedical and Health Informatics*, 24(2):568–576, 2019.

280 Simon Jégou, Michal Drozdzal, David Vazquez, Adriana Romero, and Yoshua Bengio. The one hundred layers tiramisu: Fully convolutional DenseNets for semantic segmentation. In *Proceedings of the IEEE Conference on Computer Vision and Pattern Recognition Workshops*, pages 11–19, 2017.

281 Kaiming He, Xiangyu Zhang, Shaoqing Ren, and Jian Sun. Deep residual learning for image recognition. In *Proceedings of the IEEE Conference on Computer Vision and Pattern Recognition*, pages 770–778, 2016.

282 Gao Huang, Zhuang Liu, Laurens Van Der Maaten, and Kilian Q Weinberger. Densely connected convolutional networks. In *Proceedings of the IEEE Conference on Computer Vision and Pattern Recognition*, pages 4700–4708, 2017.

283 Christian Szegedy, Wei Liu, Yangqing Jia, Pierre Sermanet, Scott Reed, Dragomir Anguelov, Dumitru Erhan, Vincent Vanhoucke, and Andrew Rabinovich. Going deeper with convolutions. In *Proceedings of the IEEE Conference on Computer Vision and Pattern Recognition*, pages 1–9, 2015.

284 Lingxi Xie and Alan Yuille. Genetic cnn. In *Proceedings of the IEEE International Conference on Computer Vision*, pages 1379–1388, 2017.

285 Karen Simonyan and Andrew Zisserman. Very deep convolutional networks for large-scale image recognition. *arXiv preprint arXiv:1409.1556*, 2014.

286 Sergey Ioffe. Batch normalization: Accelerating deep network training by reducing internal covariate shift. *arXiv preprint arXiv:1502.03167*, 2015.

287 Vinod Nair and Geoffrey E Hinton. Rectified linear units improve restricted Boltzmann machines. In *Proceedings of the 27th International Conference on Machine Learning (ICML-10)*, pages 807–814, 2010.

288 Diganta Misra. Mish: A self regularized non-monotonic activation function. *arXiv preprint arXiv:1908.08681*, 2019.

289 Dmitry Ulyanov. Instance normalization: The missing ingredient for fast stylization. *arXiv preprint arXiv:1607.08022*, 2016.

290 William M Spears and Kenneth A De Jong. An analysis of multi-point crossover. In *Foundations of Genetic Algorithms*, volume 1, pages 301–315. Elsevier, 1991.

291 Guangming Lin and Xin Yao. Analysing crossover operators by search step size. In *Proceedings of 1997 IEEE International Conference on Evolutionary Computation (ICEC'97)*, pages 107–110. IEEE, 1997.

292 Brad L Miller and David E Goldberg. Genetic algorithms, tournament selection, and the effects of noise. *Complex Systems*, 9(3):193–212, 1995.

293 Richard W Hamming. Error detecting and error correcting codes. *The Bell System Technical Journal*, 29(2):147–160, 1950.

294 John H Holland. *Adaptation in natural and artificial systems: An introductory analysis with applications to biology, control, and artificial intelligence.* MIT press, 1992.

295 Yee Leung, Yong Gao, and Zong-Ben Xu. Degree of population diversity-a perspective on premature convergence in genetic algorithms and its Markov chain analysis. *IEEE Transactions on Neural Networks*, 8(5):1165–1176, 1997.

296 Zbigniew Michalewicz. *Genetic algorithms+ data structures= evolution programs*. Springer Science & Business Media, 2013.

297 John J Grefenstette. Genetic algorithms and machine learning. In *Proceedings of the Sixth annual Conference on Computational Learning Theory*, pages 3–4, 1993.

298 Kaiming He, Xiangyu Zhang, Shaoqing Ren, and Jian Sun. Delving deep into rectifiers: Surpassing human-level performance on imagenet classification. In *Proceedings of the IEEE International Conference on Computer Vision*, pages 1026–1034, 2015.

299 Michael Zhang, James Lucas, Jimmy Ba, and Geoffrey E Hinton. Lookahead optimizer: k steps forward, 1 step back. In *Proceedings of the 33rd Conference on Neural Information Processing Systems (NeurIPS 2019)*, pages 1–12, 2019.

300 Diederik P Kingma. Adam: A method for stochastic optimization. *arXiv preprint arXiv:1412.6980*, 2014.

301 Tsung-Yi Lin, Priya Goyal, Ross Girshick, Kaiming He, and Piotr Dollár. Focal loss for dense object detection. In *Proceedings of the IEEE Conference on Computer Vision and Pattern Recognition*, pages 2980–2988, 2017.

302 Joes Staal, Michael D Abràmoff, Meindert Niemeijer, Max A Viergever, and Bram Van Ginneken. Ridge-based vessel segmentation in color images of the retina. *IEEE Transactions on Medical Imaging*, 23(4):501–509, 2004.

303 Christopher G Owen, Alicja R Rudnicka, Robert Mullen, Sarah A Barman, Dorothy Monekosso, Peter H Whincup, Jeffrey Ng, and Carl Paterson. Measuring retinal vessel tortuosity in 10-year-old children: Validation of the computer-assisted image analysis of the retina (CAIAR) program. *Investigative Ophthalmology & Visual Science*, 50(5):2004–2010, 2009.

304 AD Hoover, Valentina Kouznetsova, and Michael Goldbaum. Locating blood vessels in retinal images by piecewise threshold probing of a matched filter response. *IEEE Transactions on Medical Imaging*, 19(3):203–210, 2000.

305 Thomas Köhler, Attila Budai, Martin F Kraus, Jan Odstrčilik, Georg Michelson, and Joachim Hornegger. Automatic no-reference quality assessment for retinal fundus images using vessel segmentation. In *Proceedings of the 26th IEEE International Symposium on Computer-based Medical Systems*, pages 95–100. IEEE, 2013.

306 Qiaoliang Li, Bowei Feng, LinPei Xie, Ping Liang, Huisheng Zhang, and Tianfu Wang. A cross-modality learning approach for vessel segmentation in retinal images. *IEEE Transactions on Medical Imaging*, 35(1):109–118, 2015.

307 Paweł Liskowski and Krzysztof Krawiec. Segmenting retinal blood vessels with deep neural networks. *IEEE Transactions on Medical Imaging*, 35(11):2369–2380, 2016.

308 Roberto Vega, Gildardo Sanchez-Ante, Luis E Falcon-Morales, Humberto Sossa, and Elizabeth Guevara. Retinal vessel extraction using lattice neural networks with dendritic processing. *Computers in Biology and Medicine*, 58:20–30, 2015.

309 José Ignacio Orlando, Elena Prokofyeva, and Matthew B Blaschko. A discriminatively trained fully connected conditional random field model for blood vessel segmentation in fundus images. *IEEE Transactions on Biomedical Engineering*, 64(1):16–27, 2016.

310 Zhun Fan and Jia-Jie Mo. Automated blood vessel segmentation based on de-noising auto-encoder and neural network. In *2016 International Conference on Machine Learning and Cybernetics (ICMLC)*, volume 2, pages 849–856. IEEE, 2016.

311 Juan Mo and Lei Zhang. Multi-level deep supervised networks for retinal vessel segmentation. *International Journal of Computer Assisted Radiology and Surgery*, 12:2181–2193, 2017.

312 Zengqiang Yan, Xin Yang, and Kwang-Ting Cheng. Joint segment-level and pixel-wise losses for deep learning based retinal vessel segmentation. *IEEE Transactions on Biomedical Engineering*, 65(9):1912–1923, 2018.

313 Md Zahangir Alom, Chris Yakopcic, Mahmudul Hasan, Tarek M Taha, and Vijayan K Asari. Recurrent residual U-Net for medical image segmentation. *Journal of Medical Imaging*, 6(1):014006–014006, 2019.

314 Qiangguo Jin, Zhaopeng Meng, Tuan D Pham, Qi Chen, Leyi Wei, and Ran Su. DUNet: A deformable network for retinal vessel segmentation. *Knowledge-Based Systems*, 178:149–162, 2019.

315 Bo Wang, Shuang Qiu, and Huiguang He. Dual encoding u-net for retinal vessel segmentation. In *Medical Image Computing and Computer Assisted Intervention–MICCAI 2019: 22nd International Conference*, Shenzhen, China, October 13–17, 2019, *Proceedings, Part I 22*, pages 84–92. Springer, 2019.

316 Yicheng Wu, Yong Xia, Yang Song, Donghao Zhang, Dongnan Liu, Chaoyi Zhang, and Weidong Cai. Vessel-Net: Retinal vessel segmentation under multi-path supervision. In *Medical Image Computing and Computer Assisted Intervention–MICCAI 2019: 22nd International Conference*, Shenzhen, China, October 13–17, 2019, *Proceedings, Part I 22*, pages 264–272. Springer, 2019.

317 Lei Mou, Li Chen, Jun Cheng, Zaiwang Gu, Yitian Zhao, and Jiang Liu. Dense dilated network with probability regularized walk for vessel detection. *IEEE Transactions on Medical Imaging*, 39(5):1392–1403, 2019.

318 Lei Mou, Yitian Zhao, Huazhu Fu, Yonghuai Liu, Jun Cheng, Yalin Zheng, Pan Su, Jianlong Yang, Li Chen, Alejandro F Frangi, et al. CS2-Net: Deep learning segmentation of curvilinear structures in medical imaging. *Medical Image Analysis*, 67:101874, 2021.

319 Jason D Lohn and Silvano P Colombano. A circuit representation technique for automated circuit design. *IEEE Transactions on Evolutionary Computation*, 3(3):205–219, 1999.

320 Gregory S Hornby and Jordan B Pollack. Creating high-level components with a generative representation for body-brain evolution. *Artificial Life*, 8(3):223–246, 2002.

321 Jianjun Hu, Erik D Goodman, Shaobo Li, and Ronald Rosenberg. Automated synthesis of mechanical vibration absorbers using genetic programming. *AI EDAM*, 22(3):207–217, 2008.

322 Jianjun Hu, Kisung Seo, Shaobo Li, Zhun Fan, Ronald C Rosenber, and Erik D Goodman. Structure fitness sharing (SFS) for evolutionary design by genetic programming. In *Proceedings of the 4th Annual Conference on Genetic and Evolutionary Computation*, pages 780–787, 2002.

323 Gongyue Xu, Huafeng Ding, and Zemin Feng. Optimal design of hydraulic excavator shovel attachment based on multiobjective evolutionary algorithm. *IEEE/ASME Transactions on Mechatronics*, 24(2):808–819, 2019.

324 Jen-Chung Hsiao, Kumar Shivam, CL Chou, and Tai Yan Kam. Shape design optimization of a robot arm using a surrogate-based evolutionary approach. *Applied Sciences*, 10(7):2223, 2020.

325 Rituparna Datta and Kalyanmoy Deb. Multi-objective design and analysis of robot gripper configurations using an evolutionary-classical approach. In *Proceedings of the 13th Annual Conference on Genetic and Evolutionary Computation*, pages 1843–1850, 2011.

326 Rituparna Datta, Shikhar Pradhan, and Bishakh Bhattacharya. Analysis and design optimization of a robotic gripper using multiobjective genetic algorithm. *IEEE Transactions on Systems, Man, and Cybernetics: Systems*, 46(1):16–26, 2015.

327 Siavash Rezazadeh and Jonathan W Hurst. On the optimal selection of motors and transmissions for electromechanical and robotic systems. In *2014 IEEE/RSJ International Conference on Intelligent Robots and Systems*, pages 4605–4611. IEEE, 2014.

328 Satoshi Murata, Eiichi Yoshida, Akiya Kamimura, Haruhisa Kurokawa, Kohji Tomita, and Shigeru Kokaji. M-TRAN: Self-reconfigurable modular robotic system. *IEEE/ASME Transactions on Mechatronics*, 7(4):431–441, 2002.

329 Wei-Min Shen, Behnam Salemi, and Peter Will. Hormone-inspired adaptive communication and distributed control for CONRO self-reconfigurable robots. *IEEE Transactions on Robotics and Automation*, 18(5):700–712, 2002.

330 David Brandt, David Johan Christensen, and Henrik Hautop Lund. ATRON robots: Versatility from self-reconfigurable modules. In *2007 International Conference on Mechatronics and Automation*, pages 26–32. IEEE, 2007.

331 Jacques Denavit and Richard S Hartenberg. A kinematic notation for lower-pair mechanisms based on matrices. *Journal of Applied Mechanics*, 22(2):215–221, 1955.

332 Lelai Zhou, Shaoping Bai, and Michael Rygaard Hansen. Design optimization on the drive train of a light-weight robotic arm. *Mechatronics*, 21(3):560–569, 2011.

333 S Ivvan Valdez, Salvador Botello-Aceves, Héctor M Becerra, and Eusebio E Hernández. Comparison between a concurrent and a sequential optimization methodology for serial manipulators using metaheuristics. *IEEE Transactions on Industrial Informatics*, 14(7):3155–3165, 2018.

334 Peter Corke. *Robotics and control: Fundamental algorithms in MATLAB®*. springer Nature, 2021.

335 Jianjun Hu, Erik D Goodman, Kisung Seo, and Min Pei. Adaptive hierarchical fair competition (AHFC) model for parallel evolutionary algorithms. In *Proceedings of the 4th Annual Conference on Genetic and Evolutionary Computation*, pages 772–779, 2002.

336 Douglas Zongker, Bill Punch, and Bill Rand. lil-gp 1.0 user's manual. Technical report, Citeseer, 1996.

337 Miguel Á Sainz, Joaquim Armengol, and Josep Vehí. Fault detection and isolation of the three-tank system using the modal interval analysis. *Journal of process control*, 12(2):325–338, 2002.

338 Jian Wu, Gautam Biswas, Sherif Abdelwahed, and Eric Manders. A hybrid control system design and implementation for a three tank testbed. In *Proceedings of 2005 IEEE Conference on Control Applications, 2005. CCA 2005*, pages 645–650. IEEE, 2005.

339 Rong-Jong Wai and Ching-Hsiang Tu. Development of lyapunov-based genetic algorithm control for linear piezoelectric ceramic motor drive. *IEEE Transactions on Industrial Electronics*, 54(5):2566–2582, 2007.

340 Xinghuo Yu and Okyay Kaynak. Sliding-mode control with soft computing: A survey. *IEEE Transactions on Industrial Electronics*, 56(9):3275–3285, 2009.

341 Ahmad Saudi Samosir and Abdul Halim Mohamed Yatim. Implementation of dynamic evolution control of bidirectional dc–dc converter for interfacing ultracapacitor energy storage to fuel-cell system. *IEEE Transactions on Industrial Electronics*, 57(10):3468–3473, 2010.

342 Daniel Ashlock. Evolving finite state automata. In *Evolutionary Computation for Modeling and Optimization*, pages 143–166, Springer, 2006.

343 Christian Gagné and Marc Parizeau. Genericity in evolutionary computation software tools: Principles and case-study. *International Journal on Artificial Intelligence Tools*, 15(02):173–194, 2006.

344 Xenofon D Koutsoukos, Panos J Antsaklis, James A Stiver, and Michael D Lemmon. Supervisory control of hybrid systems. *Proceedings of the IEEE*, 88(7):1026–1049, 2000.

345 Fucheng Wang. *Design and synthesis of active and passive vehicle suspensions*. PhD thesis, University of Cambridge UK, 2001.

346 Malcolm C Smith. Achievable dynamic response for automotive active suspensions. *Vehicle System Dynamics*, 24(1):1–33, 1995.

347 Malcom C Smith and Gavin W Walker. Performance limitations and constraints for active and passive suspensions: A mechanical multi-port approach. *Vehicle System Dynamics*, 33(3):137–168, 2000.

348 Olaf Ronneberger, Philipp Fischer, and Thomas Brox. U-Net: Convolutional networks for biomedical image segmentation. In *Medical Image Computing and Computer-assisted Intervention–MICCAI 2015: 18th International Conference*, Munich, Germany, October 5–9, 2015, *Proceedings, Part III 18*, pages 234–241. Springer, 2015.

349 Peter JB Hancock. Selection methods for evolutionary algorithms. In *The Practical Handbook of Genetic Algorithms*, pages 67–92. CRC Press, 2019.

350 Joao Antonio de Vasconcelos, Jaime Arturo Ramirez, Ricardo HC Takahashi, and Rodney R Saldanha. Improvements in genetic algorithms. *IEEE Transactions on magnetics*, 37(5):3414–3417, 2001.

351 Albert Cardona, Stephan Saalfeld, Stephan Preibisch, Benjamin Schmid, Anchi Cheng, Jim Pulokas, Pavel Tomancak, and Volker Hartenstein. An integrated micro-and macroarchitectural analysis of the drosophila brain by computer-assisted serial section electron microscopy. *PLoS Biology*, 8(10):e1000502, 2010.

352 Yutong Yuan, Zhun Fan, Xiaomin Zhu, Li Ma, Ji Ouyang, Weidong Bao, Ji Wang, and Zhaojun Wang. SUNDER: Self-organized grouping and entrapping method for swarms in multitarget environments. *Defence Technology*, 26:68–83, 2023.

353 Chen Wang, Wenxi Kuang, Minqiang Gu, and Zhun Fan. AGENT: An adaptive grouping and entrapping method for flocking systems. *Journal of Computational Design and Engineering*, 10(1):357–367, 2023.

354 Yao Yao, Cao Jun-Hua, Guo Yi, Fan Zhun, Zou An-Min, Xu Biao, and Li Ke. Autonomous control method of rotor UAVs for power inspection with renewable energy based on swarm intelligence. *Frontiers in Energy Research*, 9:697054, 2021.

355 Hyondong Oh, Ataollah Ramezan Shirazi, Chaoli Sun, and Yaochu Jin. Bio-inspired self-organising multi-robot pattern formation: A review. *Robotics and Autonomous Systems*, 91:83–100, 2017.

356 René Doursat, Hiroki Sayama, and Olivier Michel. A review of morphogenetic engineering. *Natural Computing*, 12:517–535, 2013.

357 Meng Wu, Xiaomin Zhu, Li Ma, Weidong Bao, Zhun Fan, and Yaochu Jin. Multi-robot target entrapment using cooperative hierarchical gene regulatory network. *Swarm and Evolutionary Computation*, 80:101310, 2023.

358 Fan Zhun, Ma Peili, Guijie Zhu, Minchong Xie, Tianshan Chen, Xie Fei, Shi Ze, Weidong Bao, and Xiaomin Zhu. Method of swarm aggregation and control for intelligent robot based on three-dimensional gene regulatory network. *IEEE Transl. Journal of NanJing Normal University (Engineering and Technology Edition)*, 22(1):7, 2022.

359 Chen Wang, Zhaohui Shi, Minqiang Gu, Weicheng Luo, Xiaomin Zhu, and Zhun Fan. Revolutionary entrapment model of uniformly distributed swarm robots in morphogenetic formation. *Defence Technology*, 31:496–509, 2024.

360 Chen Wang, Dongliang Wang, Minqiang Gu, Huaxing Huang, Zhaojun Wang, Yutong Yuan, Xiaomin Zhu, Wu Wei, and Zhun Fan. Bioinspired environment exploration algorithm in swarm based on levy flight and improved artificial potential field. *Drones*, 6(5):122, 2022.

361 Shuai Zhang, Mingyong Liu, Xiaokang Lei, Yunke Huang, and Feihu Zhang. Multi-target trapping with swarm robots based on pattern formation. *Robotics and Autonomous Systems*, 106:1–13, 2018.

362 Ron Milo, Shai Shen-Orr, Shalev Itzkovitz, Nadav Kashtan, Dmitri Chklovskii, and Uri Alon. Network motifs: Simple building blocks of complex networks. *Science*, 298(5594):824–827, 2002.

363 Peter M Bowers, Shawn J Cokus, David Eisenberg, and Todd O Yeates. Use of logic relationships to decipher protein network organization. *Science*, 306(5705):2246–2249, 2004.

364 Hyondong Oh and Yaochu Jin. Evolving hierarchical gene regulatory networks for morphogenetic pattern formation of swarm robots. In *2014 IEEE Congress on Evolutionary Computation (CEC)*, pages 776–783. IEEE, 2014.

365 William B Langdon. Size fair and homologous tree genetic programming crossovers. *Genetic Programming and Evolvable Machines*, 1(1/2):95–119, 2000.

366 Yutong Yuan, Zhun Fan, Xiaomin Zhu, Meng Wu, Li Ma, Taosheng Fang, Zhaojun Wang, Weidong Bao, Yun Zhou, Huangke Chen, et al. TH-GRN model based collective tracking in confined environment. In *Advances in Swarm Intelligence: 10th International Conference, ICSI 2019*, Chiang Mai, Thailand, July 26–30, 2019, *Proceedings, Part II 10*, pages 33–43. Springer, 2019.

367 Hyondong Oh and Yaochu Jin. Adaptive swarm robot region coverage using gene regulatory networks. In *Advances in Autonomous Robotics Systems: 15th Annual Conference, TAROS 2014*, Birmingham, UK, September 1–3, 2014. Proceedings 15, pages 197–208. Springer, 2014.

368 Zhun FAN, Fuzan SUN, Peili MA, Wenji LI, Ze SHI, Zhaojun WANG, Guijie ZHU, Ke LI, and Bin XIN. Stigmergy-based swarm robots for target search and trapping. *Transactions of Beijing Institute of Technology*, 42(2):158–167, 2022.

369 Dongliang Wang, Wu Wei, Xinmei Wang, Yong Gao, Yanjie Li, Qiuda Yu, and Zhun Fan. Formation control of multiple mecanum-wheeled mobile robots with physical constraints and uncertainties. *Applied Intelligence*, 1–20, 2022.

370 Glenn Jocher, Alex Stoken, Jirka Borovec, Ayush Chaurasia, Liu Changyu, Adam Hogan, Jan Hajek, Laurentiu Diaconu, Yonghye Kwon, Yann Defretin, et al. ultralytics/yolov5:v5.0-YOLOv5-P6 1280 models, AWS, Supervise. ly and YouTube integrations. *Zenodo*, 2021.

371 Muhammad Abdul Haseeb, Jianyu Guan, Danijela Ristic-Durrant, and Axel GrÃ¤ser. DisNet: A novel method for distance estimation from monocular camera. *10th Planning, Perception and Navigation for Intelligent Vehicles (PPNIV18), IROS*, 2018.

372 Brent A Griffin and Jason J Corso. Depth from camera motion and object detection. In *Proceedings of the IEEE/CVF Conference on Computer Vision and Pattern Recognition*, pages 1397–1406, 2021.

373 Yaochu Jin, Hongliang Guo, and Yan Meng. A hierarchical gene regulatory network for adaptive multirobot pattern formation. *IEEE Transactions on Systems, Man, and Cybernetics, Part B (Cybernetics)*, 42(3):805–816, 2012.

374 Shital Shah, Debadeepta Dey, Chris Lovett, and Ashish Kapoor. AirSim: High-fidelity visual and physical simulation for autonomous vehicles. In *Field and Service Robotics: Results of the 11th International Conference*, pages 621–635. Springer, 2018.

375 Huaxing Huang, Guijie Zhu, Zhun Fan, Hao Zhai, Yuwei Cai, Ze Shi, Zhaohui Dong, and Zhifeng Hao. Vision-based distributed multi-UAV collision avoidance via deep reinforcement learning for navigation. In *2022 IEEE/RSJ International Conference on Intelligent Robots and Systems (IROS)*, pages 13745–13752. IEEE, 2022.

376 Antonio Loquercio, Ana I Maqueda, Carlos R Del-Blanco, and Davide Scaramuzza. Dronet: Learning to fly by driving. *IEEE Robotics and Automation Letters*, 3(2):1088–1095, 2018.

377 Antonio Loquercio, Elia Kaufmann, René Ranftl, Matthias Müller, Vladlen Koltun, and Davide Scaramuzza. Learning high-speed flight in the wild. *Science Robotics*, 6(59):eabg5810, 2021.

378 Dhiraj Gandhi, Lerrel Pinto, and Abhinav Gupta. Learning to fly by crashing. In *2017 IEEE/RSJ International Conference on Intelligent Robots and Systems (IROS)*, pages 3948–3955. IEEE, 2017.

379 Fereshteh Sadeghi and Sergey Levine. CAD2RL: Real single-image flight without a single real image. *arXiv preprint arXiv:1611.04201*, 2016.

380 Aleksandra Faust, Kenneth Oslund, Oscar Ramirez, Anthony Francis, Lydia Tapia, Marek Fiser, and James Davidson. PRM-RL: Long-range robotic navigation tasks by combining reinforcement learning and sampling-based planning. In *2018 IEEE International Conference on Robotics and Automation (ICRA)*, pages 5113–5120. IEEE, 2018.

381 Francisco Leiva and Javier Ruiz-del Solar. Robust RL-based map-less local planning: Using 2D point clouds as observations. *IEEE Robotics and Automation Letters*, 5(4):5787–5794, 2020.

382 Zhaowei Ma, Chang Wang, Yifeng Niu, Xiangke Wang, and Lincheng Shen. A saliency-based reinforcement learning approach for a UAV to avoid flying obstacles. *Robotics and Autonomous Systems*, 100:108–118, 2018.

383 Jur Van Den Berg, Stephen J Guy, Ming Lin, and Dinesh Manocha. Reciprocal n-body collision avoidance. In *Robotics Research: The 14th International Symposium ISRR*, pages 3–19. Springer, 2011.

384 Javier Alonso-Mora, Paul Beardsley, and Roland Siegwart. Cooperative collision avoidance for nonholonomic robots. *IEEE Transactions on Robotics*, 34(2): 404–420, 2018.

385 Ke Guo, Dawei Wang, Tingxiang Fan, and Jia Pan. VR-ORCA: Variable responsibility optimal reciprocal collision avoidance. *IEEE Robotics and Automation Letters*, 6(3):4520–4527, 2021.

386 Julio Godoy, Ioannis Karamouzas, Stephen Guy, and Maria Gini. Implicit coordination in crowded multi-agent navigation. In *Proceedings of the AAAI Conference on Artificial Intelligence*, volume 30, 2016.

387 Yu Fan Chen, Miao Liu, Michael Everett, and Jonathan P How. Decentralized non-communicating multiagent collision avoidance with deep reinforcement learning. In *2017 IEEE International Conference on Robotics and Automation (ICRA)*, pages 285–292. IEEE, 2017.

388 Yu Fan Chen, Michael Everett, Miao Liu, and Jonathan P How. Socially aware motion planning with deep reinforcement learning. In *2017 IEEE/RSJ International Conference on Intelligent Robots and Systems (IROS)*, pages 1343–1350. IEEE, 2017.

389 Pinxin Long, Tingxiang Fan, Xinyi Liao, Wenxi Liu, Hao Zhang, and Jia Pan. Towards optimally decentralized multi-robot collision avoidance via deep reinforcement learning. In *2018 IEEE International Conference on Robotics and Automation (ICRA)*, pages 6252–6259. IEEE, 2018.

390 Tingxiang Fan, Pinxin Long, Wenxi Liu, and Jia Pan. Distributed multi-robot collision avoidance via deep reinforcement learning for navigation in complex scenarios. *The International Journal of Robotics Research*, 39(7):856–892, 2020.

391 Renaud Bastien and Pawel Romanczuk. A model of collective behavior based purely on vision. *Science Advances*, 6(6):eaay0792, 2020.

392 Denis Yarats, Amy Zhang, Ilya Kostrikov, Brandon Amos, Joelle Pineau, and Rob Fergus. Improving sample efficiency in model-free reinforcement learning from images. In *Proceedings of the AAAI Conference on Artificial Intelligence*, volume 35, pages 10674–10681, 2021.

393 Juntong Lin, Xuyun Yang, Peiwei Zheng, and Hui Cheng. End-to-end decentralized multi-robot navigation in unknown complex environments via deep reinforcement learning. In *2019 IEEE International Conference on Mechatronics and Automation (ICMA)*, pages 2493–2500. IEEE, 2019.

394 Tuomas Haarnoja, Aurick Zhou, Pieter Abbeel, and Sergey Levine. Soft actor-critic: Off-policy maximum entropy deep reinforcement learning with a stochastic actor. In *International Conference on Machine Learning*, pages 1861–1870. PMLR, 2018.

395 Partha Ghosh, Mehdi SM Sajjadi, Antonio Vergari, Michael Black, and Bernhard Schölkopf. From variational to deterministic autoencoders. *arXiv preprint arXiv:1903.12436*, 2019.

396 Yanlin Zhou, Fan Lu, George Pu, Xiyao Ma, Runhan Sun, Hsi-Yuan Chen, and Xiaolin Li. Adaptive leader-follower formation control and obstacle avoidance via deep reinforcement learning. In *2019 IEEE/RSJ International Conference on Intelligent Robots and Systems (IROS)*, pages 4273–4280. IEEE, 2019.

397 Alex Kendall, Jeffrey Hawke, David Janz, Przemyslaw Mazur, Daniele Reda, John-Mark Allen, Vinh-Dieu Lam, Alex Bewley, and Amar Shah. Learning to drive in a day. In *2019 International Conference on Robotics and Automation (ICRA)*, pages 8248–8254. IEEE, 2019.

398 Fabian Schilling, Fabrizio Schiano, and Dario Floreano. Vision-based drone flocking in outdoor environments. *IEEE Robotics and Automation Letters*, 6(2):2954–2961, 2021.

399 Huy X Pham, Hung M La, David Feil-Seifer, and Luan V Nguyen. Autonomous UAV navigation using reinforcement learning. *arXiv preprint arXiv:1801.05086*, 2018.

400 Melrose Roderick, James MacGlashan, and Stefanie Tellex. Implementing the deep q-network. *arXiv preprint arXiv:1711.07478*, 2017.

401 Ruoxuan Li, Hang Shang, Minghui Shen, and Yueyi Xiao. Vision slam-based UAV obstacle avoidance system. In *AIP Conference Proceedings*, volume 3144. AIP Publishing, 2024.

402 Rongbing Li, Jianye Liu, Ling Zhang, and Yijun Hang. Lidar/mems imu integrated navigation (slam) method for a small UAV in indoor environments. In *2014 DGON Inertial Sensors and Systems (ISS)*, pages 1–15. IEEE, 2014.

403 Abdullah Yusefi, Akif Durdu, and Cemil Sungur. Orb-slam-based 2D reconstruction of environment for indoor autonomous navigation of UAVs. *Avrupa Bilim ve Teknoloji Dergisi*, pages 466–472, 2020.

404 Yao Yao, Jun-Hua Cao, Yi Guo, Zhun Fan, Bing Li, Biao Xu, and Ke Li. Adaptive coverage control for multi-USV system in complex environment with unknown obstacles. *International Journal of Distributed Sensor Networks*, 17(6):1550147 7211021525, 2021.

405 Akshay Walvekar, Yash Goel, Anuj Jain, Sohom Chakrabarty, and Anil Kumar. Vision based autonomous navigation of quadcopter using reinforcement learning. In *2019 IEEE 2nd International Conference on Automation, Electronics and Electrical Engineering (AUTEEE)*, pages 160–165. IEEE, 2019.

406 Kjell Kersandt, Guillem Muñoz, and Cristina Barrado. Self-training by reinforcement learning for full-autonomous drones of the future. In *2018 IEEE/AIAA 37th Digital Avionics Systems Conference (DASC)*, pages 1–10. IEEE, 2018.

407 Gaofei Han, Qingling Wu, Boxi Wang, Che Lin, Jiafan Zhuang, Wenji Li, Zhifeng Hao, and Zhun Fan. Deep reinforcement learning based multi-UAV collision avoidance with causal representation learning. In *2024 10th International Conference on Big Data and Information Analytics (BigDIA)*, 2024.

408 Bernhard Schölkopf, Francesco Locatello, Stefan Bauer, Nan Rosemary Ke, Nal Kalchbrenner, Anirudh Goyal, and Yoshua Bengio. Toward causal representation learning. *Proceedings of the IEEE*, 109(5):612–634, 2021.

409 Johann Brehmer, Pim De Haan, Phillip Lippe, and Taco S Cohen. Weakly supervised causal representation learning. *Advances in Neural Information Processing Systems*, 35:38319–38331, 2022.

410 Jovana Mitrovic, Brian McWilliams, Jacob Walker, Lars Buesing, and Charles Blundell. Representation learning via invariant causal mechanisms. *arXiv preprint arXiv:2010.07922*, 2020.

411 Mengyue Yang, Furui Liu, Zhitang Chen, Xinwei Shen, Jianye Hao, and Jun Wang. CausalVAE: Disentangled representation learning via neural structural causal models. In *Proceedings of the IEEE/CVF Conference on Computer Vision and Pattern Recognition*, pages 9593–9602, 2021.

412 Kun Zhang, Shaoan Xie, Ignavier Ng, and Yujia Zheng. Causal representation learning from multiple distributions: A general setting. *arXiv preprint arXiv:2402.05052*, 2024.

413 Weiqi Zhao, Dian Tang, Xin Chen, Dawei Lv, Daoli Ou, Biao Li, Peng Jiang, and Kun Gai. Disentangled causal embedding with contrastive learning for recommender system. In *Companion Proceedings of the ACM Web Conference 2023*, pages 406–410, 2023.

414 Seyed Sajad Mousavi, Michael Schukat, and Enda Howley. Deep reinforcement learning: An overview. In *Proceedings of SAI Intelligent Systems Conference (IntelliSys)* 2016: *Volume 2*, pages 426–440. Springer, 2018.

415 Chao Wang, Jian Wang, Yuan Shen, and Xudong Zhang. Autonomous navigation of UAVs in large-scale complex environments: A deep reinforcement learning approach. *IEEE Transactions on Vehicular Technology*, 68(3):2124–2136, 2019.

416 Chao Wang, Jian Wang, Jingjing Wang, and Xudong Zhang. Deep-reinforcement-learning-based autonomous UAV navigation with sparse rewards. *IEEE Internet of Things Journal*, 7(7):6180–6190, 2020.

417 Tuomas Haarnoja, Aurick Zhou, Kristian Hartikainen, George Tucker, Sehoon Ha, Jie Tan, Vikash Kumar, Henry Zhu, Abhishek Gupta, and Pieter Abbeel. Soft actor-critic algorithms and applications. *arXiv preprint arXiv:1812.05905*, 2018.

418 Nicklas Hansen and Xiaolong Wang. Generalization in reinforcement learning by soft data augmentation. In *2021 IEEE International Conference on Robotics and Automation (ICRA)*, pages 13611–13617. IEEE, 2021.

419 Roberta Raileanu, Maxwell Goldstein, Denis Yarats, Ilya Kostrikov, and Rob Fergus. Automatic data augmentation for generalization in reinforcement learning. *Advances in Neural Information Processing Systems*, 34:5402–5415, 2021.

420 Jie Hu, Li Shen, and Gang Sun. Squeeze-and-excitation networks. In *Proceedings of the IEEE Conference on Computer Vision and Pattern Recognition*, pages 7132–7141, 2018.

421 Petar Stojanov, Zijian Li, Mingming Gong, Ruichu Cai, Jaime Carbonell, and Kun Zhang. Domain adaptation with invariant representation learning: What transformations to learn? *Advances in Neural Information Processing Systems*, 34:24791–24803, 2021.

422 Kun Zhang, Mingming Gong, and Bernhard Schölkopf. Multi-source domain adaptation: A causal view. In *Proceedings of the AAAI Conference on Artificial Intelligence*, volume 29, 2015.

423 Dong Zhang, Hanwang Zhang, Jinhui Tang, Xian-Sheng Hua, and Qianru Sun. Causal intervention for weakly-supervised semantic segmentation. *Advances in Neural Information Processing Systems*, 33:655–666, 2020.

424 York Hagmayer, Steven A Sloman, David A Lagnado, and Michael R Waldmann. Causal reasoning through intervention. In *Causal Learning: Psychology, Philosophy, and Computation*, pages 86–100, 2007.

425 Brendan Alvey, Derek T, Anderson,, Andrew Buck, Matthew Deardorff, Grant Scott, and James M Keller. Simulated photorealistic deep learning framework and workflows to accelerate computer vision and unmanned aerial vehicle research. In *Proceedings of the IEEE/CVF International Conference on Computer Vision*, pages 3889–3898, 2021.

426 Michael Tschannen, Olivier Bachem, and Mario Lucic. Recent advances in autoencoder-based representation learning. *arXiv preprint arXiv:1812.05069*, 2018.

427 Zhun Fan, Chong Li, Ying Chen, Jiahong Wei, Giuseppe Loprencipe, Xiaopeng Chen, and Paola Di Mascio. Automatic crack detection on road pavements using encoder-decoder architecture. *Materials*, 13(13):2960, 2020.

428 Qirui Ji, Jiangmeng Li, Jie Hu, Rui Wang, Changwen Zheng, and Fanjiang Xu. Rethinking dimensional rationale in graph contrastive learning from causal perspective. In *Proceedings of the AAAI Conference on Artificial Intelligence*, volume 38, pages 12810–12820, 2024.

429 Hongxing Wei, Youdong Chen, Jindong Tan, and Tianmiao Wang. Sambot: A self-assembly modular robot system. *IEEE/ASME Transactions on Mechatronics*, 16(4):745–757, 2010.

430 Kyle Gilpin and Daniela Rus. Modular robot systems. *IEEE Robotics & Automation Magazine*, 17(3):38–55, 2010.

431 Mei Wang and Weihong Deng. Deep visual domain adaptation: A survey. *Neurocomputing*, 312:135–153, 2018.

432 Jindong Wang, Cuiling Lan, Chang Liu, Yidong Ouyang, Tao Qin, Wang Lu, Yiqiang Chen, Wenjun Zeng, and S Yu Philip. Generalizing to unseen domains:

A survey on domain generalization. *IEEE Transactions on Knowledge and Data Engineering*, 35(8):8052–8072, 2022.

433 Haoyue Bai, Fengwei Zhou, Lanqing Hong, Nanyang Ye, S-H Gary Chan, and Zhenguo Li. NAS-OoD: Neural architecture search for out-of-distribution generalization. In *Proceedings of the IEEE/CVF International Conference on Computer Vision*, pages 8320–8329, 2021.

434 Yu-Wei Wen, Sheng-Hsuan Peng, and Chuan-Kang Ting. Two-stage evolutionary neural architecture search for transfer learning. *IEEE Transactions on Evolutionary Computation*, 25(5):928–940, 2021.

435 Yanxi Li, Zhaohui Yang, Yunhe Wang, and Chang Xu. Adapting neural architectures between domains. *Advances in Neural Information Processing Systems*, 33:789–798, 2020.

436 Nan Guo, Ke Gu, Junfei Qiao, and Hantao Liu. Active vision for deep visual learning: A unified pooling framework. *IEEE Transactions on Industrial Informatics*, 18(10):6610–6618, 2021.

437 Junji Ito, Cristian Joana, Yukako Yamane, Ichiro Fujita, Hiroshi Tamura, Pedro E Maldonado, and Sonja Grün. Latency shortening with enhanced sparseness and responsiveness in v1 during active visual sensing. *Scientific Reports*, 12(1):6021, 2022.

438 Yuhui Xu, Lingxi Xie, Wenrui Dai, Xiaopeng Zhang, Xin Chen, Guo-Jun Qi, Hongkai Xiong, and Qi Tian. Partially-connected neural architecture search for reduced computational redundancy. *IEEE Transactions on Pattern Analysis and Machine Intelligence*, 43(9):2953–2970, 2021.

439 Xiaodong Lan and Mac Schwager. Rapidly exploring random cycles: Persistent estimation of spatiotemporal fields with multiple sensing robots. *IEEE Transactions on Robotics*, 32(5):1230–1244, 2016.

440 Henry Carrillo, Philip Dames, Vijay Kumar, and José A Castellanos. Autonomous robotic exploration using occupancy grid maps and graph slam based on Shannon and Rényi entropy. In *2015 IEEE International Conference on Robotics and Automation (ICRA)*, pages 487–494. IEEE, 2015.

441 Yue Meng, Wei Wang, Hao Han, and Jingxuan Ban. A visual/inertial integrated landing guidance method for UAV landing on the ship. *Aerospace Science and Technology*, 85:474–480, 2019.

442 Jibin Zheng, Tianyuan Yang, Hongwei Liu, Tao Su, and Liangtian Wan. Accurate detection and localization of unmanned aerial vehicle swarms-enabled mobile edge computing system. *IEEE Transactions on Industrial Informatics*, 17(7):5059–5067, 2020.

443 German I Parisi, Ronald Kemker, Jose L Part, Christopher Kanan, and Stefan Wermter. Continual lifelong learning with neural networks: A review. *Neural networks*, 113:54–71, 2019.

444 Xiaocong Du, Zheng Li, Jingbo Sun, Frank Liu, and Yu Cao. Evolutionary NAS in light of model stability for accurate continual learning. In *2021 International Joint Conference on Neural Networks (IJCNN)*, pages 1–8. IEEE, 2021.

445 Qiang Gao, Zhipeng Luo, Diego Klabjan, and Fengli Zhang. Efficient architecture search for continual learning. *IEEE Transactions on Neural Networks and Learning Systems*, 34(11):8555–8565, 2022.

446 Martin Mundt, Iuliia Pliushch, and Visvanathan Ramesh. Neural architecture search of deep priors: Towards continual learning without catastrophic interference. In *Proceedings of the IEEE/CVF Conference on Computer Vision and Pattern Recognition*, pages 3523–3532, 2021.

Index

Modular Design Automation of Intelligent Robot Systems, First Edition. Zhun Fan et al.

c

d

j

k

l

m

Printed in the USA/Agawam, MA
February 3, 2026

900760.016